Mukesh Kumar Awasthi, Dhananjay Yadav, J. Paulo Davim (Eds.)

Flow Dynamics and Heat Transfer

Advanced Mechanical Engineering

Edited by
J. Paulo Davim

Volume 16

Flow Dynamics and Heat Transfer

Computational Fluid Dynamics, Thermal Analysis, Numerical Simulations

Edited by
Mukesh Kumar Awasthi, Dhananjay Yadav
and J. Paulo Davim

DE GRUYTER

Editors
Dr. Mukesh Kumar Awasthi
Department of Mathematics
Babasaheb Bhimrao Ambedkar University
Vidya Vihar, Raebareli Road
Lucknow 226025
Uttar Pradesh
India
mukeshiitr.kumar@gmail.com

Dr. Dhananjay Yadav
Department of Mathematics
University of Nizwa
P.O. Box 33, PC 616
Nizwa
Oman
dhananjayadav@gmail.com

Prof. Dr. J. Paulo Davim
Department of Mechanical Engineering
University of Aveiro
Campus Santiago
3810-193 Aveiro
Portugal
pdavim@ua.pt

ISBN 978-3-11-166059-2
e-ISBN (PDF) 978-3-11-166167-4
e-ISBN (EPUB) 978-3-11-166405-7
ISSN 2367-3796

Library of Congress Control Number: 2025936675

Bibliographic information published by the Deutsche Nationalbibliothek
The Deutsche Nationalbibliothek lists this publication in the Deutsche Nationalbibliografie; detailed bibliographic data are available on the internet at http://dnb.dnb.de.

Cover image: Sky_blue/iStock/thinkstock
Typesetting: Integra Software Services Pvt. Ltd.
Printing and binding: CPI books GmbH, Leck

www.degruyter.com
Questions about General Product Safety Regulation:
productsafety@degruyterbrill.com

Contents

Aim and scope

The book *Flow Dynamics and Heat Transfer: Computational Fluid Dynamics, Thermal Analysis, Numerical Simulations* aims to provide a comprehensive exploration of the fundamental and advanced concepts in fluid flow and heat transfer, focusing on computational techniques and real-world applications. It serves as a valuable resource for researchers, engineers, and students involved in understanding and optimizing thermal-fluid systems across various domains, including mechanical, aerospace, chemical, and environmental engineering. The book bridges theoretical principles with computational and experimental approaches, offering a holistic view of emerging trends and advancements in the field.

The initial chapters introduce the core principles of flow dynamics and heat transfer, laying the groundwork for understanding fluid behavior and thermal energy exchange. **Chapter 1** provides a historical and conceptual overview, emphasizing the interdependence of flow dynamics and heat transfer in engineering applications. **Chapter 2** explores compressible fluid flow and heat transfer, highlighting key phenomena such as shock waves, boundary layers, and energy transfer in high-speed applications. **Chapter 3** shifts the focus to non-Newtonian fluid behavior, detailing analytical and numerical approaches to study complex fluid interactions under different physical conditions.

Building on these foundations, the book delves into various heat transfer mechanisms and their significance in industrial applications. **Chapter 4** examines forced and natural convection, discussing theoretical models and computational techniques for optimizing heat transfer performance. **Chapter 5** introduces numerical methods for solving coupled partial differential equations in heat transfer problems with uncertain parameters, addressing challenges in modeling and simulation. **Chapter 6** extends this discussion to porous media, where uncertainties in field variables play a crucial role in determining flow and thermal characteristics.

The interplay between fluid dynamics and computational modeling is explored in greater detail in later chapters. **Chapter 7** presents direct numerical simulations for thermal-fluid systems, providing insights into transient flow behavior and thermodynamic properties. **Chapter 8** investigates the dynamics of shock-accelerated gas interfaces, focusing on the impact of Mach numbers on vorticity and wave interactions. **Chapter 9** studies thermal convection in partially ionized plasma, using both linear and nonlinear analyses to evaluate the role of collisional frequency and compressibility in fluid stability.

Applications of flow dynamics in microscale and nanoscale systems are also a significant focus of this book. **Chapter 10** analyzes the effects of Joule heating, external magnetic fields, and radiation on electroosmotic flow in microchannels, contributing to the development of microfluidic devices and biomedical applications. **Chapter 11** explores the oscillating water column energy device in a two-layer fluid system, emphasizing its potential for wave energy harvesting. **Chapter 12** presents a data-driven

https://doi.org/10.1515/9783111661674-203

approach for predicting thermal conductivity in nanoparticle-enhanced nanofluids, demonstrating the role of machine learning in advancing heat transfer research.

The final chapters emphasize optimization techniques and industrial applications, showcasing advancements in engineering solutions for heat and mass transfer systems. **Chapter 13** discusses industrial applications of flow dynamics and heat transfer, covering topics such as heat exchanger efficiency, turbine cooling, and nanofluid-based thermal regulation. **Chapter 14** focuses on optimization methods, including gradient-based approaches and evolutionary algorithms, to enhance heat transfer performance. **Chapter 15** further expands on advanced optimization techniques, such as genetic algorithms and artificial intelligence, to improve the efficiency and sustainability of heat transfer systems.

By integrating fundamental theories, computational methodologies, and practical applications, this book aims to equip readers with the knowledge and tools needed to tackle complex fluid flow and heat transfer challenges. The inclusion of case studies, numerical simulations, and experimental validations ensures a balanced perspective, making this book a valuable reference for both academic and industrial professionals.

Editors
Mukesh Kumar Awasthi
Dhananjay Yadav
Paulo J. Davim

Preface

The book *Flow Dynamics and Heat Transfer: Computational Fluid Dynamics, Thermal Analysis, Numerical Simulations* offers an in-depth exploration of both fundamental and advanced aspects of fluid flow and heat transfer, combining theoretical principles with numerical modeling techniques. Given the critical role of these fields in engineering and scientific research, this book provides a structured approach to comprehending essential concepts and their real-world applications. Featuring contributions from experts across multiple disciplines, it presents the latest advancements and computational strategies for analyzing complex thermal-fluid systems.

Chapter 1 introduces the fundamental principles of flow dynamics and heat transfer, discussing their historical development, governing equations, and computational approaches such as CFD. **Chapter 2** explores compressible fluid flow and heat transfer, highlighting key phenomena like shock waves, boundary layer behavior, and numerical modeling techniques. **Chapter 3** delves into the study of non-Newtonian fluid flow and heat transfer, analyzing the behavior of Reiner-Rivlin fluids between rotating disks under the influence of a magnetic field. **Chapter 4** focuses on heat transfer in forced and natural convection, presenting theoretical models and computational techniques for optimizing heat transfer performance.

Chapter 5 investigates the numerical study of coupled partial differential equations in heat transfer problems with imprecisely defined parameters, employing fuzzy finite element methods to handle uncertainties. **Chapter 6** extends this discussion by exploring the effects of uncertain field variables in a porous cavity using fuzzy modeling techniques. **Chapter 7** examines thermal-fluid systems through direct numerical simulations, predicting thermodynamic properties and transient flow behavior in laboratory reactors. **Chapter 8** analyzes the dynamics of shock-accelerated V-shaped gas interfaces, emphasizing the influence of Mach numbers on interface deformation and vorticity generation.

Chapter 9 presents nonlinear and linear analyses of partially ionized plasma, studying thermal convection under different boundary conditions using energy and normal mode methods. **Chapter 10** investigates the thermo-fluid characteristics of electroosmotic flow in a hydrophobic microchannel, examining the effects of Joule heating, magnetic fields, and radiation on fluid transport. **Chapter 11** explores oscillating water column energy devices in a two-layer fluid system, analyzing hydrodynamic forces and wave interactions for renewable energy applications. **Chapter 12** applies machine learning techniques to predict the thermal conductivity ratio of nanoparticle-enhanced nanofluids, demonstrating the potential of AI in thermal analysis.

Chapter 13 highlights industrial applications of flow dynamics and heat transfer, integrating CFD techniques with experimental validation to optimize energy and manufacturing processes. **Chapter 14** explores optimization techniques in flow dynamics and heat transfer, discussing classical and modern approaches such as genetic

https://doi.org/10.1515/9783111661674-204

algorithms and particle swarm optimization. **Chapter 15** extends the discussion on optimization by introducing advanced methods for enhancing heat transfer efficiency in various engineering applications.

This book serves as a valuable resource for researchers, engineers, and students in the fields of fluid mechanics, heat transfer, and computational modeling. By covering a diverse range of topics, from fundamental theories to emerging computational techniques, the book provides a well-rounded perspective on the latest advancements in the field. As industries continue to rely on computational simulations and data-driven models, the insights presented in this book will contribute to the development of more efficient and sustainable thermal-fluid systems.

Editors
Mukesh Kumar Awasthi
Dhananjay Yadav
Paulo J. Davim

Acknowledgments

The completion of this book, *Flow Dynamics and Heat Transfer: Computational Fluid Dynamics, Thermal Analysis, Numerical Simulations*, has been made possible through the dedicated efforts of numerous individuals and institutions. We extend our deepest gratitude to all the contributing authors for their expertise, commitment, and invaluable contributions. Their extensive research and scholarly work have enriched this book with cutting-edge insights into flow dynamics, heat transfer, and computational techniques. Their collaboration and willingness to share their knowledge have played a crucial role in shaping the content and ensuring its relevance to researchers, engineers, and industry professionals.

We sincerely appreciate the unwavering support of **De Gruyter, Germany**, for providing us with the opportunity to publish this book. The editorial and production teams have been instrumental in guiding us through the publishing process, ensuring the highest standards of quality and precision. Their professionalism, patience, and valuable feedback have greatly enhanced the clarity and structure of the book, making it accessible to a wide audience in the scientific and engineering communities.

A special note of appreciation goes to our families and friends, whose patience and encouragement have been a constant source of motivation. Their understanding and support during long hours of research, writing, and revisions have been invaluable in bringing this book to fruition. Without their unwavering belief in our work, this endeavor would not have been possible.

We are also grateful to the broader scientific and engineering communities whose research and innovations have paved the way for advancements in flow dynamics and heat transfer. The numerous studies, experimental investigations, and computational models that have emerged over the years have provided a strong foundation for this book. We acknowledge the pioneers in the field whose fundamental discoveries continue to inspire new research and applications.

Finally, we hope that this book serves as a valuable resource for students, researchers, and professionals working in the domains of computational fluid dynamics, thermal analysis, and numerical simulations. We look forward to future advancements and collaborations that will further enhance our understanding of complex fluid flow and heat transfer phenomena. It is our sincere aspiration that this book will contribute to the growing body of knowledge and inspire future research in these fields.

Editors
Mukesh Kumar Awasthi
Dhananjay Yadav
Paulo J. Davim

https://doi.org/10.1515/9783111661674-205

About editors

Dr. Mukesh Kumar Awasthi has holds a Ph.D. on "Viscous Correction for the Potential Flow Analysis of Capillary and Kelvin-Helmholtz Instability." He is working as an Assistant Professor in the Department of Mathematics at Babasaheb Bhimrao Ambedkar University, Lucknow, India. Dr. Awasthi is specialized in the mathematical modeling of flow problems. He has taught courses of fluid mechanics, discrete mathematics, partial differential equations, abstract algebra, mathematical methods, and measure theory to postgraduate students. He has acquired excellent knowledge in the mathematical modeling of flow problems and he can solve these problems analytically as well as numerically. He has a good grasp of the subjects like viscous potential flow, electro-hydrodynamics, magneto-hydrodynamics, heat, and mass transfer. He has excellent communication skills and leadership qualities. He is self-motivated and responds to suggestions in a more convincing manner. Dr. Awasthi has qualified National Eligibility Test (NET) conducted on all India level in the year 2008 by the Council of Scientific and Industrial Research (CSIR) and got Junior Research Fellowship (JRF) and Senior Research Fellowship (SRF) for doing research. He has published more than 135 research publications (journal articles/books/book chapters/conference articles) with Elsevier, Taylor & Francis, Springer, Emerald, World Scientific, and many other national and international journals and conferences. Also, he has published 15 books. He is also a series editor of *Artificial Intelligence and Machine Learning for Intelligent Engineering Systems* published by CRC Press (Taylor & Francis). He has attended many symposia, workshops, and conferences in mathematics as well as fluid mechanics. He has been awarded the "Research Award" consecutively four times from 2013 to 2016 by the University of Petroleum and Energy Studies, Dehradun, India. He has also received the start-up research fund for his project "Nonlinear Study of the Interface in Multilayer Fluid System" from University Grant Commission, New Delhi. He is also listed among the top 2% of influential researchers in the world, as prepared by Stanford University, based on Scopus data for the years 2022 and 2023. His Orcid is *0000-0002-6706-5226*; Google Scholar web link is *https://scholar.google.co.in/citations?user=Dj3ktGAAAAAJ*; and research gate web link is *https://www.researchgate.net/profile/Mukesh-Awasthi-2*.

Dr. Dhananjay Yadav received his Ph.D. degree from Department of Mathematics, Indian Institute of Technology (IIT) Roorkee, India in 2013 and post-graduation (M.Sc.) in Mathematics from Deen Dayal Upadhyaya University, Gorakhpur, India in 2007. Currently, he is working as an Associate Professor in the Department of Mathematics at University of Nizwa, Oman. Prior to his appointment to University of Nizwa, Oman, he had worked as Principal Research Scientist at Athabasca University, Canada; Yonsei University, South Korea; and Jeju National University, South Korea. He is a leading expert in CO_2 capture, storage and oil recovery, computational sustainability and environmental analytics, fluid mechanics, numerical analysis, hydrodynamic and hydromagnetic stability, nanofluids, and fluid flow in porous media. He has published more than 100 research articles (high impact factor) in various reputed international journals. He has been listed in the top 2% influential researchers in the world prepared by Stanford University, based on Scopus data.

Dr. J. Paulo Davim is a Full Professor at the University of Aveiro, Portugal. He is also distinguished as Honorary Professor in several universities/colleges/institutes in China, India, and Spain. He received his Ph.D. in Mechanical Engineering, MSc in Mechanical Engineering (materials and manufacturing processes), and Mechanical Engineering degree (5 years), from the University of Porto (FEUP); the Aggregate title (Full Habilitation) from the University of Coimbra; and the D.Sc. (Higher Doctorate) from London Metropolitan University. He is Fellow (FIET) of IET-London, Fellow (FIMF) of IMF-Birmingham, and Eur Ing by Engineers Europe FEANI-Brussels. He is also a Senior Specialist Engineer by the Portuguese Order of Engineers with an MBA and specialist titles in Industrial Management as well as in Metrology. He has more than 35 years of teaching and research experience in manufacturing, materials, mechanical and industrial engineering, with special emphasis in machining, tribology, and surface engineering. He has

https://doi.org/10.1515/9783111661674-206

also interest in design, management, sustainability, Industry 5.0, engineering education, and higher education for sustainability. He has guided a large numbers of postdoc, Ph.D., and master's students as well as has coordinated and participated in several financed research projects. He has received several scientific awards and honors. He has worked as evaluator of projects for European Research Council and other international research agencies as well as examiner of Ph.D. thesis for many universities in different countries. He is the Editor in Chief of several international journals, book Series Editor, and Scientific Advisory for many conferences. ORCID: https://orcid.org/0000-0002-5659-3111

List of contributing authors

Atul Kumar Shukla
Department of Mathematics, Ganga Singh College, Chapra, Bihar, India

Mukesh Kumar Awasthi
Department of Mathematics, Babasaheb Bhimrao Ambedkar University, Lucknow, Uttar Pradesh, India

Dhananjay Yadav
Department of Mathematical and Physical Sciences, University of Nizwa, Nizwa, Oman

Suleiman A. Wali
Department of Chemical Engineering, Modibbo Adama University (MAU), Yola, Adamawa State, Nigeria

Abdulhalim Musa Abubakar
Department of Chemical Engineering, Modibbo Adama University (MAU), Yola, Adamawa State, Nigeria

Hayatuddeen Abubakar
Department of Chemical Engineering, Modibbo Adama University (MAU), Yola, Adamawa State, Nigeria

Semiu Adebayo Kareem
Department of Chemical Engineering, Modibbo Adama University (MAU), Yola, Adamawa State, Nigeria

Mohamed Ellouze
Physics Department, Faculty of Sciences of Sfax, University of Sfax, Tunisia

Reshu Gupta
Applied Science Cluster (Mathematics), UPES, Dehradun, Uttarakhand, India

Hemant Yadav
Mechanical Department, Roorkee Institute of Technology, Roorkee, Uttarakhand, India

Sudipta Priyadarshini
Department of Basic Science and Humanities, Vignan's Institute of Information Technology (A), Vishakhapatnam, Andhra Pradesh, India

Sukanta Nayak
Department of Mathematics, School of Advanced Sciences, VIT-AP University, Amaravati, Andhra Pradesh, India

Alok Dhaundiyal
Centre for Energy Research, Konkoly utca. 29-33,1121 Budapest, Hungary

Laszlo Toth
Institute of Technology, Hungarian University of Agriculture and Life Sciences, Hungary

Satyvir Singh
Institute for Applied and Computational Mathematics, RWTH Aachen University, Germany

Sunil
Department of Mathematics and Scientific Computing, National Institute of Technology Hamirpur, Himachal Pradesh, India

Vishal Chandel
Department of Mathematics and Scientific Computing, National Institute of Technology Hamirpur, Himachal Pradesh, India

Ashok K. Barik
Department of Mechanical Engineering, National Institute of Technology, Jamshedpur, Jharkhand, India

Prafulla K. Swain
Department of Mechanical Engineering, GITAM Institute of Technology, Visakhapatnam, Andhra Pradesh, India

Mohamed M. Awad
Mechanical Power Engineering Department, Faculty of Engineering, Mansoura University, Egypt

https://doi.org/10.1515/9783111661674-207

Mohammad Hassan
Mathematics and Scientific Computing Department, Madan Mohan Malaviya University of Technology, Gorakhpur, Uttar Pradesh, India

Lham Tashi
Department of Mathematics, North Eastern Regional Institute of Science and Technology, Arunachal Pradesh, India

Himanshu Upreti
School of Engineering and Technology, BML Munjal University, Gurugram, Haryana, India

Ankita Pandey
Department of Mathematics, GLA University, Mathura, Uttar Pradesh, India

Alok Kumar Pandey
Department of Mathematics, Graphic Era (Deemed to be University), Dehradun, Uttarakhand, India

Afsana Mammadova
College of Mechanics and Engineering Sciences, Hohai University, Nanjing, Jiangsu, China

Yashar Aryanfar
Department of Chemical Engineering, National Cheng Kung University, Taiwan
Thermo-Fluids Research Group, Khazar University, Azerbaijan

Ali Keçebaş
Department of Energy Systems Engineering, Technology Faculty, Muğla Sıtkı Koçman University, Turkey

Jorge Luis García Alcaraz
Department of Industrial Engineering and Manufacturing, Universidad Autónoma de Ciudad Juárez, Mexico

P. Balakrishnan
Department of Mechanical Engineering, SRM Institute of Science and Technology, Tamil Nadu, India

S. Senthilraja
Department of Mechatronics Engineering, SRM Institute of Science and Technology, Tamil Nadu, India

R. Gangadevi
Department of Mechatronics Engineering, SRM Institute of Science and Technology, Tamil Nadu, India

M. Baskaran
Department of Mechatronics Engineering, KS Rangasamy College of Technology, Tamil Nadu, India

N. Saravanan
Department of Mechanical Engineering, Erode Sengunthar Engineering College, Tamil Nadu, India

Atul Kumar Shukla, Mukesh Kumar Awasthi*, and Dhananjay Yadav

Chapter 1 Introduction to flow dynamics and heat transfer

Abstract: This chapter provides a comprehensive introduction to the fundamental principles of flow dynamics and heat transfer, exploring their historical development, key concepts, and applications in engineering and science. Flow dynamics, or fluid mechanics, examines the behavior of fluids under various conditions, focusing on properties like density, viscosity, and compressibility, as well as flow regimes such as laminar and turbulent flow. Heat transfer, on the other hand, deals with the mechanisms of thermal energy exchange, including conduction, convection, and radiation. The chapter highlights the interdependence of these fields, particularly in applications like forced and natural convection, boiling, and condensation. It also discusses the governing equations, such as the Navier-Stokes equations and Fourier's law, and introduces computational fluid dynamics (CFD) as a powerful tool for simulating complex fluid flow and heat transfer problems. The chapter concludes with a discussion of emerging trends, including the integration of AI and machine learning in CFD, and the importance of multiphysics and multiscale modeling in modern engineering.

Keywords: flow dynamics, heat transfer, computational fluid dynamics (CFD), Navier-Stokes equations, convection

1.1 Overview of flow dynamics and heat transfer

1.1.1 Definition and importance

Flow dynamics and heat transfer are fundamental concepts in engineering and physics that describe the behavior of fluids and the transfer of thermal energy [1, 2]. Fluid dynamics, or fluid mechanics, focuses on the motion of fluids, whether they are liquids or gases. It explores how fluids behave under various conditions, such as pressure, temperature, and velocity. Key concepts include fluid properties like density, viscosity, and compressibility, as well as flow regimes like laminar and turbulent flow. Conservation laws of mass, momentum, and energy are essential in analyzing fluid

*Corresponding author: **Mukesh Kumar Awasthi**, Department of Mathematics, Babasaheb Bhimarao Ambedkar University, Lucknow, Uttar Pradesh, India, e-mail: mukeshiitr.kumar@gmail.com
Atul Kumar Shukla, Department of Mathematics, Ganga Singh College, Chapra 841301, Bihar, India
Dhananjay Yadav, Department of Mathematical and Physical Sciences, University of Nizwa, Oman

https://doi.org/10.1515/9783111661674-001

flow, and boundary layer theory helps understand the behavior of fluids near solid surfaces. Dimensional analysis is a powerful tool for scaling and analyzing flow problems.

Heat transfer, on the other hand, is concerned with the mechanisms by which thermal energy is exchanged between systems [3]. There are three primary modes of heat transfer: conduction, convection, and radiation. Conduction occurs within solid materials due to temperature differences, while convection involves heat transfer between a solid surface and a moving fluid. Radiation, the third mode, transfers heat through electromagnetic waves. Key concepts in heat transfer include thermal conductivity, heat transfer coefficient, and radiative properties like emissivity and absorptivity. The two fields of flow dynamics and heat transfer are deeply intertwined. For instance, forced convection involves using fluid flow to enhance heat transfer from a surface, while natural convection relies on buoyancy forces driven by temperature differences. Boiling and condensation, phase change processes, also involve both heat transfer and fluid flow. Computational fluid dynamics (CFD) is a powerful tool for simulating and analyzing complex fluid flow and heat transfer problems [4].

1.1.2 Historical development and milestones

The historical development of flow dynamics and heat transfer is intertwined with the advancement of human civilization. Early civilizations, such as Egyptian and Greek, made significant contributions to the understanding of fluid flow through their irrigation systems, aqueducts, and the work of Archimedes on buoyancy. The Renaissance period saw a surge in scientific inquiry, with figures like Leonardo da Vinci and Galileo Galilei making substantial advancements. Da Vinci's detailed studies of fluid flow laid the foundation for future research, while Galileo's experiments on falling bodies contributed to the understanding of gravity and its impact on fluid motion. The eighteenth century marked a pivotal moment with the development of Bernoulli's principle by Daniel Bernoulli. This principle, which relates fluid pressure, velocity, and height, became a cornerstone of fluid mechanics. Additionally, the work of Leonhard Euler led to the formulation of the Euler equations, which describe the motion of inviscid fluids.

The nineteenth century witnessed further breakthroughs, with Claude-Louis Navier and George Gabriel Stokes independently developing the Navier-Stokes equations, which account for the effects of viscosity. These equations are fundamental to modern fluid dynamics and are used to model a wide range of fluid flow phenomena. In the field of heat transfer, Joseph Fourier's work on heat conduction in the early nineteenth century was groundbreaking. His mathematical framework, known as Fourier's law, provides the foundation for analyzing heat transfer through solids. The development of thermodynamics in the nineteenth century, particularly the work of Sadi Carnot, further enhanced the understanding of heat transfer processes. The twentieth century saw a rapid acceleration in the development of both flow dynamics and heat transfer. Ludwig Prandtl's introduction of boundary layer theory revolutionized the understanding of fluid flow near solid surfaces. The advent of CFD in the latter half of the twentieth century enabled the numerical simulation of complex fluid flow and heat transfer problems, leading to significant advancements in various engineering fields.

Today, flow dynamics and heat transfer continue to be active areas of research, with applications in fields ranging from aerospace and automotive engineering to climate science and biomedical engineering. As technology advances, we can expect further breakthroughs in these fields, leading to innovative solutions to global challenges.

1.1.3 Applications in engineering and science

The applications of flow dynamics and heat transfer are vast and diverse. In aerospace engineering, these principles are crucial for designing aircraft and spacecraft. Mechanical engineers rely on them for designing engines [7], heat exchangers, and heating, ventilation, and air conditioning (HVAC) systems. Civil engineers use these concepts to analyze fluid flow in rivers, canals, and pipes [5, 6]. Chemical engineers apply them in designing chemical reactors and separation processes. Even in biomedical engineering, understanding blood flow and heat transfer in the human body is essential. By comprehending the fundamentals of flow dynamics and heat transfer, engineers and scientists can design more efficient, reliable, and sustainable systems.

1.2 Fundamental principles of flow dynamics

1.2.1 Fluid properties

1.2.1.1 Density, viscosity, and compressibility

Density refers to how much mass is packed into a given volume of a substance. It essentially measures how tightly packed the matter is. Viscosity is a measure of a fluid's resistance to flow, determining how easily it flows. High viscosity means a fluid flows slowly (like honey), while low viscosity means it flows easily (like water). Compressibility measures how much a substance's volume changes under pressure. Gases are highly compressible, while liquids are generally considered incompressible.

1.2.1.2 Surface tension and capillarity

Surface tension is a property of liquids that arises from the cohesive forces between molecules at the liquid-air interface. These cohesive forces pull the surface molecules inward, creating a tension that acts like a thin elastic film. This tension is responsible for phenomena like the formation of spherical droplets and the ability of insects to walk on water. Capillarity, also known as capillary action, is the ability of a liquid to flow in narrow spaces against gravity. This phenomenon occurs due to the interplay between adhesive forces (between the liquid and the solid surface) and cohesive forces within the liquid. When a liquid is placed in a narrow tube, the adhesive forces pull the liquid up the sides of the tube, while the cohesive forces hold the liquid together. This combined effect results in the liquid rising above its normal level, defying gravity. Capillary action is essential in many natural processes, such as the movement of water through plant stems and the absorption of liquids by paper towels.

1.2.2 Governing equations

1.2.2.1 Continuity equation

The continuity equation in fluid mechanics is a fundamental principle that states that the mass flow rate of a fluid remains constant throughout a pipe or duct (Figure 1.1), regardless of changes in cross-sectional area. In simpler terms, it means that the amount of fluid entering a section of pipe must equal the amount of fluid leaving that section.

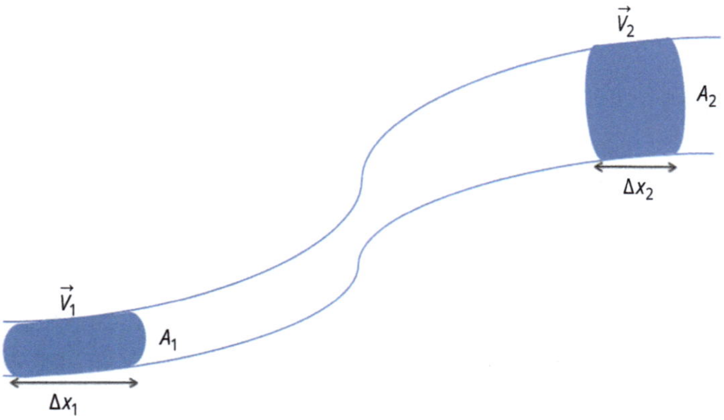

Figure 1.1: Flow through pipe.

Mathematically, the continuity equation is expressed as follows:

$$A_1\vec{V}_1 = A_2\vec{V}_2 \tag{1.1}$$

where A_1 and A_2 are the cross-sectional areas at two points in the pipe and $\vec{V}_1$ and $\vec{V}_2$are the fluid velocities at those two points.

This equation implies that if the cross-sectional area decreases, the fluid velocity must increase to maintain a constant mass flow rate. Conversely, if the area increases, the velocity must decrease. This principle has numerous applications in engineering, such as designing pipelines, nozzles, and other fluid flow systems.

1.3 Momentum equation (Navier-Stokes equations)

The Navier-Stokes equations are a set of partial differential equations that describe the motion of viscous fluids. They are fundamental in fluid mechanics and are used to model a wide range of fluid flow phenomena, from simple pipe flow to complex turbulent flows.

The equations are derived from Newton's second law of motion applied to fluid elements, taking into account the forces acting on the fluid, including pressure forces, viscous forces, and external forces. The Navier-Stokes equations are often coupled with the continuity equation, which expresses the conservation of mass.

Mathematically,

$$\rho\left(\frac{\partial\vec{V}}{\partial t} + (\vec{V}\cdot\nabla)\vec{V}\right) = \rho g - \nabla p + \nabla\cdot\left\{\mu\left[\nabla\vec{V} + \left(\nabla\vec{V}\right)^T - \frac{2}{3}\left(\nabla\cdot\vec{V}\right)I\right]\right\} + \nabla\left[\varsigma\left(\nabla\cdot\vec{V}\right)\right] \tag{1.2}$$

While the Navier-Stokes equations are relatively simple to write down, solving them analytically is challenging, except for a few special cases. For most practical applications, numerical methods are employed to solve the equations using powerful computers. This field of study is known as CFD.

Despite their complexity, the Navier-Stokes equations remain a cornerstone of fluid mechanics, providing a powerful tool for understanding and predicting the behavior of fluids in various engineering applications.

1.4 Energy equation

The energy equation in fluid mechanics is a statement of the conservation of energy principle applied to fluid flow. It states that the total energy of a fluid flow remains constant along a streamline, neglecting any energy losses due to friction or other factors.

The general form of the energy equation is

$$\frac{P_1}{\rho g} + \frac{\vec{V}_1^2}{2g} + z_1 + h_p = \frac{P_2}{\rho g} + \frac{\vec{V}_2^2}{2g} + z_2 + h_L \tag{1.3}$$

where P_1 and P_2 are the pressures at two points along the streamline, ρ is the density of the fluid, g is the acceleration due to gravity, and $\vec{V}_1$ and $\vec{V}_2$ are the velocities of the fluid at the two points. z_1 and z_2 are the elevations of the two points. $h_p$is the energy added to the fluid and z_2 is the energy lost due to friction and other factors.

This equation essentially states that the sum of pressure energy, kinetic energy, potential energy, and added energy at one point in the flow is equal to the sum of these energies at another point, accounting for energy losses.

The energy equation is a powerful tool for analyzing a wide range of fluid flow problems including pipe flow, open channel flow, and turbomachinery.

1.5 Classification of flows

1.5.1 Laminar and turbulent flows

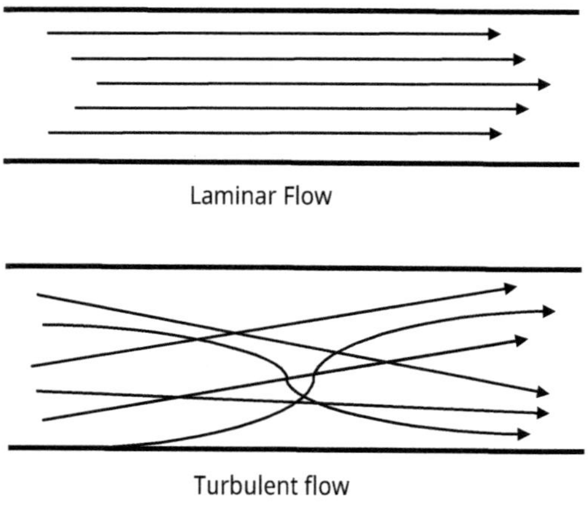

Figure 1.2: Classification of laminar and turbulent flows.

Fluid flow can be classified into two primary types (Figure 1.2): laminar and turbulent [10]. In laminar flow, fluid particles move in smooth, parallel layers, much like playing cards sliding across each other. This type of flow is typically observed at low velocities or in fluids with high viscosity. It is characterized by its predictability and low energy dissipation due to minimal friction. On the other hand, turbulent flow is chaotic and irregular. Fluid particles move in random, swirling patterns, leading to significant mixing and energy dissipation. This type of flow is often observed at high velocities or in fluids with low viscosity. Turbulent flow is much more complex to model and predict due to its inherent randomness. The transition from laminar to turbulent flow is influenced by the Reynolds number, a dimensionless quantity that balances inertial and viscous forces. As the Reynolds number increases, the flow becomes more prone to transition to turbulence. Understanding the distinction between laminar and turbulent flows is crucial in various engineering applications, from designing efficient pipelines to optimizing aerodynamic performance.

1.5.2 Compressible and incompressible flows

The distinction between compressible and incompressible flows lies in the fluid's density [8, 9]. In compressible flows, the density of the fluid changes significantly during the flow process. This is typically observed in high-speed flows, such as those involving aircraft or rockets, where the fluid's velocity approaches the speed of sound. The compressibility effects become significant when the flow Mach number exceeds approximately 0.3. In contrast, incompressible flows are those where the density remains essentially constant throughout the flow. This assumption simplifies the governing equations of fluid mechanics, making them easier to solve. Most everyday flows, such as water flowing through pipes or air flowing around a car at low speeds, can be considered incompressible. While no fluid is truly incompressible, the assumption of incompressibility is often valid for many practical applications. However, when dealing with high-speed flows or flows involving significant pressure or temperature changes, the effects of compressibility must be taken into account.

1.5.3 Internal and external flows

In fluid mechanics, flows are broadly categorized into two types: internal and external.

Internal flows occur within confined boundaries, such as pipes, ducts, or channels [11]. The fluid is completely enclosed by solid surfaces, and the flow is primarily driven by pressure differences [12]. Examples of internal flows include water flowing through pipes, air flowing through ventilation ducts, and blood flowing through blood vessels.

Key characteristics of internal flows include:

- **Fully developed flow:** After a certain distance, the flow profile becomes fully developed, meaning the velocity profile remains constant along the pipe length.
- **Pressure-driven flow:** The flow is driven by a pressure gradient along the pipe.
- **Boundary layer development:** A boundary layer forms near the pipe wall, where the fluid velocity gradually decreases to zero at the wall.

External flows [13] occur when a fluid flows over a solid surface that is not completely enclosed. The fluid is not confined by solid boundaries on all sides. Examples of external flows include air flowing over an airplane wing, water flowing around a ship hull, and wind blowing over a building.

Key characteristics of external flows:

- **Boundary layer development:** A boundary layer forms near the solid surface, where the fluid velocity gradually decreases to zero at the wall.[7]
- **Pressure and viscous forces:** The flow is influenced by both pressure forces and viscous forces.
- **Wake region:** A wake region forms behind the object, where the flow is disturbed and turbulent.

1.6 Fundamentals of heat transfer

1.6.1 Modes of heat transfer

1.6.1.1 Conduction

Conduction (Figure 1.3) is a mode of heat transfer that occurs within a solid material due to a temperature difference [14, 15]. When one part of a material is heated, the thermal energy is transferred to the neighboring atoms or molecules through vibrations and collisions. This process continues throughout the material until thermal equilibrium is reached. The rate of heat conduction depends on various factors, including the material's thermal conductivity, the temperature difference[1]across the material, and the material's geometry. Materials with high thermal conductivity, such as metals, conduct heat efficiently, while materials with low thermal conductivity, such as insulators, conduct heat poorly. Conduction is a fundamental principle in many engineering applications, such as heat exchangers, thermal insulation, and electronic cooling. Understanding the principles of conduction is essential for designing efficient and energy-saving systems.

1.6.1.2 Convection

Convection is a mode of heat transfer that involves the movement of a fluid, such as a liquid or gas [16]. It occurs when a portion of the fluid is heated, causing its density to decrease. The less dense fluid rises, while the cooler, denser fluid sinks to take its place. This creates a continuous circulation of the fluid, transferring heat from the warmer regions to the cooler regions. There are two main types of convection: natural convection and forced convection. Natural convection occurs due to density differences caused by temperature variations within the fluid [17]. For example, the rising of warm air from a radiator or the cooling of a hot cup of coffee are examples of natural convection. Forced convection involves the use of external forces, such as fans or pumps, to circulate the fluid. This enhances the rate of heat transfer compared to natural convection. Forced convection is widely used in applications like air conditioning systems, heat exchangers, and electronic cooling. Understanding the principles of convection is crucial in various engineering applications, including thermal management in electronic devices, HVAC systems, and industrial processes. Heat transfer through convection is commonly utilized by researchers in studying two-phase flow within 2D and 3D configurations [34–36].

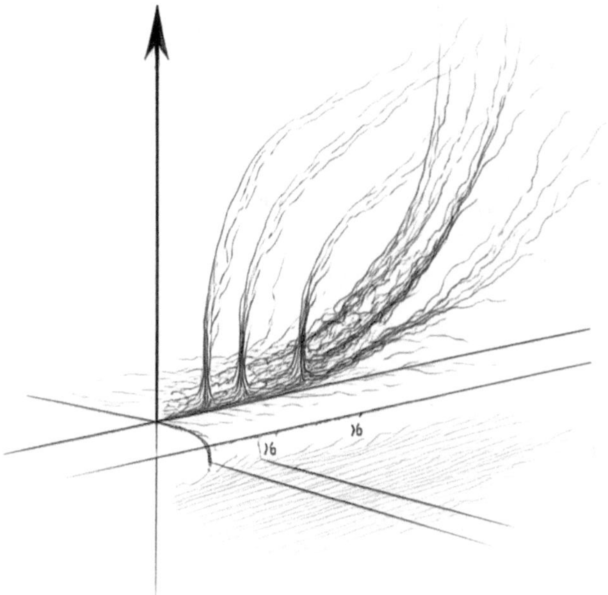

Figure 1.3: Heat convection.

1.6.1.3 Radiation

Radiation is a mode of heat transfer that occurs through electromagnetic waves [18]. Unlike conduction and convection, radiation does not require a medium to propagate. This means that heat can be transferred through a vacuum, as demonstrated by the Sun warming the Earth. All objects emit electromagnetic radiation, the intensity of which depends on their temperature. As an object's temperature increases, it emits more radiation at shorter wavelengths. This is why hot objects glow red-hot or even white-hot. The rate of heat transfer by radiation depends on several factors, including the temperature difference between the objects, their emissivity (a measure of how efficiently an object emits radiation), and their surface area. Radiation plays a significant role in various natural and engineering processes. It is responsible for the Earth's climate, the heating of buildings, and the operation of solar panels. Understanding the principles of radiation is essential for designing efficient and energy-saving systems.

1.6.2 Governing equations of heat transfer

1.6.2.1 Fourier's law of heat conduction

Fourier's law of heat conduction is a fundamental principle in heat transfer that describes the rate at which heat is conducted through a material [19]. It states that the rate of heat transfer through a material is proportional to the negative temperature gradient and the area perpendicular to the heat flow.

Mathematically, Fourier's law is expressed as:

The diagram illustrates Fourier's law of heat conduction, which states that the rate of heat transfer through a material is proportional to the negative gradient of the temperature and the area perpendicular to the direction of heat transfer:

$$Q = -kA\left(\frac{dT}{dx}\right) \tag{1.4}$$

where Q is the rate of heat transfer, K is the thermal conductivity of the material, A is the cross-sectional area perpendicular to the heat flow, and $\frac{dT}{dx}$ is the temperature gradient.

The negative sign indicates that heat flows from a region of higher temperature to a region of lower temperature.

1.7 Newton's law of cooling

Newton's law of cooling states that the rate of heat loss of a body is directly proportional to the temperature difference between the body and its surroundings [20].

Mathematically, it can be expressed as follows:

$$\frac{dT}{dt} = -k(T - T_o) \tag{1.5}$$

where dT/dt is the rate of change of temperature of the body with respect to time, k is a constant of proportionality, known as the cooling coefficient, and T is the temperature of the body at time t, and T_o is the temperature of the surroundings.

This law is particularly useful in determining the cooling rate of objects, such as hot coffee or a warm object exposed to a cooler environment. By understanding this law, engineers and scientists can analyze and predict the thermal behavior of various systems.

1.8 Stefan-Boltzmann law for radiation

The Stefan-Boltzmann law describes the total amount of electromagnetic radiation emitted by a black body at a given temperature [21]. It states that the total energy radiated per unit surface area per unit time (also known as the radiant exitance) is directly proportional to the fourth power of the black body's absolute temperature.

Mathematically, the Stefan-Boltzmann law is expressed as

$$P = \sigma A \varepsilon T^4$$

where P is the total power radiated, σ is the Stefan-Boltzmann constant (approximately P represents the total radiant power emitted by a body due to thermal radiation), A is the surface area of the black body, and ε is the emissivity of the surface (a dimensionless number between 0 and 1, representing the efficiency of radiation).

A black body is an idealized physical body that absorbs all incident electromagnetic radiation, regardless of frequency or angle of incidence. While a perfect black body does not exist in reality, many objects can be approximated as black bodies under certain conditions.

The Stefan-Boltzmann law is essential in various fields, including physics, astronomy, and engineering. It helps us understand the energy balance of stars, the heat transfer in thermal systems, and the design of energy-efficient buildings and devices.

1.9 Heat transfer in fluids

1.9.1 Heat conduction in stationary fluids

Heat conduction in stationary fluids occurs primarily through molecular interactions [22]. When a temperature difference exists within a fluid, the molecules with higher kinetic energy in the hotter regions collide with neighboring molecules, transferring some of their energy. This process continues, leading to a net transfer of energy from the hotter to the colder regions.

1.9.1.1 Key factors influencing heat conduction in stationary fluids

- **Thermal conductivity:** The thermal conductivity of a fluid determines its ability to conduct heat.[4]Fluids with higher thermal conductivity, such as metals, conduct heat more efficiently than fluids with lower thermal conductivity, like gases.
- **Temperature gradient:** The rate of heat conduction is directly proportional to the temperature gradient, which is the change in temperature per unit distance. A larger temperature difference results in a higher rate of heat transfer.
- **Fluid properties:** Fluid properties such as density, viscosity, and specific heat capacity can influence heat conduction. For example, denser fluids generally have higher thermal conductivities.

1.9.1.2 Applications of heat conduction in stationary fluids

- **Thermal insulation:** Materials with low thermal conductivity are used to reduce heat loss or gain, such as insulation in buildings and pipes [23].
- **Heat exchangers:** Heat exchangers rely on heat conduction to transfer thermal energy between two fluids [24].
- **Electronic cooling:** Heat sinks and thermal paste are used to dissipate heat generated by electronic components through conduction.

1.10 Convective heat transfer in moving fluids

Convective heat transfer is a mechanism of heat transfer that involves the combined effects of conduction and fluid flow [25]. When a fluid, such as air or water, flows over a surface with a different temperature, heat is transferred between the surface and the fluid.

1.10.1 Types of convective heat transfer

1. **Forced convection:** In forced convection, the fluid flow is driven by an external force, such as a pump or fan. Air cooling of electronic components with a fan and water cooling in car radiators. The rate of heat transfer in forced convection is significantly higher than in natural convection due to the increased fluid velocity.
2. **Natural convection:** In natural convection, the fluid flow is driven by buoyancy forces caused by density differences resulting from temperature variations. The rising of hot air from a radiator and the cooling of a hot cup of coffee are examples of natural convection. Natural convection is slower than forced convection but still plays a significant role in many heat transfer processes.

1.10.2 Key factors affecting convective heat transfer

- **Fluid properties:** Density, viscosity, and thermal conductivity of the fluid influence the rate of heat transfer.
- **Flow velocity:** Higher fluid velocity leads to increased heat transfer rates.
- **Surface geometry:** The shape and orientation of the surface affect the flow pattern and heat transfer.
- **Temperature difference:** A larger temperature difference between the surface and the fluid results in a higher rate of heat transfer.

Heat exchangers that are used in various industries to transfer heat between fluids, electronic cooling, HVAC systems, and meteorology are some applications of convective heat transfer.

1.11 Interdependence of flow dynamics and heat transfer

1.11.1 Coupled fluid flow and thermal analysis

Coupled fluid flow and thermal analysis is a complex field that involves the simultaneous consideration of fluid mechanics and heat transfer principles. In many engineering applications, these two phenomena are intricately linked, and understanding their interplay is crucial for accurate design and optimization.

Fluid flow and heat transfer involve several key mechanisms. Convective heat transfer occurs when fluids flow over surfaces, exchanging heat with the surface through convection. The rate of heat transfer is influenced by factors such as fluid velocity, temperature differences, and fluid properties. In forced convection, the fluid flow is driven by external means like pumps or fans, while in natural convection, the flow arises from buoyancy forces caused by temperature differences within the fluid. Coupling mechanisms further impact these processes. Temperature-dependent fluid properties, such as density, viscosity, and thermal conductivity, vary with temperature, influencing both the flow field and heat transfer rates. Heat generation within the fluid, from sources like chemical reactions or electrical resistance, also affects temperature distribution and flow patterns. Additionally, phase-change phenomena, such as boiling and condensation, can significantly alter fluid flow and heat transfer dynamics. CFD simulations can be used to model and analyze complex fluid flow and heat transfer problems. By coupling fluid flow solvers with thermal analysis tools, engineers can accurately predict temperature distributions, heat fluxes, and pressure drops in various systems.

Coupled fluid flow and thermal analysis find applications across various fields. In electronic cooling, these analyses are essential for designing efficient cooling systems to prevent overheating in electronic devices. In automotive engineering, they are used to optimize engine cooling systems, heat exchangers, and aerodynamic designs. Aerospace engineering relies on such analyses to evaluate heat transfer and fluid flow in aircraft and spacecraft components. Energy systems benefit from these techniques in the design of efficient heat exchangers, boilers, and turbines. Additionally, in biomedical engineering, they are applied to model blood flow and heat transfer processes within the human body, contributing to advancements in medical technology and treatments.

1.11.2 Importance of coupled fluid flow and thermal analysis in multiphysics problems

Coupled fluid flow and thermal analysis is a critical aspect of multiphysics simulations, where multiple physical phenomena interact and influence each other. This coupling is essential for accurately modeling and predicting the behavior of complex systems in various engineering fields. The importance can be understood from the following points:

1. **Real-world accuracy:** Many real-world systems involve the simultaneous occurrence of fluid flow and heat transfer. By coupling these two phenomena, simulations can more accurately replicate real-world conditions.
2. **Optimized design:** Understanding the interplay between fluid flow and heat transfer enables engineers to optimize designs for better performance and efficiency. For example, in electronic cooling, efficient heat dissipation can significantly improve the performance and lifespan of devices.
3. **Predicting system behavior:** Coupled simulations can help predict the behavior of systems under various operating conditions, allowing for early identification of potential issues and optimization of design parameters.

4. **Reducing experimental costs:** By conducting virtual simulations, engineers can reduce the need for costly physical experiments, saving time and resources.
5. **Enabling innovation:** Coupled simulations can help explore innovative designs and materials that would be difficult or impossible to test experimentally.

1.11.3 Challenges in modeling and simulation

Modeling of fluid flow and heat transfer involves challenges across multiple areas. Complex physics like turbulence, multiphase flows, and phase-change phenomena are difficult to model due to their chaotic and intricate nature. Accurate simulations require accounting for material properties that vary with temperature and may have uncertainties. Defining realistic boundary and initial conditions are critical for accuracy, especially in complex geometries. Numerical methods face issues like instability, mesh quality, and high computational costs. Validation and verification, including experimental comparison and grid independence studies, are essential for reliability.

1.12 Mathematical modeling of flow dynamics and heat transfer

1.12.1 Conservation laws

Conservation laws are fundamental principles in physics that govern the behavior of matter and energy. In fluid mechanics, these laws are applied to understand and analyze the flow of fluids. The three primary conservation laws in fluid mechanics are:

- **Conservation of mass**
  This law states that mass cannot be created or destroyed. In fluid flow, it means that the mass flowing into a control volume must equal the mass flowing out plus any accumulation within the control volume [26]. Mathematically, it is expressed as:

  $$\frac{\partial \rho}{\partial t} + \nabla \cdot (\rho \vec{V}) = 0$$

  where ρ is the fluid density and $\vec{V}$ is the fluid velocity vector.
- **Conservation of momentum**
  This law is based on Newton's second law of motion. It states that the rate of change of momentum of a fluid element is equal to the sum of the forces acting on it [27]. The momentum equation is a vector equation, representing the balance of forces in three dimensions:

$$\rho\left(\frac{\partial \vec{V}}{\partial t}+\left(\vec{V}\cdot\nabla\right)\vec{V}\right)=-\nabla p+\nabla\cdot\tau$$

where p is the pressure and τ is the viscous stress tensor.

– **Conservation of energy**
  This law states that energy cannot be created or destroyed, only converted from one form to another. In fluid flow, it accounts for the balance of kinetic energy, potential energy, and internal energy:

$$\rho\left(\frac{\partial E}{\partial t}+\left(\vec{V}\cdot\partial\right)E\right)=-\nabla\vec{V}+\tau\nabla\vec{V}$$

where E is the specific internal energy and $\vec{V}$ is the heat flux vector.

1.12.2 Dimensionless numbers in flow and heat transfer

Dimensionless numbers are ratios of physical quantities that are used to characterize fluid flow and heat transfer phenomena. They help scale and compare different flow regimes and heat transfer processes. Here are some of the most important dimensionless numbers:

1.12.2.1 Reynolds number (Re)

Reynolds number is a dimensionless quantity that characterizes the ratio of inertial forces to viscous forces in a fluid flow [28]. It is defined as follows:

$$\mathrm{Re}=\frac{\rho\vec{V}d}{\mu}$$

wher$d$ is the fluid density, $\vec{V}$ is the fluid velocity, d is the characteristic length, and μ is the dynamic viscosity of the fluid.

A high Reynolds number indicates turbulent flow, while a low Reynolds number indicates laminar flow.

1.12.2.2 Prandtl number (Pr)

Prandtl number is a dimensionless quantity that characterizes the relative importance of momentum diffusion (viscosity) to thermal diffusion (thermal conductivity) in a fluid. It is defined as follows:

$$\mathrm{Pr}=\frac{\upsilon}{a}$$

where υ is the kinematic viscosity of the fluid and α is the thermal diffusivity of the fluid.

Prandtl number helps to predict the relative thickness of the velocity and thermal boundary layers in a fluid flow.

1.12.2.3 Nusselt number (Nu)

Nusselt number is a dimensionless quantity that characterizes the enhancement of heat transfer due to convection relative to conduction. It is defined as follows:

$$\mathrm{Nu} = \frac{hL}{k}$$

where h is the convective heat transfer coefficient, L is the characteristic length and k is the thermal conductivity of the fluid.

A high Nusselt number indicates efficient convective heat transfer.

These dimensionless numbers are widely used in engineering and scientific research to analyze and predict the behavior of fluid flow and heat transfer systems. By understanding these numbers, engineers can design more efficient and effective systems.

1.12.3 Assumptions and simplifications in models

In fluid mechanics and heat transfer, models are often simplified to make the analysis tractable. These simplifications involve making certain assumptions about the physical system. Some common assumptions and simplifications include:

- **Incompressible flow**
  This assumption is valid for most liquids and gases at low Mach numbers. Incompressibility simplifies the governing equations, making them easier to solve.
- **Steady-state flow**
  This assumption assumes that the flow conditions do not change with time. Steady-state flow leads to time-independent equations, reducing the complexity of the problem.
- **One-dimensional flow**
  This assumption reduces the problem to a one-dimensional analysis, simplifying the governing equations. One-dimensional models require less computational effort compared to multidimensional models.
- **Negligible viscous effects**
  This assumption neglects the effects of viscosity, leading to simpler equations. This assumption is valid for high-speed flows where viscous effects are minimal.

- **Constant properties**
  This assumption simplifies the analysis by assuming that fluid properties like density, viscosity, and thermal conductivity are constant. Constant property assumptions lead to simpler equations and easier solutions.
- **Negligible body forces**
  This assumption is valid for many engineering applications where gravity forces are small compared to other forces. Neglecting body forces simplifies the governing equations.
- **Ideal gas law**
  This assumption relates pressure, temperature, and density for ideal gases. Ideal gas behavior is a good approximation for many gases at low pressures and high temperatures.

1.13 Simulation techniques for flow dynamics and heat transfer

1.13.1 Computational fluid dynamics (CFD)

1.13.1.1 Key steps in CFD simulations

In CFD, the simulation process follows a systematic workflow. First, geometry creation and meshing involve creating the problem's geometry using CAD software and discretizing it into a mesh of elements, where mesh quality is crucial for accuracy and convergence. Next, physics modeling defines the problem's physical aspects, including governing equations (e.g., Navier-Stokes for fluid flow, energy equations for heat transfer), turbulence models, and material properties like density and viscosity.

Boundary and initial conditions specify the conditions at domain boundaries (e.g., inlet, outlet, and walls) and initial variables (e.g., velocity, pressure, and temperature). In solver setup, a numerical method (finite difference, finite volume, or finite element) is selected, and solver settings like time steps and convergence criteria are configured. In solution and convergence, the solver iteratively solves the equations until the solution meets accuracy criteria, ensuring reliability.

Finally, post-processing involves extracting data (e.g., velocity, pressure, temperature), visualizing results using contour and vector plots, and analyzing insights into fluid flow and heat transfer. This structured approach ensures accurate and meaningful CFD simulations.

1.13.1.2 Numerical schemes and solvers

Numerical schemes and solvers are essential tools in CFD to approximate the solution of the governing equations, which are often complex differential equations.

1.13.1.3 Numerical schemes

Numerical schemes are algorithms used to discretize the governing equations into algebraic equations. Some common numerical schemes used in CFD include:
- **Finite difference method (FDM):**
  - Approximates derivatives using Taylor series expansions.
  - Simple to implement but limited to structured grids.
- **Finite volume method (FVM):**
  - Conserves physical quantities like mass, momentum, and energy over control volumes.
  - More flexible than FDM and can handle complex geometries.
- **Finite element method (FEM):**
  - Discretizes the domain into elements and approximates the solution using basis functions.
  - Well-suited for complex geometries and boundary conditions.

Solvers are algorithms used to solve the system of algebraic equations resulting from the discretization process. Some common solvers include:
- **Iterative solvers:**
  - **Jacobi method:** Simple but slow convergence.
  - **Gauss-Seidel method:** Faster than Jacobi but still iterative.
  - **Successive over-relaxation (SOR):** Accelerates convergence by introducing a relaxation factor.
- **Direct solvers:**
  - **LU decomposition:** Solves the system of equations directly by factorizing the coefficient matrix.
  - Efficient for smaller systems but can be computationally expensive for large systems.

Grid quality, convergence criteria, boundary conditions, and turbulence modeling are some key considerations in numerical simulations.

1.13.2 Thermal analysis simulations

- **Methods for conduction, convection, and radiation**

1.13.2.1 Conduction

Analytical methods for conduction are:
- **Fourier's law:** For simple geometries like slabs, cylinders, and spheres.
- **Heat conduction equation:** For more complex geometries and transient conditions.

Numerical methods for the conduction are:
- **FDM:** Discretizes the domain into grid points and approximates derivatives.
- **FEM:** Discretizes the domain into elements and solves the weak form of the governing equations.
- **FVM:** Conserves physical quantities over control volumes.

1.13.2.2 Convection

Some analytical methods for convection are:
- **Similarity solutions:** For specific cases like boundary layer flows and free convection.
- **Perturbation methods:** For small perturbations from known solutions.

Some numerical methods for the convection are:
- **CFD:** Solves the Navier-Stokes equations coupled with the energy equation.
- **Boundary layer methods:** Simplifies the governing equations for boundary layer flows.

1.13.2.3 Radiation

Some analytical methods for radiation are:
- **View factor method:** Calculates the radiative exchange between surfaces.
- **Radiation heat transfer coefficient:** Simplifies the radiative heat transfer equation for simple geometries.

Some numerical methods for the radiation:
- **Monte Carlo ray tracing:** Tracks the paths of individual photons to simulate radiation heat transfer.
- **FEM:** Solves the radiative transfer equation.

1.13.3 Emerging simulation approaches

1.13.3.1 AI and machine learning in CFD and heat transfer

Artificial intelligence (AI) and machine learning (ML) have emerged as powerful tools to revolutionize the fields of CFD and heat transfer. By leveraging these technologies, researchers and engineers can significantly enhance the accuracy, efficiency, and predictive capabilities of simulations [29].

ML accelerates simulations through reduced order modeling (ROM), creating simplified models that reduce computational cost, and surrogate modeling, which predicts results for a wider range of inputs using limited high-fidelity simulations.ML enhances accuracy with data-driven modeling, identifying patterns from large datasets, and physics-informed neural networks (PINNs), which embed physical laws into neural networks to better capture complex phenomena; it optimizes design by refining parameters through design optimization and generating optimal geometries with inverse design to achieve desired performance targets. ML enables real-time prediction for adaptive control and creates digital twins by integrating CFD simulations with sensor data to monitor and optimize system performance. The integration of AI and ML with CFD is driving innovations in industries like aerospace, automotive, energy, and healthcare, enabling efficient and sustainable solutions to complex engineering challenges.

1.13.3.2 Hybrid analytical-numerical techniques

Hybrid analytical-numerical techniques combine the strengths of analytical and numerical methods to solve complex fluid flow and heat transfer problems. By strategically combining these approaches, it is possible to achieve more accurate, efficient, and insightful solutions.

1.13.3.3 Key hybrid techniques

1. Domain decomposition methods:

- **Domain decomposition:** Dividing the domain into simpler subdomains, each solved using an appropriate method (analytical or numerical).
- **Interface matching:** Ensuring continuity of solutions across subdomain boundaries.
- **Efficient for complex geometries:** Well-suited for problems with complex geometries that can be decomposed into simpler regions.

2. Asymptotic methods:
- **Perturbation methods:** Expanding the solution as a series in a small parameter, leading to simplified equations.
- **Matched asymptotic expansions:** Combining inner and outer solutions to capture both local and global behavior.
- **Useful for singular perturbation problems:** Effective for problems with small parameters, such as thin boundary layers.

3. Spectral methods:
- **Fourier spectral methods:** Using Fourier series to represent the solution.
- **Chebyshev spectral methods:** Using Chebyshev polynomials to represent the solution.
- **High accuracy for smooth solutions:** Well-suited for problems with smooth solutions and periodic boundary conditions.

4. Hybrid numerical-analytical methods:
- **FEM with analytical solutions:** Using analytical solutions for simple subdomains and numerical methods for complex regions.
- **Boundary element method:** Converting volume integrals into boundary integrals, reducing the dimensionality of the problem.

1.14 Case studies and applications

1.14.1 Industrial applications

1.14.1.1 Cooling systems, HVAC, and power plants

Cooling systems are essential for removing excess heat from various applications, from electronic devices to large industrial facilities [30]. They can be air-cooled, liquid-cooled, or evaporative. HVAC systems, on the other hand, regulate indoor temperature, humidity, and air quality [31]. They consist of heating, cooling, and ventilation components.

Power plants generate electricity by converting different forms of energy into electrical energy. Thermal power plants use fossil fuels, nuclear power plants use nuclear fission, hydroelectric power plants harness the energy of flowing water, solar power plants convert sunlight into electricity, and wind power plants utilize wind energy. Cooling systems and HVAC systems are vital in power plants to maintain optimal operating conditions, including cooling turbines, generators, and control rooms.

1.14.2 Environmental and geophysical applications

1.14.2.1 Atmospheric flows, ocean currents, and climate modeling

Atmospheric flows, ocean currents, and climate modeling are all interconnected and play crucial roles in shaping our planet's climate. Atmospheric flows are driven by differences in temperature and pressure, resulting in global wind patterns like trade winds and jet streams. These winds redistribute heat and moisture around the globe, influencing weather patterns and climate.

Ocean currents, driven by wind, temperature differences, and salinity, transport heat and nutrients across the planet. They play a significant role in regulating Earth's climate by transferring heat from the tropics to the poles.

Climate models are complex computer programs that simulate the Earth's climate system, including the atmosphere, oceans, land surface, and ice. They use mathematical equations to represent the physical processes that govern the climate, such as radiation, heat transfer, and fluid dynamics. By analyzing past and present climate data, climate models can help us understand how the climate system works and predict future climate change [32].

The interaction between atmospheric flows, ocean currents, and climate models is crucial for understanding and predicting climate change.[9] For example, changes in atmospheric circulation patterns can affect ocean currents, which in turn can influence climate. By studying these complex interactions, scientists can better understand the factors that drive climate change and develop strategies to mitigate its impacts.

1.14.3 Advances in biomedical engineering

Biomedical engineering is revolutionizing healthcare through advancements in various fields. AI and ML are enhancing medical image analysis, personalized medicine, and drug discovery [33]. Regenerative medicine and tissue engineering are offering hope for tissue and organ repair. Wearable technology and implantable devices are enabling continuous health monitoring and early disease detection. Robotics and automation are improving surgical precision and rehabilitation. Finally, biomaterials and nanotechnology are driving innovations in drug delivery and medical implants. These advancements are improving patient outcomes and transforming the future of healthcare.

1.14.3.1 Blood flow and thermal regulation

Blood flow plays a crucial role in maintaining the body's temperature. The circulatory system acts as a heat distribution network, carrying warm blood from the core of the body to the extremities and returning cooler blood back to the core.

1.14.3.2 Vasodilation and vasoconstriction

Vasodilation occurs when the body temperature rises, blood vessels dilate, increasing blood flow to the skin's surface. This allows heat to be dissipated through radiation, convection, and evaporation. Vasoconstriction occur when the body temperature drops, blood vessels constrict, reducing blood flow to the skin's surface. This helps conserve heat.

1.15 Future trends and challenges

1.15.1 Advances in computational techniques

Recent years have witnessed significant advancements in computational techniques for fluid dynamics and heat transfer. High-performance computing (HPC) has enabled larger and more complex simulations, while advanced numerical methods like high-order schemes and adaptive mesh refinement have improved accuracy and efficiency. ML and AI are revolutionizing the field, enabling data-driven modeling, automated mesh generation, and uncertainty quantification. Multiphysics simulations, which couple fluid dynamics and heat transfer with other physical phenomena like structural mechanics and electromagnetics, are becoming increasingly powerful. Immersed boundary methods simplify the simulation of complex geometries, while large eddy simulation (LES) and direct numerical simulation (DNS) provide detailed insights into turbulent flows. The rise of open-source software like OpenFOAM and SU2 has democratized access to powerful CFD tools, empowering researchers and engineers worldwide. These advancements are driving innovation across various industries, from aerospace and automotive to energy and biomedical engineering.

1.15.2 Role of multiphysics and multiscale models

Multiphysics and multiscale models are becoming increasingly important in modern engineering and scientific research. These models enable the simulation of complex systems involving multiple physical phenomena, such as fluid flow, heat transfer, structural mechanics, electromagnetics, and chemistry. By coupling these different physical processes, researchers can gain a deeper understanding of the behavior of real-world systems. Multiscale modeling allows for the simulation of phenomena across multiple length and time scales, from microscopic to macroscopic. This capability is crucial for accurately capturing the effects of microscale processes on macroscale behavior. By integrating these powerful techniques, engineers and scientists can design more efficient, reliable, and innovative products and systems.

1.15.3 Integration of experimental and computational studies

The integration of experimental and computational studies has become a powerful approach in many scientific and engineering disciplines, including fluid dynamics and heat transfer. By combining the strengths of both methods, researchers can gain deeper insights into complex phenomena, validate computational models, and accelerate the design and optimization of systems. Experimental studies provide valuable data on real-world systems. They can be used to validate computational models, calibrate simulation parameters, and identify critical physical phenomena. However, experiments can be time-consuming, expensive, and limited in scope.

Computational studies offer a flexible and cost-effective way to explore a wide range of design parameters and operating conditions. They can be used to predict the behavior of systems under various scenarios, identify optimal designs, and gain insights into fundamental physical processes. However, computational models rely on assumptions and simplifications, which may limit their accuracy. By integrating experimental and computational studies, researchers can overcome the limitations of each individual approach. Experimental data can be used to validate and calibrate computational models, while computational simulations can be used to design and optimize experiments. This iterative process leads to more accurate and reliable results, accelerating the development of innovative solutions.

1.16 Summary and conclusion

This chapter has provided a comprehensive introduction to the fundamental principles of flow dynamics and heat transfer, highlighting their significance in engineering and scientific applications. It began by defining the core concepts of fluid dynamics and heat transfer, emphasizing their interdependence and the critical role they play in fields such as aerospace, mechanical, civil, and biomedical engineering. The historical development of these fields was explored, showcasing the contributions of key figures like Archimedes, Bernoulli, Fourier, and Prandtl, whose work laid the foundation for modern fluid mechanics and heat transfer studies. The chapter covered essential fluid properties such as density, viscosity, and compressibility, along with key governing equations like the continuity equation, Navier-Stokes equations, and the energy equation. It discussed the classification of flows into laminar vs. turbulent, compressible vs. incompressible, and internal vs. external flows, providing a clear understanding of the different flow regimes and their characteristics. The three primary modes of heat transfer – conduction, convection, and radiation – were explained with their governing equations, including Fourier's law of heat conduction, Newton's law of cooling, and the Stefan-Boltzmann law for radiation. The importance of coupled fluid flow and thermal analysis in multiphysics problems was emphasized, where the interaction between fluid dynamics and heat transfer is crucial for accurate modeling and simulation. Mathematical modeling of flow dynamics and heat transfer was explored, focusing on the conservation laws of mass, momentum, and energy, as well as the role of dimensionless numbers like the Reynolds number, Prandtl number, and Nusselt number in characterizing fluid flow and heat transfer phenomena. Challenges in modeling and simulation, including turbulence, multiphase flows, and phase change phenomena, were also discussed. Emerging trends in computational techniques, such as the integration of AI and ML in CFD and heat transfer simulations, were highlighted alongside the growing importance of multiphysics and multiscale models in modern engineering. The integration of experimental and computational studies was presented as a powerful approach to validate models and accelerate the design and optimization of systems. In conclusion, flow dynamics and heat transfer are foundational disciplines with wide-ranging applications in engineering and science. As technology continues to advance, the integration of computational tools, experimental methods, and innovative modeling techniques will drive further breakthroughs in these fields, enabling the development of more efficient, sustainable, and innovative solutions to global challenges. The future of flow dynamics and heat transfer lies in the continued exploration of multiphysics phenomena, the application of advanced computational methods, and the seamless integration of theory, simulation, and experimentation.

References

[1] Lakshminarayana B. (1995). Fluid Dynamics and Heat Transfer of Turbomachinery. John Wiley & Sons.

[2] Greene G. A., Cho Y. I., Hartnett J. P., and Bar-Cohen A. (Eds.) (2006). Advances in Heat Transfer. vol. 39, Elsevier.

[3] Bejan A. (2003). Heat Transfer Handbook. John Wiley and Sons, Inc.

[4] Jaluria Y. (2017). Computational Heat Transfer. Routledge.

[5] Awasthi M. K., Shukla A. K., and Yadav D. (2021). Rayleigh instability of power-law viscoelastic liquid with heat and mass transfer. International Communications in Heat and Mass Transfer, 129, 105657.

[6] Awasthi M. K., Shukla A. K., Singh A. K., and Shukla P. (2023). Interfacial instability analysis of viscous-viscoelastic fluid interface in an annular porous Medium. Journal of Porous Media, 26(11), 1–13.

[7] Broekaert S., Demuynck J., De Cuyper T., De Paepe M., and Verhelst S. (2016). Heat transfer in premixed spark ignition engines part I: Identification of the factors influencing heat transfer. Energy, 116, 380–391.
[8] Oosthuizen P. H., and Carscallen W. E. (1997). Compressible Fluid Flow. vol. 179, New York: McGraw-Hill.
[9] Panton R. L. (2024). Incompressible Flow. John Wiley & Sons.
[10] Sharp K. V., and Adrian R. J. (2004). Transition from laminar to turbulent flow in liquid filled microtubes. Experiments in Fluids, 36, 741–747.
[11] Greitzer E. M., Tan C. S., and Graf M. B. (2007). Internal flow: concepts and applications.
[12] Awasthi M. K., Shukla A. K., Agarwal S., Singh A. K., and Shukla P. (2024). Stability analysis of rivlin-ericksen fluid film with heat transfer through an annular porous medium. Srpm, 15, 25–39.
[13] Michalke A., and Hermann G. (1982). On the inviscid instability of a circular jet with external flow. Journal of Fluid Mechanics, 114, 343–359.
[14] Hahn D. W., and Özisik M. N. (2012). Heat Conduction. John Wiley & Sons.
[15] Chiu F. C. (2014). A review on conduction mechanisms in dielectric films. Advances in Materials Science and Engineering, 2014(1), 578168.
[16] Velarde M. G., and Normand C. (1980). Convection. Scientific American, 243(1), 92–109.
[17] Jaluria Y. (1980). Natural convection, heat and mass transfer. In Unknown Host Publication Title. Pergamon press.
[18] Sparrow E. M. (2018). Radiation Heat Transfer. Routledge.
[19] Coleman B. D., and Mizel V. J. (1963). Thermodynamics and departures from Fourier's law of heat conduction. Archive for Rational Mechanics and Analysis, 13, 245–261.
[20] Winterton R. H. S. (1999). Newton's law of cooling. Contemporary Physics, 40(3), 205–212.
[21] Lévesque L. (2014). Law of cooling, heat conduction and Stefan-Boltzmann radiation laws fitted to experimental data for bones irradiated by CO2 laser. Biomedical Optics Express, 5(3), 701–712.
[22] Currie I. G. (1967). The effect of heating rate on the stability of stationary fluids. Journal of Fluid Mechanics, 29(2), 337–347.
[23] Berge A., and Johansson P. Ä. R. (2012). Literature review of high performance thermal insulation.
[24] Shah R. K., and Sekulic D. P. (1998). Heat exchangers. Handbook of heat transfer, p. 3.
[25] Kakac S., Yener Y., and Pramuanjaroenkij A. (2013). Convective Heat Transfer. CRC press.
[26] Smith T. R., and Bretherton F. P. (1972). Stability and the conservation of mass in drainage basin evolution. Water Resources Research, 8(6), 1506–1529.
[27] Lentine M., Aanjaneya M., and Fedkiw R. (2011, August). Mass and momentum conservation for fluid simulation. In Proceedings of the 2011 ACM SIGGRAPH/Eurographics Symposium on Computer Animation (pp. 91–100).
[28] Subramanian R. S. (2014). Reynolds number. Department of Chemical and Biomolecular Engineering, Clarkson University, Clarkson.
[29] Panchigar D., Kar K., Shukla S., Mathew R. M., Chadha U., and Selvaraj S. K. (2022). Machine learning-based CFD simulations: A review, models, open threats, and future tactics. Neural Computing and Applications, 34(24), 21677–21700.
[30] Sarbu I., and Sebarchievici C. (2013). Review of solar refrigeration and cooling systems. Energy and Buildings, 67, 286–297.
[31] Trčka M., and Hensen J. L. (2010). Overview of HVAC system simulation. Automation in Construction, 19(2), 93–99.
[32] Schneider S. H., and Dickinson R. E. (1974). Climate modeling. Reviews of Geophysics, 12(3), 447–493.
[33] Griffith L. G., and Grodzinsky A. J. (2001). Advances in biomedical engineering. Jama, 285(5), 556–561.

[34] Awasthi M. K., Shukla A. K., and Pandey P. (2024). Rayleigh–Taylor instability in a Casson fluid layer with heat and mass transfer. Physics of Fluids, 36(10), 102132.
[35] Shukla A. K., and Awasthi M. K. (2022). Interfacial Characteristics of Power-Law Viscoelastic Fluid With Heat and Mass Transfer in Planar Configuration. Journal of Fluids Engineering, 144(10), 101303.
[36] Shukla A. K., and Awasthi M. K. (2023). Stability characteristics of planar Rivlin–Ericksen fluid interface with mass and heat transfer. Journal of Fluids Engineering, 145(3), 031302.

Suleiman A. Wali, Abdulhalim Musa Abubakar*,
Hayatuddeen Abubakar, Semiu Adebayo Kareem,
and Mohamed Ellouze

Chapter 2
Compressible fluid flow and heat transfer

Abstract: Compressible fluid flow and heat transfer represent critical areas in fluid dynamics, particularly in applications involving high-speed aerodynamics, energy systems, and industrial processes. Thus, the exploration of the fundamental principles governing compressible flows, highlighting key phenomena such as shock waves, expansion fans, and boundary layer behavior, is the goal of this study. A comprehensive overview of governing equations, including the Navier-Stokes equations and the energy equation, provides the foundation for understanding flow characteristics in compressible regimes. Modeling techniques specific to compressible flows are discussed, with emphasis on numerical methods such as finite volume and finite element approaches. Various simulation strategies are detailed, including steady-state and unsteady-state analyses, which are essential for predicting complex flow behaviors under varying conditions. Case studies illustrating practical applications, such as gas turbine performance and high-speed vehicle aerodynamics, demonstrate the significance of accurate modeling and simulation in optimizing design and efficiency. Challenges in simulating compressible flows, including numerical stability and turbulence modeling, are addressed, alongside advancements in computational capabilities, which enhance the accuracy of predictions. Furthermore, the interplay between flow dynamics and heat transfer is examined, emphasizing the role of temperature gradients in influencing fluid properties and system performance. As industries increasingly demand efficient thermal management in compressible systems, understanding these dynamics becomes imperative.

Keywords: heat transfer, turbulence modeling, compressible flows, flow dynamics, fluid properties

***Corresponding author: Abdulhalim Musa Abubakar**, Department of Chemical Engineering, Modibbo Adama University (MAU), P.M.B 2076, Yola 640261, Adamawa, Nigeria, e-mail: abdulhalim@mau.edu.ng
Suleiman A. Wali, Department of Chemical Engineering, Modibbo Adama University (MAU), P.M.B 2076, Yola 640261, Adamawa, Nigeria, e-mail: sawali@mau.edu.ng
Hayatuddeen Abubakar, Department of Chemical Engineering, Modibbo Adama University (MAU), P.M.B 2076, Yola 640261, Adamawa, Nigeria, e-mail: hayatumautech@gmail.com
Semiu Adebayo Kareem, Department of Chemical Engineering, Modibbo Adama University (MAU), P.M.B 2076, Yola 640261, Adamawa, Nigeria, e-mail: sakareem_1@mau.edu.ng
Mohamed Ellouze, Physics Department, Faculty of Sciences of Sfax, University of Sfax, LM2EM, B.P. 1171, 3000 Sfax, Tunisia, e-mail: mohamed.ellouze@fss.rnu.tn

https://doi.org/10.1515/9783111661674-002

2.1 Introduction

Oosthuizen and Carscallen [1] and Ames and Tang [2] previously introduce experts to compressible flow. Compressible flow refers to the movement of gases, in which density variations play a significant role due to changes in pressure and temperature. Unlike incompressible flow, where density is assumed constant, compressible flow occurs at high velocities, typically when the Mach number exceeds 0.3. This phenomenon is particularly relevant in aerospace applications, gas pipelines, and high-speed propulsion systems. When heat transfer is introduced into compressible flow, the interplay between thermal energy and fluid motion becomes even more complex. Heat addition or removal alters the fluid properties, affecting parameters such as velocity, pressure, and temperature distribution. For example, in supersonic flow, heating can lead to expansion, while cooling can cause compression, influencing shock waves and boundary layer behavior. Efficient design of propulsion systems can be carried out if the relationship between compressible flow and heat transfer is studied. As a result, the optimization of thermal management ensures safe operation of high-speed aircraft as also of industrial processes involving high-temperature gas flows [3]. Searches conducted in important article databases such as Google Scholar, ResearchGate, PubMed, ScienceDirect, Web of Science (WOS), Scopus, SpringerLink, and IEEE Xplore reveal no work with the exact title as this chapter. However, Figure 2.1 reveals the volume of works from 2019 to 2024 in ScienceDirect with the chapter keywords (Heat transfer and Compressible flows).

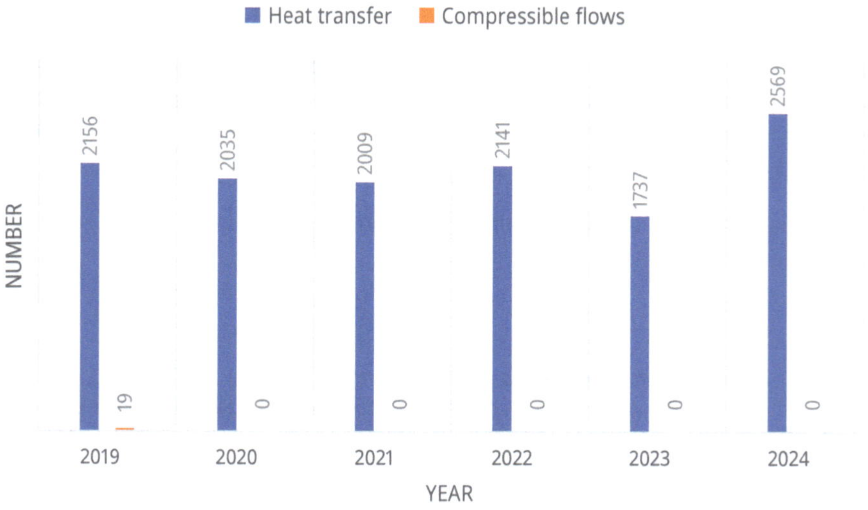

Figure 2.1: Number of publications with the study keywords.

In the 19 published works containing the keyword "compressible flows", shown in Figure 2.1, none is related to the scope of the current review. And despite a consistent volume of studies containing the keyword "heat transfer" from 2019 to 2024, none linked their study to compressible fluid flow. Looking at this, it is correct to describe this current work as innovative and contributory to the study of fluid transport thermodynamics. In the current study, equations governing compressible fluid flow and heat transfer are discussed. Boundary layer phenomenon and other flow characteristics, as briefly explained in Schreier [4], are again highlighted differently, with clear case studies across important areas of application. In addition, numerical stability issues and turbulence modeling and simulation, are comprehensively highlighted. This study helps engineers and scientists to understand how gases behave under varying pressure, temperature, and velocity conditions, which is crucial for applications in aerospace, energy systems, and industrial processes. It provides insights into optimizing propulsion systems, turbine efficiency, and cooling mechanisms in high-speed flows, thereby improving performance and safety. Advancements in this field contribute to the development of sustainable energy solutions, better thermal management in engineering designs, and innovative technologies for supersonic and hypersonic travel.

2.2 Governing equations

2.2.1 Navier-Stokes equations

When dealing with compressible fluids, the Navier-Stokes equations help explain how gases behave when subjected to changes in pressure and temperature [5, 6]. Unlike incompressible fluids, where density remains constant [7], compressible fluids expand or contract based on their thermal and pressure conditions [8]. This makes the equations particularly useful in aerodynamics, where airflows around high-speed aircraft or rockets experience drastic temperature and pressure variations. Heat transfer also plays a key role, as energy moves through conduction, convection, or radiation, affecting fluid motion and stability [9]. In practical applications, these equations guide engineers in designing better cooling systems [10], optimizing fuel combustion [11], and improving overall efficiency in various industrial and aerospace processes [12]. The Navier-Stokes equations describe the motion of fluid substances, accounting for velocity, pressure, density, and external forces [13]. For compressible fluids with heat transfer, the equations take the forms in Table 2.1.

where ρ is the density of the fluid (kg/m^3); ∇ is the nabla operator (used for gradient, divergence, and curl operations); τ is the stress tensor, representing viscous forces (Pa); t is the time (s); v is the velocity vector of the fluid (m/s); p is the pressure (Pa or N/m^2); e is the specific internal energy of the fluid (J/kg); g is the gravitational acceleration vector (m/s^2); q is the heat flux vector (W/m^2); h is the heat source term (W/m^3); R is the specific gas constant (J/kg $\cdot$ K); and T is the temperature (K).

Table 2.1: Fundamental Navier-Stokes equations for compressible fluid flow and heat transfer.

Kind	Mathematical form	Equation no.	Description	Source
Continuity equation (mass conservation)	$\frac{\partial \rho}{\partial t}+\nabla.(\rho v)=0$	(2.1)	Ensures that mass is conserved in a compressible fluid	[14]
Momentum equation (Newton's second law for fluids)	$\frac{\partial(\rho v)}{\partial t}+\nabla.(\rho vv)=-\nabla p+\nabla.\tau+\rho g$	(2.2)	Describes how forces (pressure, viscosity, and gravity) affect fluid motion	[15, 16]
Energy equation (first law of thermodynamics)	$\frac{\partial(\rho e)}{\partial t}+\nabla.(\rho ev)=-\nabla.q+\tau{:}\nabla v+\rho h$	(2.3)	Governs heat transfer and energy conservation in a moving fluid	[17–20]
Equation of state (ideal gas law)	$p=\rho RT$	(2.4)	Relates pressure, density, and temperature in a compressible fluid	[21, 22]

2.3 Energy equation for compressible flows

The energy equation for compressible flows is fundamental in understanding how energy moves and changes within a fluid, especially when the fluid's density, temperature, and pressure fluctuate [23]. Essentially, it describes how heat and work interact in a system, ensuring energy is conserved [24]. For fluids that compress or expand, like air at high speeds or gases in engines, this equation takes into account not just the internal energy of the fluid, but also how heat flows in and out and how the fluid's movement affects its energy [25, 26]. In its simplest form, it shows how the energy inside the fluid changes over time, how it moves with the fluid, and how external forces (heat transfer or viscous forces) play a role [27]. It is crucial for areas such as aerospace, combustion, and even atmospheric studies, where temperature and pressure shifts can greatly influence fluid behavior [28]. It helps engineers and scientists predict how fluids will behave under different conditions and optimize systems such as engines, turbines, and cooling mechanisms [29, 30]. For a compressible fluid, this equation accounts for internal energy, kinetic energy, and heat transfer mechanisms such as conduction, convection, and radiation [31]. Mathematically, it is expressed in accordance with eq. (2.3), where ρe represents the total energy per unit volume and $\tau{:}\nabla v$ denotes the work done by viscous forces. The term ρh accounts for any internal heat generation within the fluid.

2.4 Flow phenomena

2.4.1 Shock waves and expansion fans

In high-speed compressible fluid flows, shock waves and expansion fans play a crucial role in dictating how pressure, temperature, and velocity change within a system [32]. Supersonic and hypersonic flows are the primary causes of this phenomenon [33–36]. Shock waves are sudden and nearly discontinuous changes in flow properties occur when a supersonic flow encounters an obstruction or a rapid change in geometry [37]. These waves represent a strong compression of the fluid, where abrupt or gradual changes in flow properties are observed due to variations in pressure, density, and temperature as the velocity decreases [38]. Energy dissipation as heat due to irreversible losses occurs, because the process happens instantaneously [39]. Mathematically, shock waves are governed by the Rankine-Hugoniot relations in eqs. (2.5)–(2.8) in Table 2.2, which describe how pressure, temperature, and velocity vary across the shock front [40]. There are two primary types of shock waves: normal shocks occurring perpendicular to the flow direction, causing a drastic reduction in velocity while increasing temperature and pressure [41], and oblique shocks that form at an angle, leading to more complex changes in flow direction and thermodynamic properties [42–46].

Table 2.2: Derived relations for a normal shock.

Property	Formula	Equation no.	Description	Reference
Density ratio (compression ratio)	$\frac{\rho_2}{\rho_1} = \frac{(\gamma+1)M_1^2}{(\gamma-1)M_1^2+2}$	(2.5)	Density increases across the shock due to compression	[47]
Pressure ratio	$\frac{p_2}{p_1} = 1 + \frac{2\gamma}{\gamma+1}\left(M_1^2-1\right)$	(2.6)	Pressure rises significantly as the shock strength increases	[48]
Temperature ratio	$\frac{T_2}{T_1} = \frac{p_2}{p_1}\frac{\rho_1}{\rho_2}$	(2.7)	Temperature increases due to compression effects	[49]
Mach number after the shock	$M_2^2 = \frac{(\gamma-1)M_1^2+2}{2\gamma M_1^2-(\gamma-1)}$	(2.8)	The Mach number decreases after passing through the shock wave	[48]

where ρ_1 and ρ_2 are densities of the fluid before and after the shock, respectively (kg/m^3); p_1 and p_2 are pressures of the fluid before and after the shock, respectively (Pa); T_1 and T_2 are temperatures of the fluid before and after the shock, respectively (K); M_1 and M_2 are Mach numbers before and after the shock (dimensionless); γ is the ratio of specific heats (C_p/C_v), typically 1.4 for air; and R is the specific gas constant (J/kg · K).

In terms of heat transfer, shock waves cause localized temperature spikes that affect thermal loads in high-speed aircraft [50, 51], rockets [52], and gas turbines [53]. Advanced cooling techniques to prevent material failure are a major concern in aerospace engineering as a result of the strong compressive heating in shock waves [54, 55]. In contrast to shock waves, expansion fans occur when a supersonic flow encounters a sudden expansion, such as when passing around a convex corner [56, 57]. Unlike shocks, expansion fans cause a smooth and gradual decrease in pressure [58], temperature [59], and density [60], while increasing the velocity of the fluid [61], as mentioned in Table 2.3. This process follows the Prandtl-Meyer expansion theory, which describes how a fluid accelerates while cooling down when moving through an expansion wave [2]. Table 2.3 summarizes their differences.

Table 2.3: Comparison and engineering applications.

Feature	Shock waves	References	Expansion fans	References
Nature	Sudden, compressive	[62]	Gradual, expansive	[63]
Effect on pressure	Increases sharply	[64]	Decreases smoothly	[65]
Effect on temperature	Increases due to compression	[66]	Decreases due to expansion	[67]
Effect on velocity	Decreases	[68]	Increases	[69]
Energy loss	High (due to entropy increase)	[70]	Low (nearly isentropic)	[71]
Heat transfer concern	Strong thermal load due to heating	[72, 73]	Cooling effect due to expansion	[74, 75]

Unlike shock waves, expansion fans are isentropic, meaning they do not generate entropy or significant heat losses [76]. Thus, it makes them beneficial in applications where controlled expansion is needed, such as in supersonic nozzles [77], wind tunnels [78, 79], and space propulsion systems [80].

2.4.2 Boundary layer behavior in compressible flows

When a fluid moves past a solid surface, a thin region called the boundary layer forms near the surface [81]. In this region, the velocity gradually increases from zero at the wall (due to friction) to match the speed of the free-stream flow [82–84]. However, when dealing with compressible flows, especially at high speeds, things become more complicated because density, temperature, and pressure can change significantly within the boundary layer [85–87]. One major factor that sets compressible boundary layers apart from their incompressible counterparts is how density varies with temperature [88]. At high velocities, such as in supersonic and hypersonic flows, the air near the surface gets heated due to friction and compression effects [89, 90]. This heating lowers the density of the air close to the wall, which alters the way the velocity changes across the boundary layer [91]. Another critical issue is the interaction between shock waves and the boundary layer [92–94]. When a shock wave – a sudden jump in pressure and temperature – hits the boundary layer, it can disrupt the smooth flow of air, sometimes causing flow separation [95, 96]. This separation increases drag and can make vehicles unstable, which is a serious concern in aerospace engineering [97, 98]. Engineers must carefully design surfaces to minimize these effects, ensuring smoother and more efficient airflow [99]. Heat transfer is another key aspect. At high speeds, the temperature difference between the air and the surface becomes extreme [93]. It leads to intense heating, which is why spacecraft reentering the atmosphere need heat shields to survive the harsh conditions [100–105]. The same principles apply to high-speed jets and rockets, where managing heat is just as important as controlling aerodynamics [106, 107].

2.5 Modeling techniques

Two widely used numerical approaches for Navier-Stokes equations, energy conservation equations, and species transport equations problems are the finite volume method (FVM) [108] and the finite element method (FEM) [109], each offering distinct advantages, depending on the application and computational requirements. The FVM is widely applied in computational fluid dynamics (CFD) due to its inherent conservation properties [110]. In this method, the computational domain is divided into discrete control volumes, and the governing equations are integrated over each volume to ensure mass, momentum, and energy conservation [111]. The fluxes at the control volume boundaries are computed using numerical schemes [110, 112], such as Godunov's method [113], upwind schemes [114], or flux-splitting techniques [115], which are particularly useful for capturing shock waves [116] and discontinuities in compressible flow [117]. The advantage of FVM in compressible flow is its ability to handle sharp gradients [118], shock waves [119], and expansion waves efficiently [120]; as such, becoming a preferred method in aerospace applications [121], high-speed aerodynamics [122], and supersonic flow simulations [123]. Additionally, in heat transfer problems, FVM ensures accurate representation of conductive, convective, and radiative heat transfer mechanisms – well-suited for thermal analysis in turbulent flows [124]. FEM, on the other hand, provides a more flexible approach, especially for problems involving complex geometries and anisotropic materials [125]. FEM discretizes the domain into smaller elements and approximates the solution using basis functions (such as polynomial functions) over each element [126]. In compressible fluid flow problems, FEM is often used with stabilization techniques called the streamline-upwind Petrov-Galerkin (SUPG) method to handle numerical instabilities associated with high-speed flows [127, 128]. While, traditionally, more common in structural mechanics, FEM has been successfully applied to heat transfer problems [129], particularly when dealing with heterogeneous materials [130, 131], phase-change processes [132, 133], and coupled fluid–structure interactions [134, 135]. In heat transfer analysis, FEM is advantageous for solving conduction-dominated problems [136], as it allows precise modeling of thermal gradients in materials with varying properties [137–139]. Table 2.4 lists a third technique called finite difference method (FDM).

Table 2.4: Comparison of modeling techniques for compressible flows.

Aspect	FVM	FEM	FDM	Emphasis
Discretization approach	Control volumes with flux balancing	Elements with shape functions	Grid points with direct differentiation	[140–142]
Governing equations	Integral form of conservation laws	Weak form of conservation laws	Differential form of governing equations	[143–145]
Conservation properties	Naturally conservative	Requires special formulations for conservation	Not inherently conservative	[146–149]
Geometry handling	Works well with structured/ unstructured meshes	Highly flexible for complex geometries	Best suited for structured grids	[150–152]
Accuracy in shock capture	High flux-based methods and upwind schemes	Requires stabilization techniques (SUPG and DG)	Struggles with sharp discontinuities	[153–155]
Computational cost	Moderate to high, depending on scheme	High due to matrix assembly and inversion	Low to moderate	[156–158]
Application in compressible flow	Preferred for high-speed flows, shock waves, turbulence	Used in aerodynamics, but less common in CFD	Less preferred for high-speed flows	[159–161]
Heat transfer modeling	Efficient for convection and radiation problems	Best for conduction-dominated problems	Suitable for simple heat transfer cases	[162–164]
Numerical stability	Stable with proper flux schemes	Requires stabilization techniques	Can be unstable for complex problems	[165–167]
Ease of implementation	Moderate (flux evaluation required)	Complex (matrix operations needed)	Simple for structured grids	[168–170]
Parallelization potential	Highly parallelizable	Challenging, due to matrix dependencies	Easily parallelizable	[171–173]

FDM is less common for compressible flows [174], though it is simple and efficient for structured grids but struggles with shock waves and conservation properties [175]. FDM is conceptually straightforward, as it directly approximates derivatives using Taylor series expansions, compared to the more complex integral-based FVM or matrix-based FEM [176]. High-resolution simulations in simple domains employ FDM scheme, because it allows faster computations, particularly in time-dependent problems [177]. As data dependencies are minimal, compared to FEM's global matrix operations, FDM's structured nature boosts the efficiency of parallel computing for explicit methods [178]. In problems with rectangular or Cartesian grids [179], FDM provides highly accurate solutions with minimal computational overhead, whereas FVM and FEM require additional geometric transformations or unstructured meshing [180]. Due to its straightforward derivation and implementation (including in Excel) [181], FDM is widely used in teaching, research, and for verifying results, before applying more complex methods [182].

2.6 Simulation strategies

During the analysis of compressible fluid flow and heat transfer, one of the fundamental considerations is whether the system operates under steady-state or unsteady-state conditions [183]. Steady-state analysis assumes that all properties, such as velocity, pressure, and temperature, remain constant over time [184]. That way, it becomes computationally efficient and easier to solve, as it simplifies simulations by focusing only on spatial variations [185]. It is particularly useful for engineering applications where conditions stabilize over time, such as heat exchangers [186], supersonic nozzles [187], and aerodynamic surfaces experiencing consistent flow patterns [188]. Conversely, unsteady-state analysis accounts for variations in fluid properties over time through the capture of transient behaviors that are crucial for understanding dynamic systems [189]. This type of simulation is essential when dealing with rapid changes in flow conditions, such as shock waves [190], turbulence [191, 192], and pulsating heat transfer phenomena [193]. In compressible flows, transient effects play a significant role in engine combustion, aerodynamic stall, and weather modeling scenarios, where flow properties continuously evolve [194]. Even though unsteady-state simulations proffer deeper insights into real-world behaviors, they require significantly more computational power and complex numerical methods to resolve time-dependent equations accurately [195, 196].

2.7 Case studies

2.7.1 Gas turbine performance optimization

Since gas turbines rely on compressible fluid flow and heat transfer, their performance is influenced by several factors, including pressure ratios, temperature levels, and aerodynamic design. One of the most effective ways to enhance efficiency is by increasing the turbine inlet temperature, while maintaining material integrity. Another key aspect of optimization is reducing energy losses due to aerodynamics and heat transfer inefficiencies. Carefully designing turbine blades with improved profiles and advanced thermal coatings helps minimize flow separation and pressure drops. CFD simulations, also brought in Table 2.5, will allow engineers to analyze airflow patterns and make necessary adjustments to turbine geometries.

Table 2.5: Optimization of gas turbine performance, considering compressible fluid flow and heat transfer.

Topic	Goal	Reference
Gas turbine performance characteristics	Understanding the performance metrics of gas turbines, including power output and efficiency, is essential for optimization	[197]
Coupled flow network and CFD analysis	Integrating flow network models with CFD enhances the design of cooling structures in turbine blades	[198]
Thermo-fluid dynamics in aircraft gas turbines	Analyzing thermo-fluid dynamics is crucial for the design and performance assessment of aircraft gas turbines	[199]
Heat transfer and fluid flow modeling	Modeling heat transfer and fluid flow in the hot sections of gas turbines aids in performance optimization	[200]
Optimization of blade cooling passages	Geometric optimization of cooling passages in turbine blades enhances heat transfer and overall efficiency	[201]
Impact of specific heat ratio	Investigating the effect of specific heat ratio on compressible flow performance in turbines is vital for accurate modeling	[202]
Heat transfer rate and fluid flow analysis	Analyzing heat transfer rates and fluid flow characteristics influences the design and optimization of gas turbines	[203]
Internal cooling channel heat transfer	Studying heat transfer characteristics of internal cooling channels with pin-fins in turbine blades leads to improved cooling efficiency	[204]
Robust optimization under uncertainties	Optimizing gas turbine performance, considering geometrical and operational uncertainties, ensures reliable operation	[205]
Laval nozzle dynamics	Understanding the dynamics of Laval nozzles is crucial for optimizing compressible fluid flow in turbines	[206]

Integrating heat exchangers such as recuperators and regenerators can help recover waste heat. Consequently, it will help in improving fuel efficiency by preheating compressed air before it enters the combustion chamber. Material advancements have also played a significant role in boosting gas turbine efficiency. The use of high-temperature-resistant alloys and ceramic matrix composites has allowed turbines to operate at higher temperatures without compromising structural integrity.

2.7.2 High-speed vehicle aerodynamics

Air is the most commonly studied fluid in this field (Table 2.6). Other compressible fluids, such as carbon dioxide (CO_2), helium, and hydrogen, play significant roles in specialized applications, including planetary entry, high-speed propulsion, and experimental aerodynamics.

Table 2.6: Compressible fluid flow and heat transfer in high-speed vehicle aerodynamics.

Compressible fluid used	Heat transfer type	Objective	Case study
Air	Convective heat transfer	Review computational methods for predicting aerodynamic heating during high-speed atmospheric flight	[207]
Air	Turbulent boundary layer heat transfer	Assess the effects of heat transfer and Mach number on high-speed turbulent boundary layers	[208]
Air	Transitional flow heat transfer	Develop a heat flux correlation for high-speed flow in the transitional regime	[209]
Air	Aerothermodynamic correlations	Provide correlations for predicting heat transfer in high-speed flows	[210]
Air	Viscous flow and structural interaction	Investigate the interaction between high-speed compressible viscous flow and structural response	[211]

Carbon dioxide, for instance, is a key fluid in studying aerodynamic heating for vehicles entering the Martian atmosphere [212]. Due to its higher molecular weight and specific heat properties, compared to air, CO_2 affects convective and radiative heat transfer differently, necessitating modified computational models for heat flux prediction. Studies on CO_2 flow focus on understanding hypersonic boundary layer transition and its impact on thermal loads experienced by entry vehicles [213, 214]. Similarly, helium, with its low molecular weight and high thermal conductivity, is often used in experimental aerodynamics to simulate rarefied flow conditions [215]. Its ability to mimic high-speed, low-density environments makes it useful in studying heat transfer phenomena in the upper atmosphere and space vehicle reentry scenarios. Hydrogen, another important compressible fluid, is extensively utilized in high-speed propulsion systems, particularly in scramjets and rocket engines [216]. As a cryogenic fuel, hydrogen undergoes complex convective and radiative heat transfer interactions, especially during combustion and shock-induced flow regions [217, 218]. The high enthalpy and diffusivity of hydrogen require advanced thermal management strategies to prevent excessive heating and ensure efficient energy conversion. Supercritical hydrogen is additionally investigated for its role in regenerative cooling systems, where it absorbs heat from engine walls before combustion, thus enhancing thermal efficiency [219, 220].

2.7.3 Heat exchangers in energy systems

Heat exchanger devices are widely used in industries such as power generation, refrigeration, chemical processing, and aerospace engineering, where effective thermal management is critical. In high-speed aerodynamics and compressible flow applications, heat exchangers play a key role in cooling systems for jet engines, scramjets, and hypersonic vehicles. They experience extreme temperatures due to aerodynamic heating and combustion processes, making thermal regulation a crucial factor in their performance and longevity. The choice of heat exchanger depends on factors such as fluid properties, pressure conditions, and the primary mode of heat transfer – whether conduction, convection, or radiation. Several types of heat exchangers are commonly used across various applications, as detailed in Table 2.7.

Table 2.7: Key studies on compressible flow and heat transfer in heat exchangers.

Work	Highlight	Application	Reference
Rayleigh flow in heat exchangers	Examines the behavior of compressible flow with heat addition or removal in ducts	Relevant for designing heat exchangers in propulsion systems where heat is added to compressible fluids	[221]
High-speed compressible flow in microchannel heat exchangers	Investigates the performance of heat exchangers utilizing high-speed compressible air flow in microchannels	Applicable in compact heat exchanger designs for electronics cooling and aerospace applications	[222]
Modeling compressible fluid flow in heat exchangers	Focuses on CFD modeling of compressible fluid flow and heat transfer in heat exchangers	Useful for optimizing heat exchanger designs in power plants and chemical processing industries	[223]

Table 2.7 (continued)

Work	Highlight	Application	Reference
Thermal and fluid flow analysis in heat exchangers	Analyzes heat transfer and fluid flow characteristics in heat exchangers handling compressible fluids	Essential for improving efficiency in systems like gas turbines and refrigeration units	[224]
Fundamentals of heat exchangers	Provides an overview of heat exchanger types, functions, and operational mechanisms	Applicable across various energy systems for efficient thermal management	[225]
Shell-and-tube heat exchangers: basic calculations	Covers design calculations and operational considerations for shell and tube heat exchangers	Widely used in power generation and petrochemical industries	[226]
Heat exchanger design and performance	Discusses design principles and performance assessment of heat exchangers	Critical in power plants and HVAC systems to maintain energy efficiency	[227]
Mechanisms of heat transfer in heat exchangers	Explores conduction and convection mechanisms in heat exchangers	Important for understanding heat transfer processes in various applications	[228]

Shell-and-tube heat exchangers, known for their durability, are commonly found in power plants, where they handle high-pressure fluids with ease. Plate heat exchangers, which have a compact design and enhanced surface area, are often preferred in situations where space is limited but efficient heat transfer is necessary. In aerospace applications, regenerative heat exchangers are particularly useful, as they utilize high-temperature exhaust gases to preheat incoming air or fuel, improving the overall thermal efficiency and reducing energy losses. The materials used in heat exchanger construction have also evolved, with ceramic composites and superalloys becoming increasingly popular due to their ability to withstand extreme temperatures and mechanical stress. The performance of a heat exchanger is influenced by factors such as fluid flow arrangement (parallel, counterflow, or crossflow), turbulence levels, and phase-change processes like condensation and evaporation. To enhance efficiency, researchers and engineers focus on optimizing surface geometries using advanced cooling techniques, and incorporating nanofluids with superior thermal conductivity. In aerospace applications, innovative approaches such as transpiration cooling are being explored to handle extreme heat loads more effectively.

2.8 Numerical stability issues and turbulence modeling

When studying compressible fluid flow, numerical stability issues are a common challenge [229], particularly when using computational methods such as CFD for simulating complex fluid behavior [230]. Issues of these kinds arise from the discretization of the governing fluid equations, which can lead to inaccuracies and instability in the solution, if not handled properly [231, 232]. The primary causes of instability include improper grid resolution [233], inadequate time-stepping methods [234], and the misapplication of boundary conditions [235]. Numerical oscillations, also known as "spurious" solutions, can occur when the spatial and temporal resolutions are too coarse, or the numerical scheme is not well-suited to the flow physics [236]. For compressible flows, these issues are especially critical because they can affect the accuracy of shock capturing and the prediction of high-speed phenomena (especially shock waves and expansion fans) [237]. In accordance with Argyropoulos and Markatos [238], turbulence modeling further complicates the numerical simulation of compressible flows. Turbulence is inherently chaotic, and accurately capturing the effects of turbulence on fluid dynamics is essential for high-fidelity simulations [239]. There are several approaches to turbulence modeling, ranging from the k-ε (k-epsilon) and k-ω (k-omega) simple models (Table 2.8), which are popular due to their relative simplicity and computational efficiency [240], to more advanced methods such as large eddy simulation and direct numerical simulation, which provide higher accuracy but at the cost of increased computational expense [241].

Table 2.8: Common turbulence models in compressible fluid flow.

Turbulence model	Approach	Applications	Advantages	Disadvantages
k-ε (k-epsilon)	RANS (Reynolds-averaged Navier-Stokes)	General engineering applications, industrial flows	Simple, efficient, good for steady-state flows	Less accurate in complex, unsteady flows
k-ω (k-omega)	RANS	Aerospace, high-speed compressible flows	Better for boundary layer prediction; works well for adverse pressure gradients	Sensitive to freestream conditions; less accurate in turbulence near walls
LES (large eddy simulation)	Direct simulation of large eddies	High-fidelity simulations of complex, turbulent flows	Accurate; captures large-scale turbulence dynamics	High computational cost; limited to relatively simple cases

Table 2.8 (continued)

Turbulence model	Approach	Applications	Advantages	Disadvantages
DNS (direct numerical simulation)	Full resolution of all turbulence scales	Very high-accuracy simulations	Provides most detailed results	Extremely computationally expensive; limited to small-scale problems

Each model has its strengths and limitations, and the choice of turbulence model often depends on the flow conditions, the level of accuracy required, and the computational resources available [242]. To ensure numerical stability, advanced numerical methods such as implicit schemes or high-order schemes are often employed, which provide better convergence properties and reduce the impact of numerical errors [243, 244].

2.9 Interplay between flow dynamics and heat transfer

Temperature gradients have a profound impact on fluid properties, altering density, viscosity, and thermal conductivity in complex ways. In gases, an increase in temperature generally leads to a rise in viscosity due to enhanced molecular activity, while in liquids, viscosity tends to decrease as intermolecular forces weaken. Changes in temperature also influence fluid density, which can drive buoyancy effects and cause variations in pressure distribution, especially in compressible flows. Thermal conductivity, another key property, varies with temperature and affects how efficiently heat is transferred within the fluid. These shifts in fluid behavior play a critical role in determining flow stability, turbulence intensity, and overall system performance in engineering applications.

2.10 Conclusion

The study of compressible fluid flow and heat transfer is essential in engineering, particularly in aerospace, energy, and industrial applications where high-speed gas dynamics play a critical role. Throughout this work, we have explored the fundamental equations governing compressible flows, analyzed key flow characteristics, and examined the influence of heat transfer on fluid behavior. The relationship between compressibility and thermal effects shapes the performance of systems like jet engines, gas turbines, and high-speed vehicles, where shock waves, boundary layers, and expansion fans determine efficiency and stability. By leveraging advanced numerical modeling techniques such as finite volume and finite element methods, engineers and scientists can better predict and optimize these complex flows. However, challenges remain, particularly in turbulence modeling and numerical stability, which require continuous advancements in computational methods and experimental validation. As industries push for more efficient thermal management and aerodynamic designs, further research in compressible fluid flow will be instrumental in shaping future innovations in propulsion, energy systems, and high-performance engineering applications.

References

[1] Oosthuizen P. H., and Carscallen W. E. (2013). Introduction to Compressible Fluid Flow, 2nd ed. CRC Press.

[2] Ames F. E., and Tang C. C. (2021). An Introduction to Compressible Flow, 2nd ed. Boca Raton: CRC Press, doi: 10.1201/9781003042945.

[3] Saad M. A. (1985). Compressible Fluid Flow, [Online]. Available, Englewood Cliffs, NJ: Prentice-Hall, Inc., https://ui.adsabs.harvard.edu/abs/1985ph . . . book . . . .S/abstract.

[4] Schreier S. (1982). Compressible Flow, United States: U.S. Department of Energy's OSTI, [Online]. Available https://www.osti.gov/biblio/6590140.

[5] Bistafa S. R. (2018). The Navier–Stokes equation from Maxwell's Theory of gases. Journal of the Brazilian Society of Mechanical Sciences and Engineering, 40(1), 1–15. doi: 10.1007/s40430-017-0957-8.

[6] Schneiderbauer S., and Krieger M. (2014). What do the Navier-Stokes equations mean?. European Journal of Physics, 35(015020), 1–26. doi: 10.1088/0143-0807/35/1/015020.

[7] Deville M. O. (2021). An Introduction to the Mechanics of Incompressible Fluids, Lausanne, Vaud, Switzerland: Springer, doi: 10.1007/978-3-031-04683-4.

[8] Som S. K. (2024). Introduction to fluid machines, and compressible flow. PSG College of Arts and Science. [Online]. Available: http://elearn.psgcas.ac.in/nptel/courses/video/112105182/lec26.pdf

[9] Goldstein M. E., and Siegel R. (1971). Analysis of heat transfer for compressible flow in two-dimensional porous media. International Journal of Heat and Mass Transfer, 14(10), 1677–1690. doi: 10.1016/0017-9310(71)90076-7.

[10] Ahmad J. U., Pandya S. A., Chan W. M., and Chaderjian N. M. (2005). Navier-Stokes simulation of air-conditioning facility of a large modern computer room, Proc. FEDSM2005 2005 ASME Fluids Eng. Div. Summer Meet. Exhib. June 19–23, 2005, Houston, TX, USA, pp. 1–6. [Online]. Available: https://www.nas.nasa.gov/assets/nas/pdf/staff/Chan_W_Navier-Stokes

[11] Manna P., Dharavath M., Sinha P. K., and Chakraborty D. (2013). Optimization of a flight-worthy scramjet combustor through CFD. Aerospace Science and Technology, 27(1), 138–146. doi: 10.1016/j.ast.2012.07.005.

[12] Cai J., Wang Y., and Yu S. (2022). The recent progress and the State-of-art applications of Navier Stokes equation. Highlights in Science, Engineering and Technology, 12, 114–120. doi: 10.54097/hset.v12i.1413.

[13] Li D. L. (2005). The Green's function of the Navier-Stokes equations for gas dynamics in $\mathbb{R}3$. Communications in Mathematical Physics, 257, 579–619. doi: 10.1007/s00220-005-1351-4.

[14] Wahba E. M. (2021). Derivation of the differential continuity equation in an introductory engineering fluid mechanics course. International Journal of Mechanical Engineering Education, 50(2). doi: 10.1177/03064190211014460.
[15] Leishman J. G. (2022). Momentum Equation. In Introduction to Aerospace Flight Vehicles, Ch. 21, Embry-Riddle Aeronautical University, doi: 10.15394/eaglepub.2022.1066.n17.
[16] Anderson J. D. (1992). Governing equations of fluid dynamics. In Computational Fluid Dynamics, Wendt J. F. (Ed.) Berlin Heidelberg: Springer-Verlag, pp. 15–51. Ch. 2, doi: 10.1007/978-3-662-11350-9_2.
[17] Jin C. (2023). Confirmation of the First Law of thermodynamics in theory and extended Bernoulli equation. Journal of Applied Mathematics and Physics, 11(2), 409–420. doi: 10.4236/jamp.2023.112023.
[18] Shanthini R. (2006). The First Law applied to steady state processes. In Thermodynamics for Beginners, Ch. 10, pp. 211–248. [Online]. Available https://www.researchgate.net/publication/326607787.
[19] Boxer G. (1987). First Law of thermodynamics – Flow processes. In Work Out Engineering Thermodynamics, 1st ed. London:Macmillan Work Out Series. Palgrave, pp. 54–67. Ch. 5, doi: 10.1007/978-1-349-09346-5_6.
[20] Mínguez J. M. (2005). The work–energy theorem and the First Law of thermodynamics. International Journal of Mechanical Engineering Education, 33(1), 1–6. doi: 10.7227/IJMEE.33.1.8.
[21] Sirignano W. A. (2018). Compressible flow at high pressure with linear Equation of State. Journal of Fluid Mechanics, 843, 244–292. doi: 10.1017/jfm.2018.166.
[22] Rapp-Kindner I., Ősz K., and Lente G. (2025). The ideal gas law: Derivations and intellectual background. ChemTexts, 11(1), 1–22. doi: 10.1007/s40828-024-00198-9.
[23] De Michele C., and Coppola G. (2023). Numerical treatment of the energy equation in compressible flows simulations. Comput Fluids, 250(105709). doi: 10.1016/j.compfluid.2022.105709.
[24] Kostic M. M. (2018). Nature of heat and thermal energy: From caloric to Carnot's reflections, to entropy, exergy, entransy and beyond. Entropy, 20(584). doi: 10.3390/e20080584.
[25] Wibolo A., Made A. A. I., and Sudirman (2021). Experimental study of air velocity effect of heat transfer in distillation condenser. International Conference on Communication Technology, [ICOMTA 2023], 1175(012014), 1–5. doi: 10.1088/1757-899X/1175/1/012014.
[26] Sanli A., Ozsezen A. N., Kilicaslan I., and Canakci M. (2008). The influence of engine speed and load on the heat transfer between gases and in-cylinder walls at fired and motored conditions of an IDI diesel engine. Applied Thermal Engineering, 28(11–12), 1395–1404. doi: 10.1016/j.applthermaleng.2007.10.005.
[27] Leishman J. G. (2022). Energy equation and Bernoulli's equation. In Introduction to Aerospace Flight Vehicles, Embry-Riddle Aeronautical University, Ch. 22, doi: 10.15394/eaglepub.2022.1066.n18.
[28] Lu W. et al. (2023). Evaluation of the effect of pressure and heat transfer on the efficiency of a batch fuel reactor, using Iron-based Oxygen Carrier with a CFD model. Fuel, 33(126266), 1–16. doi: 10.1016/j.fuel.2022.126266.
[29] Manafi H., Mansoori G. A., and Ghotbi S. (1999). Phase behavior prediction of petroleum fluids with minimum characterization data. Journal of Petroleum Science and Engineering, 22(1–3), 67–93. doi: 10.1016/S0920-4105(98)00058-8.
[30] Rahman S. (2024). Computational Fluid Dynamics for predicting and controlling fluid flow in industrial equipment. European Journal of Advances in Engineering and Technology, 11(9), 1–9. [Online]. Available http://www.ejaet.com.
[31] Rowlinson J. S., and Watson I. D. (1969). The prediction of the thermodynamic properties of fluids and fluid mixtures-I The principle of corresponding states and its extensions. Chemical Engineering Science, 24(10), 1565–1574. doi: 10.1016/0009-2509(69)80095-3.

[32] Das S., Toh H. T., Abu A. B. H., Das S., Mondal K., and Quen L. K. (2016). A preliminary numerical study of high speed compressible flow around a solid triangular prism MJJIC 2016 Kuala Lumpur, pp. 1–5., [Online]. Available: https://www.researchgate.net/publication/340886333
[33] Stemmer C., and Adams N. A. (2009). Supersonic and hypersonic boundary-layer flows. In Turbulence and Interactions, Deville M., Lê T.-H., and Sagaut P. (Eds.) Berlin, Heidelberg: Springer, pp. 77–91. doi: 10.1007/978-3-642-00262-5_4.
[34] Edisson S. (2011). Supersonic and hypersonic flows on 2D unstructured context: Part I. WSEAS Transactions on Fluid Mechanics, 4(6), 199–226. [Online]. Available https://www.researchgate.net/publication/291088065.
[35] Lee S. et al. (2024). A review of flow control strategies for supersonic/hypersonic fluid dynamics. Aerospace Research Communications, 2(13149), 1–25. doi: 10.3389/arc.2024.13149.
[36] Hui W. H. (1997). Supersonic and hypersonic flow with attached shock waves over delta wings. Proceedings of the Royal Society A, 325(1561). doi: 10.1098/rspa.1971.0168.
[37] Hussein I. A. (2009). Shock waves in compressible flow. Developments in Petroleum Science, 32, 112–145. Ch. 5, doi: 10.1016/S0376-7361(09)70009-3.
[38] Stein K., Schommer J., and Heppner B. (2012). Undergraduate studies on compressible flows and shock waves, [Online]. Available: https://www.researchgate.net/publication/258598398
[39] Zhu Q., Wu Y., Zhou W., Yang Q., and Xu X. (2024). A comprehensive study on the roles of viscosity and heat conduction in shock waves. Journal of Fluid Mechanics, 984, doi: 10.1017/jfm.2024.264.
[40] Gedalin M. (2023). Rankine-Hugoniot relations in turbulent shocks. Frontiers in Physics, 11, 1–6. doi: 10.3389/fphy.2023.1325995.
[41] Nilam V., Xavier S. M., Dondu H. B., and Avula B. B. (2024). Comprehensive analysis of normal shock wave propagation in turbulent non-ideal gas flows with analytical and neural network methods. The Physics of Fluids, 36(095120). doi: 10.1063/5.0220497.
[42] Babu V. (2021). Oblique shock waves. In Fundamentals of Gas Dynamics, Cham: Springer, pp. 103–119. doi: 10.1007/978-3-030-60819-4_7.
[43] Avramenko A. A., Tyrinov A. I., Shevchuk I. V., and Dmitrenko N. P. (2023). Oblique shock wave in turbulent flow. Journal of Non-Equilibrium Thermodynamics, 48(4), 389–401. doi: 10.1515/jnet-2022-0093.
[44] Vimercati D., Gori G., and Guardone A. (2018). Non-ideal oblique shock waves. Journal of Fluid Mechanics, 847, 266–285. doi: 10.1017/jfm.2018.328.
[45] Zocca M., Guardone A., Cammi G., Cozzi F., and Spinelli A. (2019). Experimental observation of oblique shock waves in steady non-ideal flows. Experiments in Fluids, 60(101). doi: 10.1007/s00348-019-2746-x.
[46] Bulat P., Melnikova A., Upyrev V., and Volkov K. (2021). Refraction of oblique shock wave on a tangential discontinuity. Fluids, 6(301). doi: 10.3390/fluids6090301.
[47] Wassgren C. (2016). Gas dynamics. Normal Shock Waves, 1365–1373. Ch. 13, [Online]. Available https://engineering.purdue.edu/~wassgren/teaching/ME30900/NotesAndReading.
[48] Wassgren C. (2016). Basic relations. Normal Shock Waves, 1–4. Ch. 1, [Online]. Available https://www.aerostudents.com/courses/aerodynamics-1/normalShockWaves.pdf.
[49] ANSYS. Normal shock waves (Shock-expansion theory – Lesson 2), 2020. [Online]. Available: https://innovationspace.ansys.com/courses/wp-content/uploads/sites/5/2020/12/LT3-C2-L2-Handout-Final.pdf
[50] Bulat P. V., and Volkov K. N. (2016). Shock waves oscillations in the interaction of supersonic flows with the head of the aircraft. International Journal of Environmental and Science Education, 11(12), 4976–4984. [Online]. Available http://www.ijese.net/arsiv/191.html.
[51] Gnani F., Zare-Behtash H., and Kontis K. (2016). Pseudo-shock waves and their interactions in high-speed intakes. Progress in Aerospace Sciences, 82, 36–56. doi: 10.1016/j.paerosci.2016.02.001.

[52] Afraimovich E. L., Kosogorov E. A., and Plotnikov A. V. (2002). Shock–acoustic waves generated during rocket launches and earthquakes. Cosmic Research, 40, 241–25. doi: 10.1023/A:1015925020387.

[53] Zhang Q., O'Dowd D. O., He L., Wheeler A. P. S., Ligrani P. M., and Cheong B. C. Y. (2011). Overtip shock wave structure and its impact on turbine blade tip heat transfer. Journal of Turbomachinery, 133(4), 1–8. doi: 10.1115/1.4002949.

[54] Rita A., Sivakumar A., and Dhas S. A. M. B. (2019). Influence of shock waves on structural and morphological properties of copper oxide NPs for aerospace applications. Journal of Nanostructure in Chemistry, 9, 225–230. doi: 10.1007/s40097-019-00313-0.

[55] Koen D.(2011). Investigation of novel cooling methods to enhance aerospace component manufacturing practices, Degree of Master of Science in Engineering Thesis, Department of Industrial Engineering, Faculty of Engineering, Stellenbosch University, . [Online]. Available: http://hdl.handle.net/10019.1/18058

[56] Cengel Y. A., and Boles M. A. (2008). Compressible flow. In Thermodynamics: An Engineering Approach, 6th ed. Ch. 17, McGraw-Hill, pp. 1–41. [Online]. Available https://kh.aquaenergyexpo.com/wp-content/uploads/2023/01/Chapter-17-Compressible-Flow.pdf.

[57] Wu G. et al. (2024). Effect of expansion waves on cowl shock wave and boundary layer interaction in hypersonic inlet. Propulsion and Power Research, 13(1), 80–97. doi: 10.1016/j.jppr.2024.02.001.

[58] Balaji G., Kumar B. N., Vijayarangam J., Vasudevan A., and Pandiyarajan R. (2021). Numerical investigation of expansion Fan optimization of truncated annular aerospike nozzle. Materials Proceedings, 46(Part 9), 4283–4288. doi: 10.1016/j.matpr.2021.03.124.

[59] Yen C. (2024). Wave propagation, shock waves and expansion fans in scramjet engine design. Journal of Aeronautics and Aerospace Engineering, 13(3), 1–2. doi: 10.35248/2168-9792.24.13.358.

[60] Page R. H., and Sernas V. (2012). Apparent reverse transition in an expansion fan. AIAA Journal, 8(1). doi: 10.2514/3.5641.

[61] Huo J., Yi S., Zheng W., Niu H., Lu X., and Gang D. (2022). Experimental investigation of expansion effect on shock wave boundary layer interaction near a compression ramp. Chinese Journal of Aeronautics, 35(12), 89–101. doi: 10.1016/j.cja.2022.04.014.

[62] Sasoh A. (2021). Compressible Fluid Dynamics and Shock Waves, 1st ed. Springer Singapore, doi: 10.1007/978-981-15-0504-1.

[63] Nel L., Skews B., and Naidoo K. (2015). Schlieren techniques for the visualization of an expansion fan/shock wave interaction. Journal of Visualization, 18, 469–479. doi: 10.1007/s12650-014-0253-y.

[64] Seitz M. W., and Skews B. W. (2006). Effect of compressible foam properties on pressure amplification during shock wave impact. Shock Waves, 15, 177–197. doi: 10.1007/s00193-006-0033-2.

[65] Pathan K. A., Dabeer P. S., and Khan S. A. (2019). Influence of expansion level on base pressure and reattachment length. CFD Letters, 11(5), [Online]. Available https://www.akademiabaru.com/index.php/archives/article/view/196.

[66] Zhu X.-K., Yu C.-P., Tong F.-L., and Li X.-L. (2016). Numerical study on wall temperature effects on shock wave/turbulent boundary-layer interaction. American Institute of Aeronautics and Astronautics, 55(1). doi: 10.2514/1.J054939.

[67] Luo X., Prast B., Van Dongen M. E. H., Hoeijmakers H. W. M., and Yang J. (2006). On phase transition in compressible flows: Modelling and validation. Journal of Fluid Mechanics, 548, 403–430. doi: 10.1017/S0022112005007809.

[68] Zhilin A. A., and Fedorov A. V. (1998). The shock-wave structure in a two-velocity mixture of compressible media with different pressures. Journal of Applied Mechanics and Technical Physics, 39, 166–174. doi: 10.1007/BF02468081.

[69] Dang T. Q., and Bushnell P. R. (2009). Aerodynamics of cross-flow fans and their application to aircraft propulsion and flow control. Progress in Aerospace Sciences, 45(1–3), 1–29. doi: 10.1016/j.paerosci.2008.10.002.

[70] Thornber B., Drikakis D., Williams R. J. R., and Youngs D. (2008). On entropy generation and dissipation of kinetic energy in high-resolution shock-capturing schemes. Journal of Computational Physics, 227(10), 4853–4872. doi: 10.1016/j.jcp.2008.01.035.

[71] Dashora G., Sasidharan U., and Kumar R. (2025). Unsteady flow features of an isentropic compression ramp. AIAA SCITECH 2025 Forum, doi: 10.2514/6.2025-0547.

[72] Bernardini M., Asproulias I., Larsson J., Pirozzoli S., and Grasso F. (2016). Heat transfer and wall temperature effects in shock wave turbulent boundary layer interactions. Physical Review Fluids, 1, doi: 10.1103/PhysRevFluids.1.084403.

[73] Tzou D. Y. (1989). On the thermal shock wave induced by a moving heat source. ASME Journal of Heat and Mass Transfer, 111(2), 232–238. doi: 10.1115/1.3250667.

[74] Gammeter C., Krismer F., and Kolar J. W. (2015). Weight optimization of a cooling system composed of fan and extruded-fin heat sink. IEEE Transactions on Industry Applications, 51(1), 509–520. doi: 10.1109/TIA.2014.2336977.

[75] Salman A. D., Al-Badr A. R., and Al-Hassani A. H. (2023). Experimental study of the split-type air conditioner with the variable-speed compressor, variable-speed supply fan, and electronic expansion valve. Wasit Journal of Engineering Sciences, 10(3), 57–70. doi: 10.31185/ejuow.Vol11.Iss3.458.

[76] Baumgärtner D., Otter J. J., and Wheeler A. P. S. (2021). The effect of isentropic exponent on supersonic turbine wakes. In Proceedings of the 3rd International Seminar on Non-Ideal Compressible Fluid Dynamics for Propulsion and Power, Cham: Springer, pp. 153–161. doi: 10.1007/978-3-030-69306-0_16.

[77] Khan S. A., and Rathakrishnan E. (2004). Control of suddenly expanded flows from correctly expanded nozzles. International Journal of Turbo and Jet-Engines, 21, 255–278. doi: 10.1515/TJJ.2004.21.4.255.

[78] Tichenor N., Semper M., Bowersox R., Srinvasan R., and North S. (2012). Calibration of an actively controlled expansion hypersonic wind tunnel. In 27th AIAA Aerodynamic Measurement Technology and Ground Testing Conference on 28 June 2010 – 01 July 2010 Chicago, Illinois, doi: 10.2514/6.2010-4793.

[79] Semper M., Tichenor N., Bowersox R., Srinvasan R., and North S. (2012). On the design and calibration of an actively controlled expansion hypersonic wind tunnel. In 47th AIAA Aerospace Sciences Meeting Including the New Horizons Forum and Aerospace Exposition, doi: 10.2514/6.2009-799.

[80] Tajmar M. (2003). Advanced Space Propulsion Systems, 1st ed. Springer Vienna, doi: 10.1007/978-3-7091-0547-4.

[81] Von Karman T., and Tsien H. S. (2012). Boundary layer in compressible fluids. Aerospace Science, 5(6). doi: 10.2514/8.591.

[82] Lee H., Helm C., Martín P. M., and Williams O. J. H. (2023). Compressible boundary layer velocity transformation based on a generalized form of the total stress. Physical Review Fluids, 8(074604), 1–24. doi: 10.1103/PhysRevFluids.8.074604.

[83] Baronti P. O., and Libby P. A. (2012). Velocity profiles in turbulent compressible boundary layers. Aiaaj, 4(2), 193–197. doi: 10.2514/3.3419.

[84] Freire A. P. S., Cruz D. O. A., and Pellegrini C. C. (1995). Velocity and temperature distributions in compressible turbulent boundary layers with heat and mass transfer. International Journal of Heat and Mass Transfer, 38(13), 2507–2515. doi: 10.1016/0017-9310(94)00352-V.

[85] Zienkiewicz O. C., Taylor R. L., and Nithiarasu P. (2014). Compressible high-speed gas flow. In The Finite Element Method for Fluid Dynamics, Ch. 7, Elsevier Ltd, pp. 225–281. doi: 10.1016/B978-1-85617-635-4.00007-8.

[86] Codoni D., Moutsanidis G., Hsu M.-C., Bazilevs Y., Johansen C., and Korobenko A. (2021). Stabilized methods for high-speed compressible flows: Toward hypersonic simulations. Computational Mechanics, 67, 785–809. doi: 10.1007/s00466-020-01963-6.
[87] Jiang B. (1998). High-speed compressible flows. In The Least-Squares Finite Element Method, Berlin, Heidelberg: Springer, pp. 303–328. doi: 10.1007/978-3-662-03740-9_13.
[88] Stewartson K. (1949). Correlated incompressible and compressible boundary layers. Proceedings of the Royal Society A, 200(1060), 84–100. doi: 10.1098/rspa.1949.0160.
[89] Novean M., Schetz J., Bowersox R., Novean M., Schetz J., and Bowersox R. (2012). Direct measurements of skin friction in complex supersonic flows. In 35th Aerospace Sciences Meeting and Exhibit[06–09 January 1997, doi: 10.2514/6.1997-394.
[90] Zhao X., Yi S., Mi Q., Hu Y. F., and Ding H. (2022). Skin friction reduction of hypersonic body by supersonic layer. Fluid Dyn, 57(5), 686–696. doi: 10.1134/S0015462822050123.
[91] Qu Z., Li X., Luo F., Long Y., and Chen W. (2024). Mechanisms and characteristics of wall skin friction reduction by boundary layer injection under hypervelocity inflow conditions. International Journal of Heat and Fluid Flow, 106(109269). doi: 10.1016/j.ijheatfluidflow.2023.109269.
[92] Délery J. M. (2003). The different facets of an old but always present concern: Shock-wave/boundary layer interaction. In IUTAM Symposium Transsonicum IV, Sobieczky H. (Ed.) Dordrecht: Springer, pp. 91–98. doi: 10.1007/978-94-010-0017-8_15.
[93] Sandham N. D., Shock-wave/boundary-layer interactions, 2012. [Online]. Available: https://www.sto.nato.int/publications/STO Educational Notes/RTO-EN-AVT-195/EN-AVT-195–05.pdf.
[94] Bao Y. et al. (2022). Study of shock wave/boundary layer interaction from the perspective of nonequilibrium effects. The Physics of Fluids, 34(4). doi: 10.1063/5.0085570.
[95] Liou M.-F., and Lee B. J. (2013). Mitigation of adverse effects caused by shock wave boundary layer interactions through optimal wall shaping. In 31st AIAA Appl. Aerodyn. Conf.[June 24–27, 2013 San Diego, CA], pp. 1–20. doi: 10.2514/6.2013-2653.
[96] Chen Y. (2024). Shock wave-boundary Layer interaction: A survey of recent development. Highlights in Science, Engineering and Technology, 112, 422–430. doi: 10.54097/smvame92.
[97] Bian X., Wang Q., Su X., and Yuan X. (2020). Interaction mechanisms of shock waves with the boundary layer and wakes in a highly-loaded NGV using hybrid RANS/LES. Chinese Journal of Aeronautics, 33(1), 149–160. doi: 10.1016/j.cja.2019.07.008.
[98] Huang W., Wu H., Yang Y., Yan L., and Li S. (2020). Recent advances in the shock wave/boundary layer interaction and its control in internal and external flows. Acta Astronautica, 174, 103–122. doi: 10.1016/j.actaastro.2020.05.001.
[99] Jana T., and Kaushik M. (2022). Survey of control techniques to alleviate repercussions of shock-wave and boundary-layer interactions. Advances in Aerodynamics, 4(27), 1–30. doi: 10.1186/s42774-022-00119-9.
[100] Munoth K. K. J. (2015). Ablation heat shields during satellite and missile re-entry to the Earth's atmosphere, [Online]. Available: https://www.researchgate.net/publication/280716670
[101] Fedele A., Omar S., Cantoni S., Savino R., and Bevilacqua R. (2021). Precise re-entry and landing of propellantless spacecraft. Adv. Sp. Res, 68(11), 4336–4358. doi: 10.1016/j.asr.2021.09.029.
[102] Bharghava D. S. N., Mali A. K., Jana T., and Kaushik M. (2024). Implementation of concavity over heat shield of a reentry vehicle in reducing aerodynamic heating. International Journal of Heat and Fluid Flow, 107(109413). doi: 10.1016/j.ijheatfluidflow.2024.109413.
[103] Mazzaracchio A., and Marchetti M. (2012). Effect of spacecraft aerodynamics and heat shield characteristics on optimal aeroassisted transfer. Engineering, 4(6), 307–320. doi: 10.4236/eng.2012.46040.
[104] Dhawan S., Vishal M., and Taploo A. (2015). Thermal protection for a re-entry vehicle using heat ablation process. International Journal of Science and Research, 4(11), 1293–2095. [Online]. Available https://www.researchgate.net/publication/285584215.

[105] Rajamani A. (2023). Heat shields for re-entry vehicles: A review. International Journal of Engineering Research and Technology, 12(3), 165–170. doi: 10.17577/IJERTV12IS030106.
[106] Sabnis K., and Babinsky H. (2023). A review of three-dimensional shock wave-boundary-layer interactions. Progress in Aerospace Sciences, 143(100953). doi: 10.1016/j.paerosci.2023.100953.
[107] Prasanna K. D., Ravi B., and Jeyan J. V. M. L. (2021). A review on effect of shock wave boundary layer interaction in different flows fields. Journal of Emerging Technologies and Innovative Research, 8(9), d687–d693. [Online]. Available https://www.researchgate.net/publication/354850013.
[108] Singh S., and Kashyap P. (2023). Finite Volume Method: An efficient tool for transport equation solutions. Tuijin Jishu/Journal of Propulsion Technology, 44(6), 3378–3388. [Online]. Available https://www.propulsiontechjournal.com/index.php/journal/article/download/3938.
[109] Lesaint P. (1974). Finite element methods for the transport equation. Anal Numérique, 8(2), 67–93. [Online]. Available https://www.esaim-m2an.org/articles/m2an/pdf/1974/01/m2an197408R200671.pdf.
[110] Moukalled F., Mangani L., and Darwish M. (2016). The Finite Volume Method in Computational Fluid Dynamics-An Advanced Introduction with OpenFOAM® and Matlab, 1st ed. Springer Cham, doi: 10.1007/978-3-319-16874-6.
[111] Kitamura K. (2020). Introduction: Brief review of finite volume method (FVM) in Computational Fluid Dynamics. Advancement of Shock Capturing Computational Fluid Dynamics Methods, 1–5. doi: 10.1007/978-981-15-9011-5_1.
[112] Kolditz O. (2002). Finite Volume Method. In Computational Methods in Environmental Fluid Mechanics, Berlin, Heidelberg: Springer, pp. 173–190. doi: 10.1007/978-3-662-04761-3_8.
[113] Alcrudo F., and Garcia-Navarro P. (1993). A high-resolution Godunov-type scheme in finite volumes for the 2D shallow-water equations. International Journal of Numerical Methods in Fluids, 16(6), 489–505. doi: 10.1002/fld.1650160604.
[114] Neofytou P. (2005). A 3rd order upwind finite volume method for generalised Newtonian fluid flows. Advances in Engineering Software, 36(10), 664–680. doi: 10.1016/j.advengsoft.2005.03.011.
[115] Xu C., and Amano R. S. (2001). Flux-splitting finite volume method for turbine flow and heat transfer analysis. Computational Mechanics, 27, 119–127. doi: 10.1007/s004660000219.
[116] Motheau E., and Wakefield J. (2020). Investigation of finite-volume methods to capture shocks and turbulence spectra in compressible flows. Communications on Applied Mathematics and Computation, 15(1), 1–36. doi: 10.2140/camcos.2020.15.1_.
[117] Zeng X. (2014). A high-order hybrid finite difference–finite volume approach with application to inviscid compressible flow problems: A preliminary study. Comput Fluids, 98, 91–110. doi: 10.1016/j.compfluid.2014.02.007.
[118] Li G., and Modest M. F. (2021). A hybrid finite volume/PDF Monte Carlo method to capture sharp gradients in unstructured grids. In ASME 2000 Int. Mech. Eng. Congr. Expo. Novemb. 5–10, 2000 Orlando, Florida, USA, vol. 5, pp. 225–233. doi: 10.1115/IMECE2000-1580.
[119] Alekseev I., and Kustova E. (2020). Numerical simulations of shock waves in viscous carbon dioxide flows using finite volume method. Vestnik St. Petersburg University, Mathematics, 53, 344–350. doi: 10.1134/S1063454120030024.
[120] Greaves D. (2010). Application of the finite volume method to the simulation of nonlinear water waves. Advanced Series on Ocean Engineering, 357–396. doi: 10.1142/9789812836502_0010.
[121] Oliveira F. B., and Azevedo J. L. F. (2023). Assessment of interface gradient reconstruction techniques for Finite Volume Methods in aerospace applications. In AIAA AVIATION 2023 Forum on 12–16 June 2023 San Diego, CA and Online, doi: 10.2514/6.2023-3241.
[122] Yatsukhno D. S. (2018). Application of the finite volume method for the standard ballistic model aerodynamics calculations. In 12th Int. Conf. Aerophysics Phys. Mech. Class. Quantum Syst. – APhM-2018 27–29 Novemb. 2018, Moscow, Russ. Fed, vol. 1250, pp. 1–9. doi: 10.1088/1742-6596/1250/1/012011.

[123] Deng X., Xie B., Teng H., and Xiao F. (2019). High resolution multi-moment finite volume method for supersonic combustion on unstructured grids. Applied Mathematical Modelling, 66, 404–423. doi: 10.1016/j.apm.2018.08.010.
[124] Beyers J. H. M., Harms T. M., and Kröger D. G. (2001). A finite volume analysis of turbulent convective heat transfer for accelerating radial flows. Numerical Heat Transfer, Part A: Applications, 40(2), 117–138. doi: 10.1080/104077801750468453.
[125] Okereke M., and Keates S. (2018). Finite Element applications-A Practical Guide to the FEM Process, 1st ed. Springer Cham, doi: 10.1007/978-3-319-67125-3.
[126] Rao S. S. (2011). Discretization of the domain. In The Finite Element Method in Engineering, 5th ed. ch. 2, Butterworth-Heinemann, pp. 53–73. doi: 10.1016/B978-1-85617-661-3.00002-7.
[127] Zhang X., and Xu X. (2023). The moving finite element method with streamline-upwind Petrov–Galerkin for magneto-hydrodynamic flows problems at high Hartmann numbers. Engineering Analysis with Boundary Elements, 157, 400–412. doi: 10.1016/j.enganabound.2023.09.022.
[128] Codoni D., Johansen C., and Korobenko A. (2022). A streamline-upwind Petrov–Galerkin formulation for the analysis of hypersonic flows in thermal non-equilibrium. Computer Methods in Applied Mechanics and Engineering, 398(115185). doi: 10.1016/j.cma.2022.115185.
[129] Bharatish A. (2020). Finite elements for heat transfer problems. In Heat Conduction Analysis, ch. 4, Florida: University of Florida, pp. 1–68. [Online]. Available https://web.mae.ufl.edu/nkim/IntroFEA/Chapter4.pdf.
[130] Eso R., and Arman A. (2023). The finite element method of flow and heat transfer in heterogeneous materials, WSEAS Transactions on Heat and Mass Transfer, 18, 1–7. doi: 10.37394/232012.2023.18.1.
[131] Soodmand I. et al. (2024). Heterogeneous material models for finite element analysis of the human mandible bone – A systematic review. Heliyon, 10(24), 1–13. doi: 10.1016/j.heliyon.2024.e40668.
[132] Woodfield J., Alvarez M., Gómez-Vargas B., and Ruiz-Baier R. (2019). Stability and finite element approximation of phase change models for natural convection in porous media. Journal of Computational and Applied Mathematics, 360, 117–137. doi: 10.1016/j.cam.2019.04.003.
[133] Brusche J. H., Segal A., and Urbach H. P. (2005). Finite-element model for phase-change recording. Journal of the Optical Society of America A, 22(4), 773–786. doi: 10.1364/JOSAA.22.000773.
[134] Sigrist J.-F. (2015). Fluid–structure coupling. In Fluid-Structure Interaction: An Introduction to Finite Element Coupling, ch. 5, John Wiley & Sons, Ltd, pp. 185–224. doi: 10.1002/9781118927762.
[135] Kalita A., and Rao T. V. V. L. N. (2023). An overview of fluid-structure interaction: Modelling, Finite Element Method and applications. Annals of Multidisciplinary Research, Innovation and Technology, 2(2), 93–105. [Online]. Available: https://adtu.in/files/current_issue/vol2_issue2/vol2-issue2-article014.pdf.
[136] Pentenrieder B. (2005). Finite Element solutions of heat conduction problems in complicated 3D geometries using the multigrid method, Diplomarbeit, Fakultat fur Mathematik, Technische Universitat Munchen (TUM), [Online]. Available: https://www5.in.tum.de/pub/pentenriederDA05.pdf
[137] Grigorova V., Shcheka S., Salehi F., Kamenev K. V., and Clark S. M. (2023). Use of finite element analysis for the thermal gradient determination of a neutron transparent high-pressure sample environment for neutron tomography. High Pressure Research, 43(3), 215–230. doi: 10.1080/08957959.2023.2225702.
[138] Hemanth J., and Yogesh K. B. (2018). Finite Element Analysis (FEA) and thermal gradient of a solid rectangular fin with embossing's for aerospace applications. Advances in Aerospace Sciences and Technology, 3(3), 49–60. doi: 10.4236/aast.2018.33004.
[139] Bai T., Qiao Y., Wang X., and Zhao N. (2022). Finite element method modeling of temperature gradient-induced Cu atomic thermomigration in Cu/Sn/Cu micro solder joint. Microelectronics Reliability, 129(114479). doi: 10.1016/j.microrel.2022.114479.

[140] Russell R. D., Williams J. F., and Xu X. (2007). A comparison of direct discretization of 4th order problems versus system reduction. World Scientific Connect, 195–205. doi: 10.1142/9789812709356_0011.
[141] Macri M., and De S. (2005). Towards an automatic discretization scheme for the method of finite spheres and its coupling with the finite element method. Computers and Structures, 83(17–18), 1429–1447. doi: 10.1016/j.compstruc.2004.09.019.
[142] Edwards M. G. (2002). Unstructured, control-volume distributed, full-tensor finite-volume schemes with flow based grids. Computers and Geosciences, 6, 433–452. doi: 10.1023/A:1021243231313.
[143] Nagel J. R. (2012). Solving the generalized Poisson equation using the finite-difference method (FDM). [Online]. Available: https://my.ece.utah.edu/~ece6340/LECTURES
[144] Cardoso-Ribeiro F. L., Matignon D., and Lefèvre L. (2021). A partitioned finite element method for power-preserving discretization of open systems of conservation laws. IMA Journal of Mathematical Control and Information, 38(2), 493–533. doi: 10.1093/imamci/dnaa038.
[145] Liu Y., Vinokur M., and Wang Z. J. (2006). Spectral (finite) volume method for conservation laws on unstructured grids V: Extension to three-dimensional systems. Journal of Computational Physics, 212(2), 454–472. doi: 10.1016/j.jcp.2005.06.024.
[146] Tezduyar T. E., Aliabadi S. K., Behr M., and Mittal S. (1994). Massively parallel finite element simulation of compressible and incompressible flows. Computer Methods in Applied Mechanics and Engineering, 119(1–2), 157–177. doi: 10.1016/0045-7825(94)00082-4.
[147] Aliabadi S. K., Ray S. E., and Tezduyar T. E. (1993). SUPG finite element computation of viscous compressible flows based on the conservation and entropy variables formulations. Computational Mechanics, 11, 300–312. doi: 10.1007/BF00350089.
[148] Deng X., Min Y., Mao M., Liu H., Tu G., and Zhang H. (2013). Further studies on Geometric Conservation Law and applications to high-order finite difference schemes with stationary grids. Journal of Computational Physics, 239, 90–111. doi: 10.1016/j.jcp.2012.12.002.
[149] Liu R., Wang D., Zhang X., Li W., and Yu B. (2013). Comparison study on the performances of finite volume method and finite difference method. Journal of Applied Mathematics, doi: 10.1155/2013/596218.
[150] Schönauer W., and Adolph T. (2003). FDEM: How we make the FDM more flexible than the FEM. Journal of Computational and Applied Mathematics, 158(1), 157–167. doi: 10.1016/S0377-0427(03)00461-8.
[151] Gerstmayr J., and Schöberl J. (2006). A 3D Finite Element Method for flexible multibody systems. Multibody System Dynamics, 15, 305–320. doi: 10.1007/s11044-006-9009-3.
[152] Lygidakis G. N., and Nikolos I. K. (2012). Using the Finite-Volume Method and hybrid unstructured meshes to compute radiative heat transfer in 3-D geometries. Numerical Heat Transfer, Part B: Fundamentals, 62(5), 289–314. doi: 10.1080/10407790.2012.707012.
[153] Çar K., (2024). Adaptive mesh refinement for one dimensional scalar conservation laws, Degree of Master of Science in Mathematics Thesis, Graduate School of Natural and Applied Science, Middle East Technical University, [Online]. Available: https://www.proquest.com/openview/e0f0e70ce801453ace62.
[154] Yurun F. (1997). A comparative study of the discontinuous Galerkin and continuous SUPG finite element methods for computation of viscoelastic flows. Computer Methods in Applied Mechanics and Engineering, 141(1–2), 47–65. doi: 10.1016/S0045-7825(96)01102-4.
[155] Frolkovič P., and Mikula K. (2007). Flux-based level set method: A finite volume method for evolving interfaces. Applied Numerical Mathematics, 57(4), 436–454. doi: 10.1016/j.apnum.2006.06.002.
[156] Ikegawa M., Kaiho M., and Kato C. (1994). FEM/FDM composite scheme for viscous incompressible flow analysis. Computer Methods in Applied Mechanics and Engineering, 112(1–4), 149–163. doi: 10.1016/0045-7825(94)90023-X.

[157] Gantner G., Haberl A., Praetorius D., and Schimanko S. (2021). Rate optimality of adaptive finite element methods with respect to overall computational costs. Mathematics of Computation, 90, 2011–2040. doi: 10.1090/mcom/3654.

[158] Jeon J., and Kim S. J.. (2021). FVM network to reduce computational cost of CFD simulation, arXiv, 1–20, [Online]. Available: https://www.researchgate.net/profile/Joongoo-Jeon/publication/351448497_FVM

[159] Pirozzoli S. (2011). Numerical methods for high-speed flows. Annual Review of Fluid Mechanics, 43, 163–194. doi: 10.1146/annurev-fluid-122109-160718.

[160] Rajanna M. R. et al. (2022). Finite element methodology for modeling aircraft aerodynamics: Development, simulation, and validation. Computational Mechanics, 70, 549–563. doi: 10.1007/s00466-022-02178-7.

[161] Lunati I., and Jenny P. (2006). Multiscale finite-volume method for compressible multiphase flow in porous media. Journal of Computational Physics, 216(2), 616–636. doi: 10.1016/j.jcp.2006.01.001.

[162] Özişik M. N., Orlande H. R. B., Colaço M. J., and Cotta R. M. (2017). Finite Difference Methods in Heat Transfer, 2nd ed. Boca Raton: CRC Press, doi: 10.1201/9781315121475.

[163] Haddad H., Guessasma M., and Fortin J. (2014). Heat transfer by conduction using DEM–FEM coupling method. Computational Materials Science, 81, 339–347. doi: 10.1016/j.commatsci.2013.08.033.

[164] Węcel G. (2006). BEM-FVM solution of the conjugate radiative and convective heat transfer problem. Archives of Computational Methods in Engineering, 13, 171–248. doi: 10.1007/BF02980230.

[165] Fasel H. (2006). Investigation of the stability of boundary layers by a finite-difference model of the Navier – Stokes equations. Journal of Fluid Mechanics, 78(2), 355–383. doi: 10.1017/S0022112076002486.

[166] Rispoli F., and Rafael S. G. (2006). A stabilized finite element method based on SGS models for compressible flows. 196(1–3, 652–664. doi: 10.1016/j.cma.2006.07.006.

[167] Wen Z., Jie O., Xiaodong W., Jin S., and Binxin Y. (2016). Numerical simulation of viscoelastic fluid flows using a robust FVM framework on triangular grid. Journal of Non-Newtonian Fluid Mechanics, 236, 18–34. doi: 10.1016/j.jnnfm.2016.08.003.

[168] Liu Y., Yang L., Shu C., and Zhang H. (2021). A multi-dimensional shock-capturing limiter for high-order least square-based Finite Difference-Finite Volume Method on unstructured grids. Advances in Applied Mathematics and Mechanics, 13(3), 671–700. doi: 10.4208/aamm.OA-2020-0255.

[169] Dabrowski M., Krotkiewski M., and Schmid D. W. (2008). MILAMIN: MATLAB-based finite element method solver for large problems. Geochemistry, Geophys. Geosytems, 9(4). doi: 10.1029/2007GC001719.

[170] Joshi H., Agarwal A., Puranik B., Shu C., and Agrawal A. (2010). A hybrid FVM–LBM method for single and multi-fluid compressible flow problems. International Journal of Numerical Methods in Fluids, 62(4), 403–427. doi: 10.1002/fld.2024.

[171] Kim J. W., and Sandberg R. D. (2012). Efficient parallel computing with a compact finite difference scheme. Comput Fluids, 58, 70–87. doi: 10.1016/j.compfluid.2012.01.004.

[172] Cheng Z., and Paraschivoiu M. (2003). Parallel computations of finite element output bounds for conjugate heat transfer. Finite Elements in Analysis and Design, 39(7), 581–597. doi: 10.1016/S0168-874X(02)00129-4.

[173] Afzal A., Ansari Z., Faizabadi A. R., and Ramis M. K. (2017). Parallelization strategies for Computational Fluid Dynamics software: State of the Art review. Archives of Computational Methods in Engineering, 24, 337–363. doi: 10.1007/s11831-016-9165-4.

[174] He Z., Li L., Zhang Y., and Tian B. (2018). Consistent implementation of characteristic flux-split based finite difference method for compressible multi-material gas flows. Comput Fluids, 168, 190–200. doi: 10.1016/j.compfluid.2018.04.007.

[175] An H., Ichikawa Y., Tachikawa Y., and Shiiba M. (2010). Three-dimensional finite difference saturated-unsaturated flow modeling with nonorthogonal grids using a coordinate transformation method. Water Resources Research, 46(11). doi: 10.1029/2009WR009024.
[176] Sujecki S. (2013). Application of Extended Taylor Series based Finite Difference Method in photonics. In 15th International Conference on Transparent Optical Networks (ICTON), 23–27 June 2013, Cartagena, Spain: IEEE, pp. 1–3. doi: 10.1109/ICTON.2013.6603003.
[177] Zhao M., Wang H., and Cheng A. (2018). A fast Finite Difference Method for three-dimensional time-dependent Space-fractional diffusion equations with fractional derivative boundary conditions. Journal of Scientific Computing, 74, 1009–1033. doi: 10.1007/s10915-017-0478-8.
[178] Saipol H. F. S., Alias N., and Nordin S. Z. (2024). Generalization of parallel performance for multidimensional finite difference method of PDE problem. In 3rd International Conference on Applied and Industrial Mathematics and Statistics 2022 (Icoaims2022): Mathematics and Statistics Manifestation the Excellence of Civilization, 24–26 August 2022 at Pahang, Malaysia, doi: 10.1063/5.0192112.
[179] Zhan W., Rao X., Zhao H., Zhang H., Hu S., and Dai W. (2022). Generalized finite difference method (GFDM) based analysis for subsurface flow problems in anisotropic formation. Engineering Analysis with Boundary Elements, 140, 48–58. doi: 10.1016/j.enganabound.2022.04.008.
[180] Zhen M., Qu K., and Cai J. (2021). A novel Finite Difference Method for Euler equations in 2D unstructured meshes. arXiv, 1–31. doi: 10.48550/arXiv.2102.12933.
[181] Salmasi F., and Abraham J. (2023). On the Finite Differences Method using Microsoft Excel. In Research Highlights in Mathematics and Computer Science, vol. 6, ch. 11, B P International, pp. 140–186. doi: 10.9734/bpi/rhmcs/v6/5540A.
[182] Makauskas P. (2022). Comparison of FDM, FVM with NN for solving the forward problem, Master's Final Degree Project (Thesis), Faculty of Mathematics and Natural Sciences, Kauno Technologijos Universitetas, [Online]. Available: https://epubl.ktu.edu/object/elaba:132245383/
[183] Leishman J. G. (2022). Types of fluid flows. In Introduction to Aerospace Flight Vehicles, ch. 14, Embry-Riddle Aeronautical University, doi: 10.15394/eaglepub.2022.1066.n11.
[184] Cheon H. L., Fyhn H., Hansen A., Wilhelmsen Ø., and Sinha S. (2023). Steady-state two-phase flow of compressible and incompressible fluids in a capillary tube of varying radius. Transport in Porous Media, 147, 15–33. doi: 10.1007/s11242-022-01893-2.
[185] Uma M. (2023). Flow pattern of compressible fluids in steady state. International Journal of Science and Research, 12(1), 1233–1237. doi: 10.21275/SR231010132523.
[186] Jain P., Vijay S. V., Asati Y., and Habibullah H. (2020). Compressible Flow in a Heat Exchanger Tubing, arXiv e-prints, pp. 1–55. doi: 10.13140/RG.2.2.31499.92964.
[187] Louisos W. F., and Hitt D. L. (2008). Supersonic micro-nozzles. In Encyclopedia of Microfluidics and Nanofluidics, Li D. (Ed.) 1st ed. Boston, MA: Springer, pp. 1900–1911. doi: 10.1007/978-0-387-48998-8_1486.
[188] Leishman J. G. (2022). Equations of fluid motion. In Introduction to Aerospace Flight Vehicles, ch. 19, Embry-Riddle Aeronautical University, [Online]. Available: https://eaglepubs.erau.edu/introductiontoaerospaceflightvehicles.
[189] Chervyakov V. M., and Vorob'ev Y. V. (2005). Unsteady-state flow of an ideal compressible medium in channels of a rotary apparatus. Theoretical Foundations of Chemical Engineering, 39(1), 62–69. doi: 10.1007/PL00022162.
[190] Gan T., and Wang Q. (2024). Experimental investigation of unsteady nature of shock waves induced by various ramps. The Physics of Fluids, 36(12). doi: 10.1063/5.0239594.
[191] Emelyanov V., Volkov K., and Yakovchuk M. (2022). Unsteady flow simulation of compressible turbulent flow in dual-bell nozzle with movement of extendible section from its initial to working position. Acta Astronautica, 194, 514–523. doi: 10.1016/j.actaastro.2021.10.007.

[192] Barakos G., and Drikakis D. (1999). An implicit unfactored method for unsteady turbulent compressible flows with moving boundaries. Comput Fluids, 28(8), 899–922. doi: 10.1016/S0045-7930(98)00057-7.

[193] Ye Q., Zhang Y., and Wei J. (2021). A comprehensive review of pulsating flow on heat transfer enhancement. Applied Thermal Engineering, 196(117275). doi: 10.1016/j.applthermaleng.2021.117275.

[194] Thorley A. R. D., and Tiley C. H. (1987). Unsteady and transient flow of compressible fluids in pipelines – A review of theoretical and some experimental studies. International Journal of Heat and Fluid Flow, 8(1), 3–15. doi: 10.1016/0142-727X(87)90044-0.

[195] Birken P. (2022). Numerical Methods for Unsteady Compressible Flow Problems, 1st ed. Chapman and Hall, [Online]. Available: https://www.routledge.com/Numerical-Methods-for-.

[196] Schafer O. (2002). Simulation of unsteady compressible flow in turbomachinery. Progress in Computational Fluid Dynamics, an International Journal, 2(1). doi: 10.1504/PCFD.2002.003212.

[197] Brooks F. J. (2000). GE Gas Turbine Performance Characteristics, Schenectady, NY. [Online]. Available: https://www.gevernova.com/content/dam/gepower-new/global

[198] Kukutla P. R., and Prasad B. V. S. S. S. (2017). Coupled flow network model and CFD analysis for a combined impingement and film cooled gas turbine nozzle guide vane. Adv. Model. Anal, 86(1), 250–270. doi: 10.18280/mmc_b.860118.

[199] ATET. (2012). Aircraft gas turbine thermo-fluid dynamics, design and performance, In PD031360 – Edexcel BTEC Level 6 Diploma in Engineering (QCF) units T24, 1, Pearson Education Limited, 1–8. [Online]. Available: https://qualifications.pearson.com/content/dam/pdf/BTEC-Professional-Qualifications

[200] Samarasinghe T., Abeykoon C., and Turan A. (2018). Modelling of heat transfer and fluid flow in the hot section of gas turbines used in power generation: A comprehensive survey. International Journal of Energy Research, 1–20. doi: 10.1002/er.4296.

[201] Ali U., and Kamran M. S. (2020). Geometric optimization of a gas turbine blade cooling passage using CFD. International Journal of Thermal and Environmental Engineering, 17(2), 105–114. doi: 10.5383/ijtee.17.02.004.

[202] Von Backstro T. W. (2009). The effect of specific heat ratio on the performance of compressible flow turbo-machines. In ASME Turbo Expo 2008: Power for Land, SeA, and Air on June 9–13, 2008 Berlin, Germany[Paper No: GT2008-50183]-Volume 6: Turbomachinery, Parts A, B, and C, International Gas Turbine Institute, pp. 2111–2117. doi: 10.1115/GT2008-50183.

[203] Saraireh M. (2024). Heat transfer rate and fluid flow analysis with design parameters of gas turbine using Beta-clog2-LSTM. Engineering, Technology and Applied Science Research, 14(5), 16281–16289. doi: 10.48084/etasr.8152.

[204] Walker T. D., and Bonham C. (2023). Volume 7B: Heat transfer – General interest/additive manufacturing impacts on heat transfer; Internal air systems; Internal cooling. In ASME Turbo Expo 2023: Turbomachinery Technical Conference and Exposition on June 26–30, 2023, Boston, Massachusetts, USA: International Gas Turbine Institute, doi: 10.1115/GT2023-FM7B.

[205] Bashi H., Esfahanian V., Tavakoli A., Ansari M., and Kordi M. (2025). Robust optimization of a gas turbine performance under geometrical and operational uncertainties using a novel robustness criterion. Structural and Multidisciplinary Optimization, 68(12). doi: 10.1007/s00158-024-03934-4.

[206] Arsenjev S. L., Lozovitski I. B., and Sirik Y. P. (2005). Laval turbonozzle dynamics. arXiv, 1–11. doi: 10.48550/arXiv.physics/0508003.

[207] Tauber M. E. (1989). A review of high-speed, convective, heat-transfer computation methods. In National Aeronautics and Space Administration (NASA), Office of Management Scientific and Technical Information Division, [Online]. Available: https://ntrs.nasa.gov/api/citations/19890017745/downloads/19890017745.pdf.

[208] Cogo M., Baù U., Chinappi M., Bernardini M., and Picano F. (2023). Assessment of heat transfer and Mach number effects on high-speed turbulent boundary layers. Journal of Fluid Mechanics, 974 (A10). doi: 10.1017/jfm.2023.791.

[209] Singh N., and Schwartzentruber T. E. (2016). Heat flux correlation for high-speed flow in the transitional regime. Journal of Fluid Mechanics, 792, 981–996. doi: 10.1017/jfm.2016.118.

[210] Singh N., and Schwartzentruber T. E. (2017). Aerothermodynamic correlations for high-speed flow. Journal of Fluid Mechanics, 821, 421–439. doi: 10.1017/jfm.2017.195.

[211] Limtrakarn W., and Dechaumphai P. (2004). Interaction of high-speed compressible viscous flow and structure by adaptive finite element method. International Journal of the Korean Society of Mechanical Engineers, 18, 1837–1848. doi: 10.1007/BF02984332.

[212] Liu F., Xiong T., and Yuan J. (2021). Relationship between the actual entry of vehicles into the Martian atmosphere and its simulation in a wind tunnel using Mars entry vehicles of various geometries. Transactions of the Japan Society for Aeronautical and Space Sciences, 19(3), 285–294. doi: 10.2322/tastj.19.285.

[213] Rhisat F. N. (2024). Numerical investigation of hypersonic flow characteristics and surface heating of a Mars entry vehicle, Degree of Master of Science Thesis in Mechanical Engineering, Graduate School of Southern Illinois University Edwardsville, [Online]. Available: https://www.proquest.com/openview/c9d2c43ccd654941349

[214] Hollis B. R., and Liechty D. S. (2012). Transition due to heat-shield cavities on a mars entry vehicle. Journal of Spacecraft and Rockets, 43(2). doi: 10.2514/1.19669.

[215] Zhu L., Wu L., Zhang Y., and Sharipov F. (2019). Ab initio calculation of rarefied flows of helium-neon mixture: Classical vs quantum scatterings. International Journal of Heat and Mass Transfer, 145 (118765). doi: 10.1016/j.ijheatmasstransfer.2019.118765.

[216] Zhang S., Feng Y., Zhang D., Jiang Y., Qin J., and Bao W. (2016). Parametric numerical analysis of regenerative cooling in hydrogen fueled scramjet engines. International Journal of Hydrogen Energy, 41(25), 10942–10960. doi: 10.1016/j.ijhydene.2016.03.176.

[217] Hamilton B. (2024). A new method for cryogenic fluid thermal conductivity measurements and thermophysical and transport properties of hydrogen-helium mixtures. In Degree of Doctor of Philosophy in Mechanical Engineering, Massachusetts Institute of Technology, [Online]. Available: https://dspace.mit.edu/bitstream/handle/1721.1/155652/.

[218] Tamilarasan A. (2015). Numerical modelling of heat transfer and evaporation characteristics of cryogenic liquid propellant. In Degree of Master of Science in Mechanical Engineering, University of Washington, [Online]. Available: https://digital.lib.washington.edu/server/api/core/bitstreams/3379.

[219] Shanmugam A. R., and Park K. S. (2024). Flow and heat transfer of supercritical hydrogen in a regenerative cooling channel with the arc ribs of a rocket engine. Applied Thermal Engineering, 236 (Part A). doi: 10.1016/j.applthermaleng.2023.121451.

[220] Locke J. M., and Landrum D. B. (2012). Study of heat transfer correlations for supercritical hydrogen in regenerative cooling channels. Journal of Propulsion and Power, 24(1). doi: 10.2514/1.22496.

[221] Mohekar A. A., Tilley B. S., and Yakovlev V. V. (2023). Resonance, Rayleigh flows, and thermal choking: Compressible coolant states in porous electromagnetic heat exchangers. Journal of Applied Physics, 133(12). doi: 10.1063/5.0139723.

[222] Maniscalco N. I. (2014). Design and performance characterization of high power density air-cooled compact condenser heat exchangers, Degree of Doctor of Philosophy Dissertation in Mechanical Engineering in the Graduate College of the University of Illinois at Urbana-Champaign, [Online]. Available:: https://www.ideals.illinois.edu/items/50450

[223] Abeykoon C. (2021). Modelling of heat exchangers with computational fluid dynamics. In Proceedings of the 8th International Conference on Fluid Flow, Heat and Mass Transfer (FFHMT'21) Niagara Falls, Canada Virtual Conference – May 21–23, 2021, pp. 1–9. doi: 10.11159/ffhmt21.127.

[224] Satra N., and Dixit S. (2018). Thermal and fluid flow analysis of a heat exchanger: 'A comprehensive report,'. International Journal of Engineering Research and Technology, 5(11), 1012–1020. [Online]. Available: https://www.irjet.net/archives/V5/i11/IRJET-V5I11194.pdf.
[225] USDE. (1993). Heat exchanger fundamentals-Module 2. In DOE Fundamentals Handbook-Mechanical Science[Volume 1 of 2], DOE-HDBK-1018/1–93, 20585, Washington, D.C: U.S. Department of Energy FSC-6910, pp. 1–18. [Online]. Available: https://d6s74no67skb0.cloudfront.net/course-material/ME905-Heat-Exchanger-Fundamentals.pdf.
[226] Jurandir P. P. E. (2010). Shell and tube heat exchangers basic calculations. [Online]. Available: https://pdhonline.com/courses/m371/m371content.pdf
[227] El-Semary I. M. M. (Mar 04 2025). Lecture 4: Heat exchanger. Accessed: [Online] Available: https://bu.edu.eg/portal/uploads/Engineering%2CShoubra/.
[228] Levenspiel O. (1984). The three mechanisms of heat transfer: Conduction, convection, and radiation. In Engineering Flow and Heat Exchange. The Plenum Chemical Engineering Series, ch. 9, Boston, MA: Springer, pp. 161–188. doi: 10.1007/978-1-4615-6907-7_9.
[229] Brès G. A., and Colonius T. (2008). Three-dimensional instabilities in compressible flow over open cavities. Journal of Fluid Mechanics, 599, 309–339. doi: 10.1017/S0022112007009925.
[230] Pilkhwal D. S., Ambrosini W., Forgione N., Vijayan P. K., Saha D., and Ferreri J. C. (2007). Analysis of the unstable behaviour of a single-phase natural circulation loop with one-dimensional and computational fluid-dynamic models. Annals of Nuclear Energy, 34(5), 339–355. doi: 10.1016/j.anucene.2007.01.012.
[231] Behl M. (2007). Discretization methods in fluid dynamics, Indo-German Winter Academy. [Online]. Available: https://www.researchgate.net/profile/Mahesh-Nagarkar
[232] Teena R., Chaturvedi B. K., and Preeti J. (2023). Numerical stability and accuracy in CFD for unsteady flows. International Journal of Engineering Mathematics, 12(6), 169–178. [Online]. Available: https://www.indianjournals.com/ijor.aspx?target.
[233] Roache P. J. (1997). Quantification of uncertainty in Computational Fluid Dynamics. Annual Review of Fluid Mechanics, 29, 123–160. doi: 10.1146/annurev.fluid.29.1.123.
[234] Sarshar A., Tranquilli P., Pickering B., McCall A., Roy C. J., and Sandu A. (2017). A numerical investigation of matrix-free implicit time-stepping methods for large CFD simulations. Comput Fluids, 159, 53–63. doi: 10.1016/j.compfluid.2017.09.014.
[235] Han G. (2003). Rock stability under different fluid flow conditions, Degree of Doctor of Philosophy Thesis in Chemical Engineering, University of Waterloo, [Online]. Available: https://uwspace.uwaterloo.ca/items/b1f3f3e7-683a-45f3-beae-933f6391dadd
[236] Lerat A., and Peyret R. (2005). The problem of spurious oscillations in the numerical solution of the equations of gas dynamics. In Proceedings of the Fourth International Conference on Numerical Methods in Fluid Dynamics, Richtmyer R. D. (Ed.) Berlin, Heidelberg: Springer, pp. 251–256. doi: 10.1007/BFb0019759.
[237] Gao T. et al. (2023). One-dimensional turbulence modeling of compressible flows: II. Full compressible modification and application to shock–turbulence interaction. The Physics of Fluids, 35(3). doi: 10.1063/5.0137435.
[238] Argyropoulos C. D., and Markatos N. C. (2015). Recent advances on the numerical modelling of turbulent flows. Applied Mathematical Modelling, 39(2), 693–732. doi: 10.1016/j.apm.2014.07.001.
[239] Vinuesa R. (2021). High-fidelity simulations in complex geometries: Towards better flow understanding and development of turbulence models. Results in Engineering, 11(100254), 1–11. doi: 10.1016/j.rineng.2021.100254.
[240] Abd Halim M. A., Mohd N. A. R. N., Nasir N., and Bin Dahalan N. (2018). The evaluation of k-ε and k-ω turbulence models in modelling flows and performance of S-shaped diffuser. International Journal of Automotive and Mechanical Engineering, 15(2), 5161–5177. doi: 10.15282/ijame.15.2.2018.2.0399.

[241] Shishkina O., Wagner C., Kath C., Wagner S., Weinman K., and Feldmann D. (2013). Direct Numerical Simulations (DNS) and Large Eddy Simulations (LES) – Theory and practice-Course Notes. In Course Note Conference Presentation at Göttingen, Wagner C., and Klein S. (Eds.) [Online]. Available: https://elib.dlr.de/81590/.

[242] Škerlavaj A., Škerget L., Ravnik J., and Lipej A. (2011). Choice of a turbulence model for pump intakes. Proceedings of the Institution of Mechanical Engineers, Part A: Journal of Power and Energy, 225(6). doi: 10.1177/0957650911403870.

[243] Ranadive H. (2019). High-order compressible formulation for LES/PDF simulations of turbulent reacting flows, PhD Doctorate Thesis, School of Mechanical and Manufacturing Engineering, Faculty of Engineering, The University of New South Whales, doi: 10.26190/unsworks/3800.

[244] Mor-Yossef Y. (2024). Quasi-Newton positivity-preserving scheme for the Spalart-Allmaras turbulence model using unstructured grids. Journal of Computational Physics, 513(113166). doi: 10.1016/j.jcp.2024.113166.

Reshu Gupta

Chapter 3
Non-Newtonian fluid flow and heat transfer

Abstract: This chapter examines the analytical study of flow and heat transfer in a non-Newtonian Reiner-Rivlin fluid confined between two rotating disks in the presence of a magnetic field. We applied the Von Kármán similarity transformation to the higher-order partial differential equations, resulting in a system of nonlinear ordinary differential equations (ODEs) of flow and heat transfer. The system of nonlinear ODEs related to boundary conditions is solved using the differential transform method (DTM). The energy equation takes the viscous dissipation function into account. Therefore, the Eckert number is involved in the energy equation. We go into great detail about how the temperature and velocity profiles are affected by factors like Reynolds number, viscoinelastic parameter, forced parameter, Prandtl number, and magnetic field. The accuracy of DTM is demonstrated by a comparison of the findings produced by DTM with numerical results. Shear stress is computed for the top and bottom disks. Temperature profiles and velocity components can be calculated numerically and analytically using Maple software.

Keywords: Reiner-Rivlin fluid, heat transfer, magnetic field, differential transform method

3.1 Introduction

The research on the phenomenon of heat and mass transfer in fluid flow systems has numerous industrial applications, including gas turbines, processing of foods, thermal management techniques, and energy conversion processes. Its numerous applications have made it a popular field for researchers, and as a result, the phenomenon of rotating fluid flow has generated a lot of interest. It has been extensively examined by numerous scholars who have stayed committed to Von Kármán's [1] groundbreaking study on the rotating disk flow problem. He made the assumption that in a still fluid, a disk with a large radius rotates at a constant angular velocity. The fluid exits the disk radially close to it due to the absence of centrifugal force. This fluid is replaced by fluid above the disk in a downward spiraling motion known as the disk free pumping action. Many further studies on spinning disk flows have been influenced by von Kármán's work. The von Kármán problem is a limiting instance of the family of flows

Reshu Gupta, Applied Science Cluster (Mathematics), UPES, Dehradun 248001, Uttarakhand, India, e-mail: guptareshu2002@gmail.com

https://doi.org/10.1515/9783111661674-003

that arises when the ambient fluid and the disk rotate along the same axis at different angular velocities, as demonstrated by Bachelor [2]. Bödewadt [3] began by addressing another specific scenario in which the disk is stationary and the fluid at infinity is in a state of rotation. According to the results of the research of Jalili et al. [4], a higher Reynolds number typically leads to higher velocities and lower temperature distributions, while the effects of other parameters, like the forced and cross viscous parameters, on the temperature distribution and velocity profiles are more complicated and location dependent. Zhao et al. [5] discussed the flow of Ree-Eyring fluid between two rotating disks using mixed convective entropy-optimized nanomaterial magnetohydrodynamics (MHD). Porosity and velocity slip effects are taken into account. The angular frequency and stretching rates of the two revolving disks differ. The important chapter can be studied in [6–10].

The non-Newtonian fluid model, which has numerous applications in various fields, was covered by Reiner [11] and Rivlin [12]. In order to clarify the idea of "dilatancy," or a material's ability to expand its volume when disturbed, Reiner [11] proposed a class of fluids. This type of behavior seen in granular materials is comparable to the rod-climbing phenomenon brought on by churning and the expansion of non-Newtonian fluids caused by shearing. The Reiner-Rivlin (RR) model has a clear function in the dynamics of rotating-disk systems because it may be used to examine the flow behavior of geological formations, biological fluids, food products, and chemicals. Tabassum and Mustafa [13] conclude that as the parameters for wall roughness increase, the radial velocity profile rises away from the disk and falls close to it. Also, with an increase in the thermal slip parameter, the thermal boundary layer contracts and the rate of heat transfer increases. The main goal of Puspanathan et al. [14] research is to use a special RR fluid to clarify the behavior of swirling flow around a rotating disk that is radially diminishing. The first solution is determined to be stable and physically credible after a stability study is conducted to evaluate their physical relevance. On the other hand, the second option is unstable. Recent efforts in this field are available through [15–22].

MHD is the study of the properties, effects, and behavior of magnetic fields over viscous fluids. Examples of magnetofluids include electrolytes, plasmas, and salt water. Researchers and investigators are paying close attention to this topic in the modern day. In this area, a great deal of work has been done. The interplay of conducting fluid and magnetic field creates the tectonic uses of MHD in engineering, chemistry, physics, industrial tackle, and many other industries, such as pumps, bearings, MHD generators, and boundary layer control. The work of many explorations has been discussed in relation to these applications. The first person to innovate in the field of MHD was Hannes Alfén [23]. He was awarded the 1970 Nobel Prize in Physics for his contributions to the field of MHD. Applications of MHD fluid flow in distinct configurations relevant to human body parts are highly intriguing and tectonic in the field of medicine. Rashidi et al. [24] examine the significant uses of MHD in drug distribution, pulsatile flow, simple flow, and peristaltic flow. The majority of the most remarkable real-world applications are nonlinear and characterized by nonlinear equations. Nonlinear ordinary differential equations (ODEs) are present in many mathematical issues and can be resolved both analytically and quantitatively. The differential transform approach is one of the semi-analytical methods that does not call for tiny parameters. This method can be used to convert nonlinear ODEs into linear ones. Zhou was the first to apply the differential transform method (DTM) to solve the equations for electric circuits [25]. One can find recent work applying DTM through [26–29].

We examined the heat transmission and flow characteristics of non-Newtonian liquids over geometry with physical qualities in each of the aforementioned studies. With the goal of using DTM via Maple to represent non-Newtonian RR flow, continuity, momentum, and concentration are being mathematically formulated in terms of the RR fluid model. Using charts, the distribution profiles, which include velocity and temperature are carefully examined.

3.2 Mathematical formulation

Consider the upper disk (UD) is rotating uniformly about the vertical axis, causing the non-Newtonian viscoinelastic RR fluid to flow through the gap between the two disks. Figure 3.1 shows the entire structure of the cylindrical coordinate system (r, θ, z) model, with RR fluid positioned between two disks. The lower disk (LD) and UD are positioned at $z = 0$ and $z = d$. The LD is stretching at the rate of S and UD is rotating with the angular velocity Ω. The temperatures at both disks are T_1, and T_2. Figure 3.1 illustrates the vertical imposition of the magnetic field of strength B_0. The RR fluid is injected with the velocity ω_0 from UD.

The base equation suggested by Reiner [11] and Rivlin [12] for RR fluid is:

$$\tau_{ij} = -p\delta_{ij} + 2\mu_{nv} d_{ij} + 4\mu_{cv} c_{ij}$$

where

$$d_{ij} = \frac{1}{2}\left(u_{i,j} + u_{j,i}\right), \quad c_{ij} = d_{im} d_j^m$$

where τ_{ij} is the stress tensor,δ_{ij} is the Kronecker's delta, μ_{nv} is the coefficient of Newtonian viscosity, μ_{cv} is the coefficient of cross-viscosity, and d_{ij}, c_{ij} are physical components of the stress tensor.

The governing equations in the cylindrical form are:

$$\frac{\partial u}{\partial r} + \frac{u}{r} + \frac{\partial w}{\partial z} = 0 \tag{3.1}$$

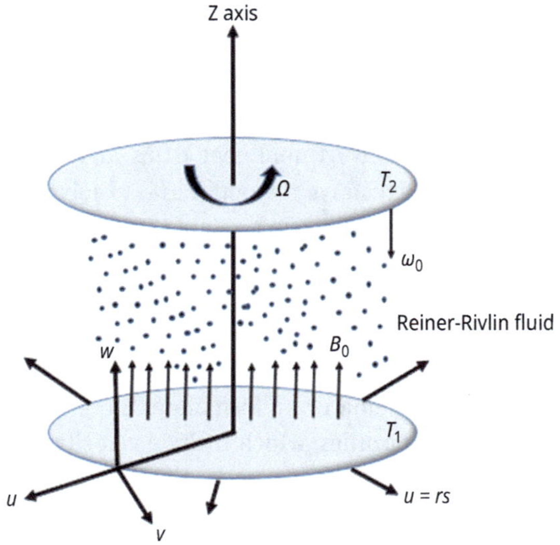

Figure 3.1: Model description.

$$u\frac{\partial u}{\partial r}-\frac{v^2}{r}+w\frac{\partial u}{\partial z}=\frac{1}{\rho}\left(\frac{\partial \tau_{rr}}{\partial r}+\frac{\partial \tau_{rz}}{\partial z}+\frac{\tau_{rr}-\tau_{\theta\theta}}{r}-\sigma {B_0}^2 u\right) \tag{3.2}$$

$$u\frac{\partial v}{\partial r}+\frac{uv}{r}+w\frac{\partial v}{\partial z}=\frac{1}{\rho}\left(\frac{\partial \tau_{r\theta}}{\partial r}+\frac{\partial \tau_{\theta z}}{\partial z}+\frac{2\tau_{r\theta}}{r}-\sigma {B_0}^2 v\right) \tag{3.3}$$

$$u\frac{\partial w}{\partial r}+w\frac{\partial w}{\partial z}=\frac{1}{\rho}\left(\frac{\partial \tau_{rz}}{\partial r}+\frac{\partial \tau_{zz}}{\partial z}+\frac{\tau_{rz}}{r}\right) \tag{3.4}$$

The energy equation for the steady flow excluding the external force is:

$$\left(u\frac{\partial T}{\partial t}+w\frac{\partial T}{\partial z}\right)=\frac{k}{\rho c_v}\left(\frac{\partial^2 T}{\partial t^2}+\frac{1}{r}\frac{\partial T}{\partial r}+\frac{\partial^2 T}{\partial z^2}\right)+\Phi \tag{3.5}$$

$$\Phi=\tau_{rr}d_{rr}+\tau_{\theta\theta}d_{\theta\theta}+\tau_{zz}d_{zz}+2(\tau_{r\theta}d_{r\theta}+\tau_{rz}d_{rz}+\tau_{\theta z}d_{\theta z}) \tag{3.6}$$

where ρ is the density, σ is the conductivity, τ_{rr}, $\tau_{\theta\theta}$, τ_{zz}, $\tau_{r\theta}$, $\tau_{\theta z}$, τ_{rz} are the components of the stress tensor, c_v is the specific heat at fixed volume, k is the thermal conductivity, T is the temperature, Φ is the viscous dissipation function.

The associated boundary conditions are:

$$\text{At } z=0\text{: } u=rs,\quad v=0,\quad w=0,\quad T=T_1$$

$$\text{At } z=d\text{: } u=0,\quad v=r\Omega,\quad w=-w_0,\quad T=T_2 \tag{3.7}$$

The appropriate similarity transformation is:

$$u = r\Omega f'(\eta), \quad v = r\Omega g(\eta), \quad w = -2d\Omega f(\eta), \quad T^* = \frac{T - T_1}{T_2 - T_1}, \quad \eta = \frac{z}{d} \tag{3.8}$$

The suggested form of temperature distribution is:

$$T = T_0 + \frac{\nu\Omega}{c_v}\left(\phi(\eta) + \xi^2\psi(\eta)\right), \quad \xi = \frac{r}{d} \tag{3.9}$$

Solving eqs. (3.2)–(3.4) with the help of eq. (3.8), we get the following transformed equations:

$$f'''' + 2R_o\left(gg' + ff'''\right) - 2R_o\tau_3\left(f'f'''' + 3g'g'' + 2f''f'''\right) - M^2 f'' = 0 \tag{3.10}$$

$$g'' + 2R_o\tau_3\left(f''g' - f'g''\right) + 2R_o\left(fg' - f'g\right) - M^2 g = 0 \tag{3.11}$$

Substituting eq. (3.9) into eq. (3.5) and comparing the coefficients of ξ^2 and free from ξ^2, we get:

$$\psi'' + R_oP_r\left(2f\psi' - 2\psi f' + f''^2 + g'^2\right) - 3R_o^2\tau_3P_r\left(f'f''^2 + f'g'^2\right) = 0 \tag{3.12}$$

$$\phi'' + 4\psi - 24R_o^2\tau_3P_rf'^3 + 2P_rR_o\left(6f'^2 - f\phi'\right) = 0 \tag{3.13}$$

where $R_o\left[= \frac{\rho d^2\Omega}{\mu_{nv}}\right]$ is the Reynolds number, $P_r\left[= \frac{\mu_{nv}c_v}{k}\right]$ is the Prandtl number, $\tau_3\left[= \frac{\mu_{cv}}{\rho d^2}\right]$ is the cross-viscosity parameter, $M\left[= \frac{B_o{}^2\sigma d^2}{\mu_{nv}}\right]$ is the magnetic field parameter.

The dimensionless form the temperature distribution is:

$$T^* = \frac{T - T_1}{T_2 - T_1} = E_r\left(\phi(\eta) + \xi^2\psi(\eta)\right) \tag{3.14}$$

$E_r\left[= \frac{\rho c_v(T_2 - T_1)}{\mu_{nv}\Omega}\right]$ is the Eckert number

The transformed boundary conditions are:

$$f'(0) = S\left[= \frac{s}{\Omega}\right]; \; g(0) = 0; \; f(0) = 0; \quad \phi(0) = 0; \; \psi(0) = 0$$

$$f'(1) = 0; \; g(1) = 1; \; f(1) = A\left[= \frac{w_0}{2d\Omega}\right]; \quad \phi(1) = \frac{1}{E_c}; \; \psi(1) = 0 \tag{3.15}$$

The skin friction coefficient on the LD and UD is calculated by:

$$\tau_{rz} = \mu_{nv}\left(\frac{\partial u}{\partial z}\right)_{z=0} = \frac{r\Omega}{d}f''(0), \tau_{\theta z} = \mu_{nv}\left(\frac{\partial v}{\partial z}\right)_{z=0} = \frac{r\Omega}{d}g'(0)$$

The Nusselt number on the LD and the UD is evaluated by:

$$Nu_a = \frac{T_a d}{(T_1 - T_2)k}, \; Nu_b = \frac{T_b d}{(T_1 - T_2)k}$$

where

$$T_a = \frac{1}{\pi\left(\xi^2 - {\xi_0}^2\right)} \int_{\xi_0}^{\xi} 2\pi\xi H_{fa} d\xi, \; T_b = \frac{1}{\pi\left(\xi^2 - {\xi_0}^2\right)} \int_{\xi_0}^{\xi} 2\pi\eta\xi H_{fb} d\xi$$

where heat flux

$$H_{fa} = -\frac{k}{d}\left(\frac{\partial T^*}{\partial \eta}\right)_{\eta=0}, \quad H_{fb} = -\frac{k}{d}\left(\frac{\partial T^*}{\partial \eta}\right)_{\eta=1}$$

3.2.1 Application of DTM

Using the theorem from Gupta et al. [27], the transformed form of eqs. (3.10)–(3.13) are:

$$\begin{aligned}
&(k+4)(k+3)(k+2)(k+1)F(k+4) + 2R_o\left(\sum_{l=0}^{k}(l+1)G(l+1)G(k-l)\right.\\
&\left. + \sum_{l=0}^{k}(l+3)(l+2)(l+1)F(l+3)F(k-l)\right)\\
&- 2R_o\tau_3\left(\sum_{l=0}^{k}(l+4)(l+3)(l+2)(l+1)F(l+4)(k-l+1)F(k-l+1)\right.\\
&+ 3\sum_{l=0}^{k}(l+2)(l+1)G(l+2)(k-l+1)G(k-l+1)\\
&\left. + 2\sum_{l=0}^{k}(l+3)(l+2)(l+1)F(l+3)(k-l+2)(k-l+1)F(k-l+2)\right)\\
&- M^2(k+2)(k+1)F(k+2) = 0
\end{aligned} \tag{3.16}$$

$$\begin{aligned}
&(k+2)(k+1)G(k+2) + 2R_o\tau_3\left(\sum_{l=0}^{k}(l+2)(l+1)F(l+2)(k-l+1)G(k-l+1)\right.\\
&\left. - \sum_{l=0}^{k}(l+2)(l+1)G(l+2)(k-l+1)F(k-l+1)\right) + 2R_o\left(\sum_{l=0}^{k}(l+1)G(l+1)F(k-l)\right.\\
&\left. - \sum_{l=0}^{k}(l+1)F(l+1)G(k-l)\right) - M^2G(k) = 0
\end{aligned} \tag{3.17}$$

$$(k+2)(k+1)\Psi(k+2)+R_oP_r\left(2\sum_{l=0}^{k}(l+1)\Psi(l+1)F(k-l)\right.$$

$$-2\sum_{l=0}^{k}(l+1)F(l+1)\Psi(k-l)+\sum_{l=0}^{k}(l+2)(l+1)F(l+2)(k-l+2)(k-l+1)F(k-l+2)$$

$$\left.+\sum_{l=0}^{k}(l+1)G(l+1)(k-l+1)G(k-l+1)\right)$$

$$-3R_o{}^2\tau_3P_r\left(\sum_{l_1=0}^{k}\sum_{l=0}^{l_1}(l+2)(l+1)F(l+2)(l_1-l+2)(l_1-l+1)F(l_1-l+2)(k-l+1)F(k-l+1)\right.$$

$$\left.+\sum_{l_1=0}^{k}\sum_{l=0}^{l_1}(l+1)G(l+1)(l_1-l+1)G(l_1-l+1)(k-l_1+1)F(k-l_1+1)+\right)=0 \quad (3.18)$$

Transformed Boundary Conditions are:

$$F(0)=0,\quad F(1)=S,\quad F(2)=a,\quad F(3)=b,\quad G(0)=0$$

$$G(1)=c,\quad \Phi(0)=0,\quad \Psi(0)=0,\quad \Phi(1)=d,\quad \Psi(1)=e \quad (3.19)$$

Using BCs from eq. (3.19) together with eq. (3.15), we get values of constants a, b, c, d, e.

3.3 Results and discussions

Figure 3.2 shows the behavior of the AV for different values of the Reynolds number. It states that AV has increasing nature from LD to the UD. It is also clear from this figure that AV increased with increasing Reynolds number because the effect of inertial forces increases, causing the velocity to increase. Figure 3.3 displays the nature of the RV for various values of the Reynolds number. RV rises near the LD and falls thereafter. Also, RV rises on raising the values of the Reynolds number while its behavior is reversed after nearing the UD. According to Figure 3.4, TV decreases with an increase in the Reynolds number. Figure 3.5 reflects the influence of the Reynolds number on temperature. It rises rapidly but falls near the UD. It has a maximum temperature near the UD. Also, it rises with the increasing value of the Reynolds number. Tables 3.1 and 3.2 present the data of the shear stress on the LD and UD.

Figures 3.6–3.9 reflect the nature of the AV, RV, TV, and temperature influenced by A. Figure 3.6 shows the increasing nature of AV while RV increases near the UD and decreases thereafter in Figure 3.7. Both velocities rise on rising the values of A. On the other side, Figure 3.8 displays that TV falls with increased A. According to Figure 3.9, temperature rises with A.

Table 3.1: Data of shear stress on LD.

R	*M*	τ_3	*A*	$f''(0)$		$g'(0)$	
				DTM	NM	DTM	NM
0.01	5	10	0.5	0.8115118	0.8115124	0.0596619	0.0596624
0.02	5	10	0.5	0.8566179	0.8566184	0.0521425	0.0521432
0.03	5	10	0.5	0.9060861	0.9060865	0.0448659	0.0448665
0.05	2	10	0.5	1.1832178	1.1832185	0.4357358	0.4357364
0.05	3	10	0.5	1.1510145	1.1510153	0.1968176	0.1968184
0.05	4	10	0.5	1.0859457	1.0859464	0.0805795	0.0805798
0.05	5	6	0.5	0.9075205	0.9075213	0.0450436	0.0450444
0.05	5	7	0.5	0.9337341	0.9337348	0.0414612	0.0414621
0.05	5	8	0.5	0.9612171	0.9612177	0.0379607	0.0379614
0.05	5	10	0.2	− 1.6702846	− 1.6702855	0.0458277	0.0458283
0.05	5	10	0.3	− 0.7547030	− 0.7547037	0.0411041	0.0411047

Table 3.2: Data of shear stress on UD.

R	*M*	τ_3	*A*	$f''(1)$		$g'(1)$	
				DTM	NM	DTM	NM
0.01	5	10	0.5	− 3.0593193	− 3.0593195	5.1602042	5.1602048
0.02	5	10	0.5	− 2.7894151	− 2.7894155	5.3172531	5.3172538
0.03	5	10	0.5	− 2.4933878	− 2.4933881	5.4686116	5.4686121
0.05	2	10	0.5	− 1.6066751	− 1.6066757	2.6029375	2.6029382
0.05	3	10	0.5	− 1.6183761	− 1.6183770	3.6199398	3.6199404
0.05	4	10	0.5	− 1.6970922	− 1.6970927	4.6716819	4.6716829
0.05	5	6	0.5	− 2.4927079	− 2.4927085	5.4583766	5.4583774
0.05	5	7	0.5	− 2.3352437	− 2.3352444	5.5332146	5.5332154
0.05	5	8	0.5	− 2.1708801	− 2.1708807	5.6053984	5.6053990
0.05	5	10	0.2	0.2183122	0.2183127	5.1076921	5.1076926
0.05	5	10	0.3	− 0.4868069	− 0.4868076	5.3068072	5.3068078
0.05	5	10	0.4	− 1.1671870	− 1.1671875	5.5172735	5.5172738

The influence of S on the velocity and temperature profile is displayed in Figures 3.10–3.13. Figure 3.10 shows the rising nature of the AV. Also, it increases by increasing the values of S. Figure 3.11 displays that RV increases with near the UD while decreases with S thereafter. No significant changes in RV are shown in Figure 3.12. Also, the temperature falls with increased values of S in Figure 3.13.

The influence τ_3 of on the velocity and temperature profile is displayed in Figures 3.14–3.17. In Figure 3.14, AV increases with the increasing values of τ_3. RV also has similar behavior, but its behavior is opposite near the UD (Figure 3.15). TV falls with more τ_3 (Figure 3.16). Figure 3.17 reflects the rising nature of temperature with τ_3.

The influence M of on the velocity and temperature profile is displayed in Figures 3.18–3.21. In Figure 3.18, AV falls with the increasing values of M. RV also has a

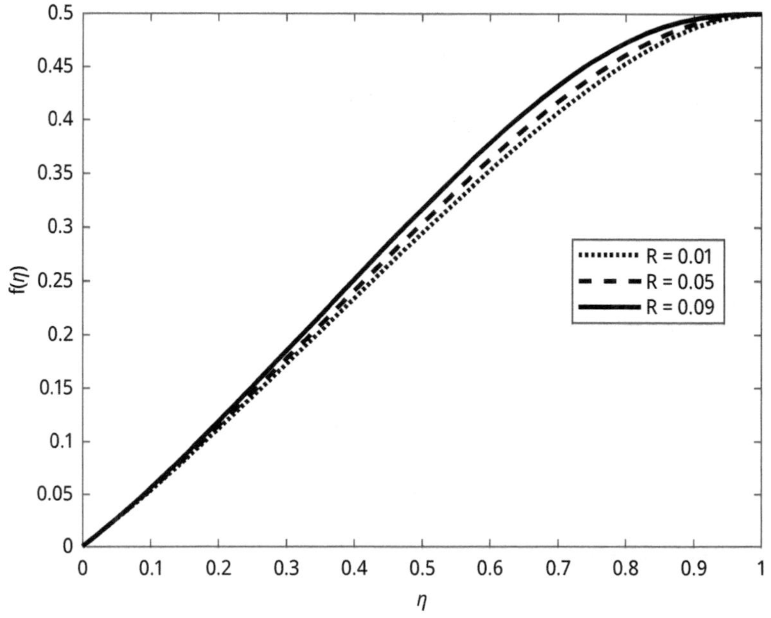

Figure 3.2: AV nature with respect to *R*.

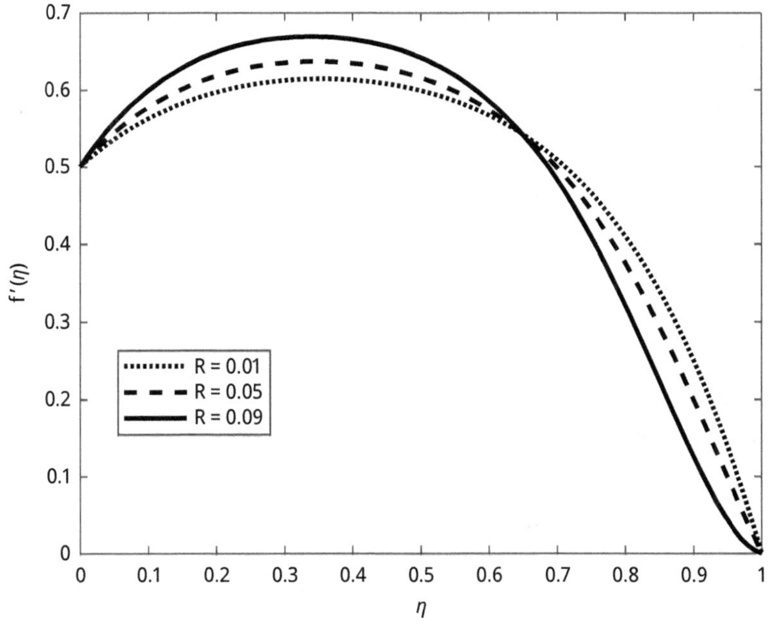

Figure 3.3: RV nature with respect to *R*.

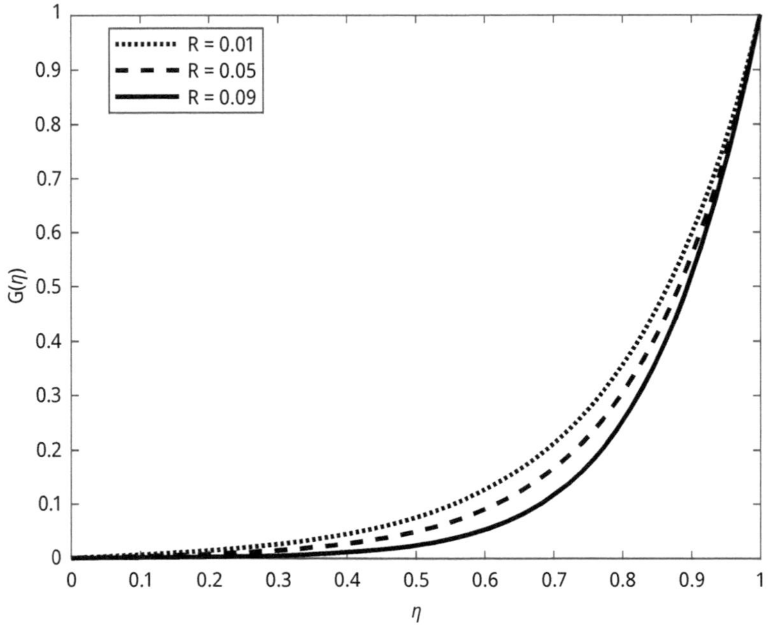

Figure 3.4: TV nature with respect to *R*.

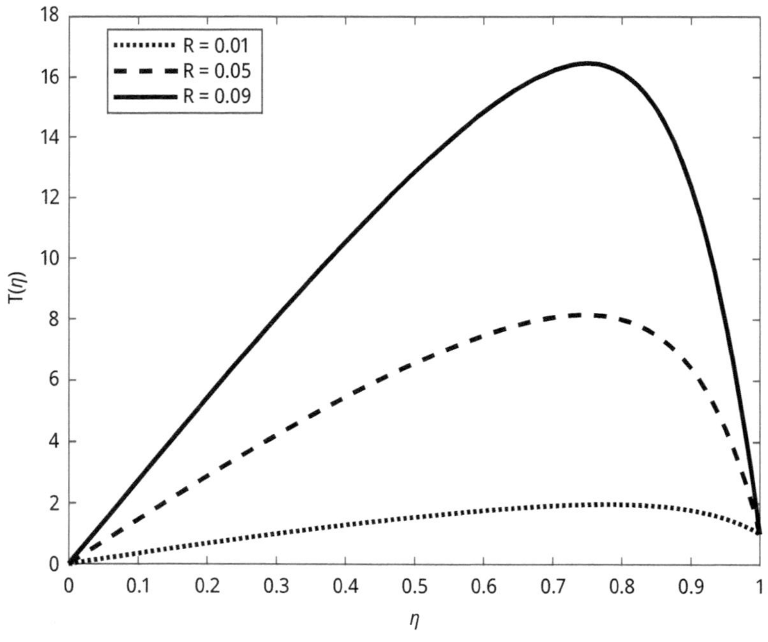

Figure 3.5: Temperature nature with respect to *R*.

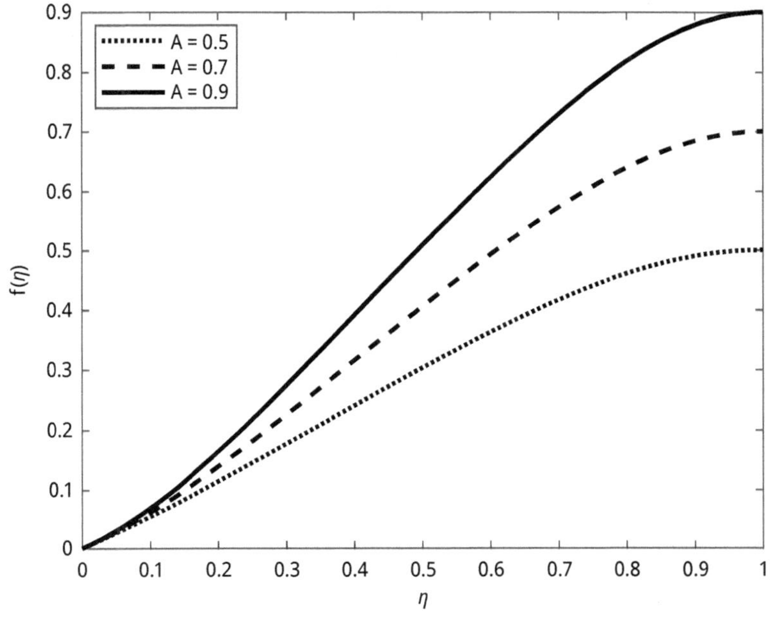

Figure 3.6: AV nature with respect to *A*.

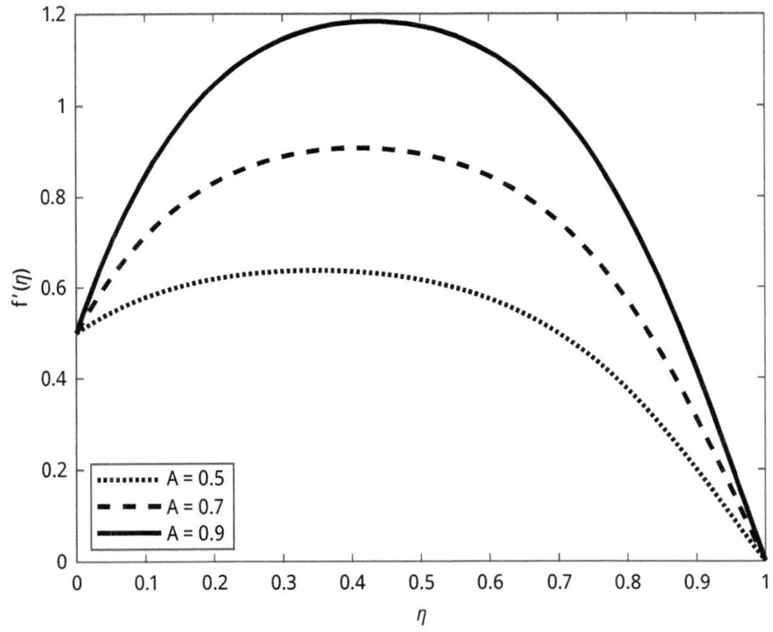

Figure 3.7: RV nature with respect to *A*.

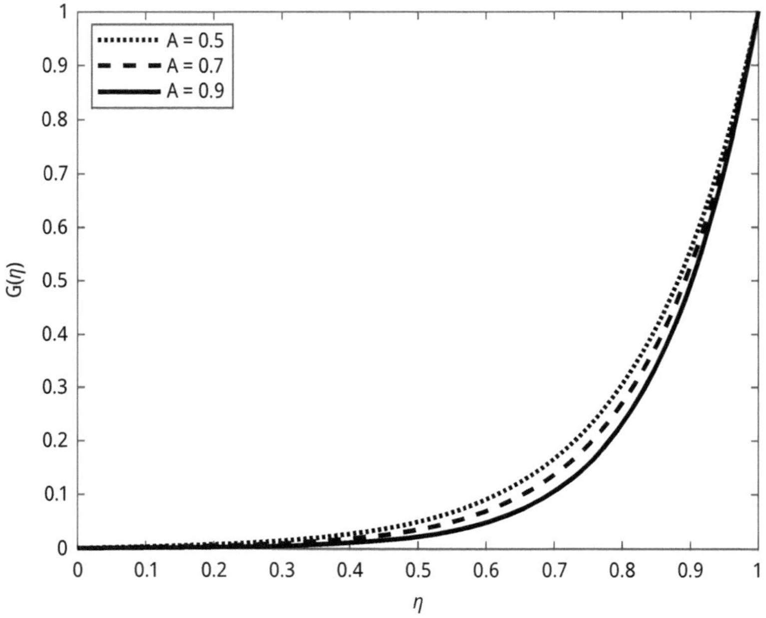

Figure 3.8: TV nature with respect to *A*.

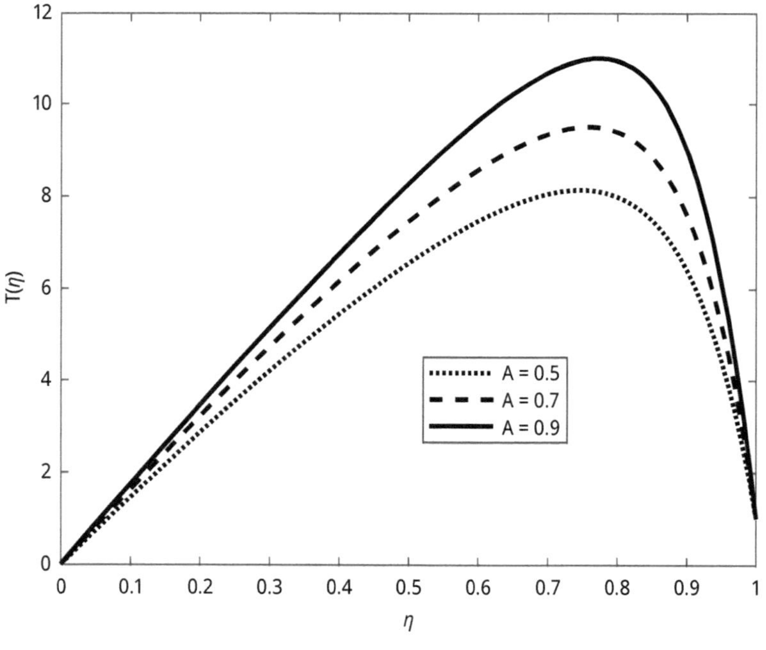

Figure 3.9: Temperature nature with respect to *A*.

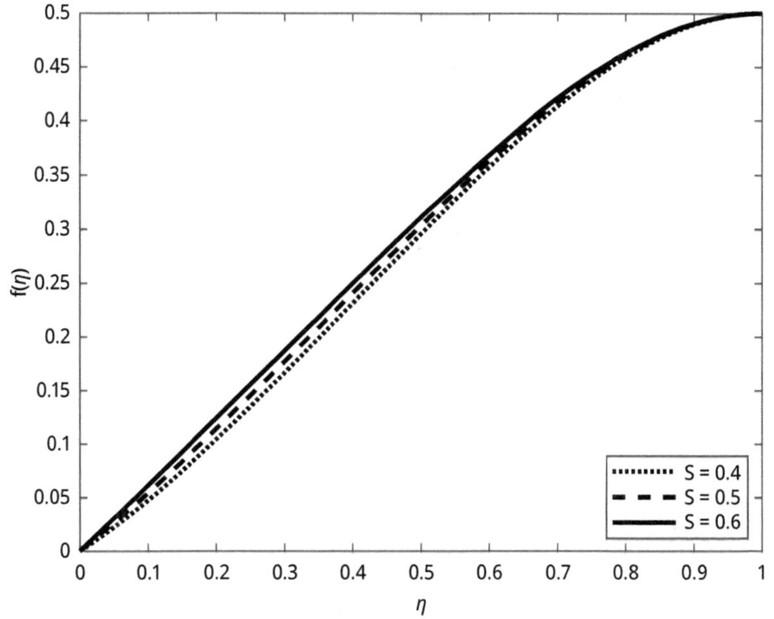

Figure 3.10: AV nature with respect to *S*.

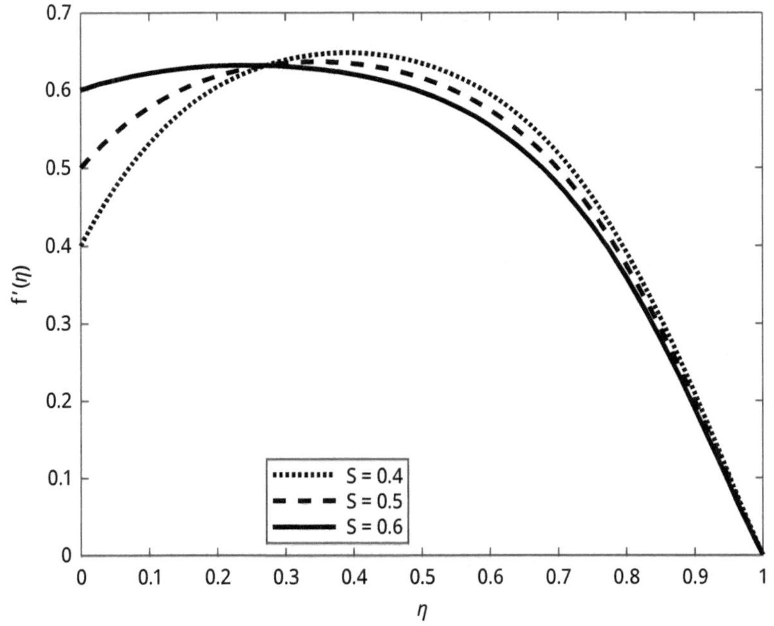

Figure 3.11: RV nature with respect to *S*.

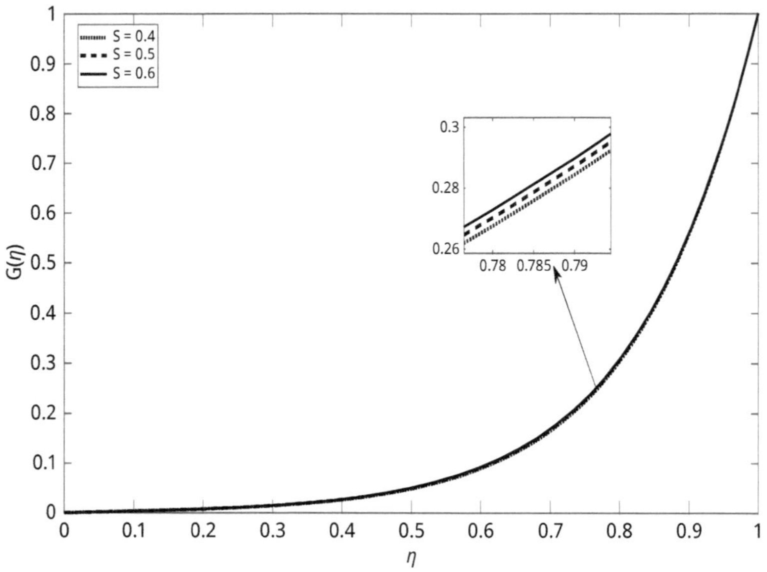

Figure 3.12: TV nature with respect to *S*.

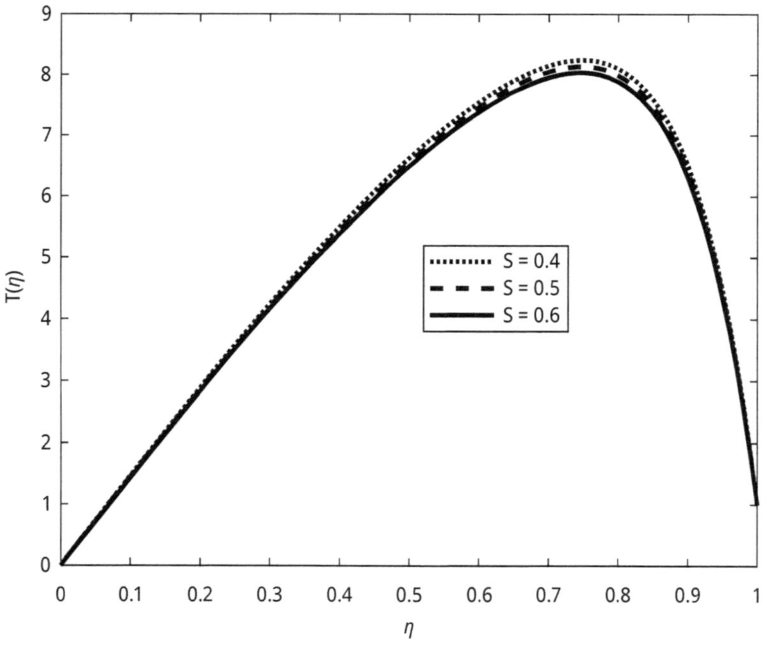

Figure 3.13: Temperature nature with respect to *S*.

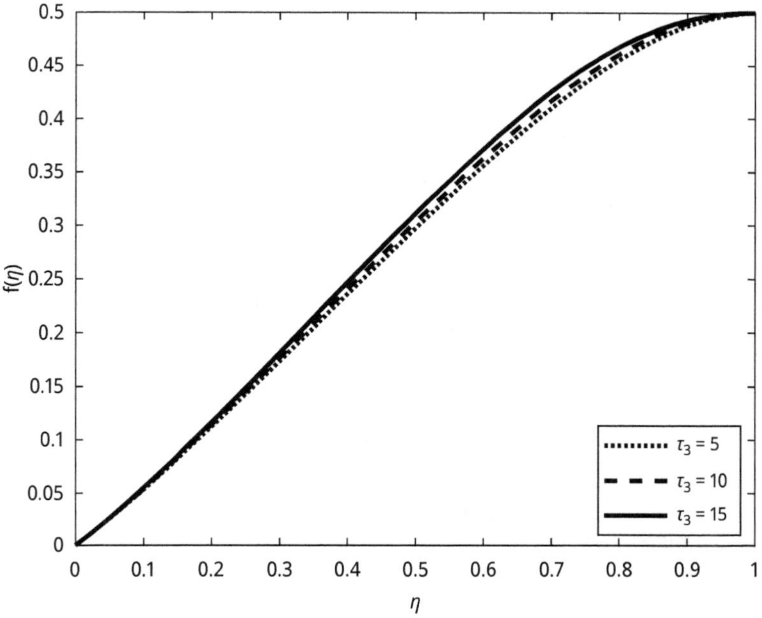

Figure 3.14: AV nature with respect to τ_3.

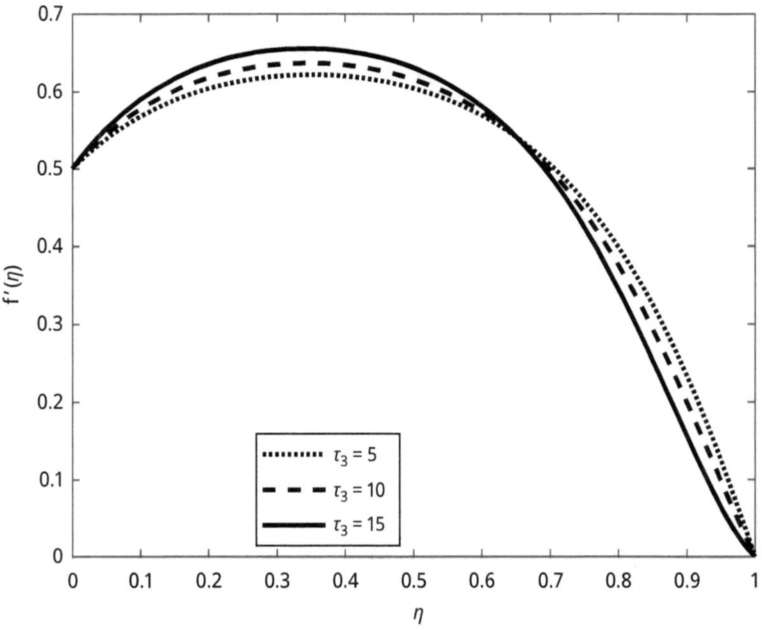

Figure 3.15: RV nature with respect to τ_3.

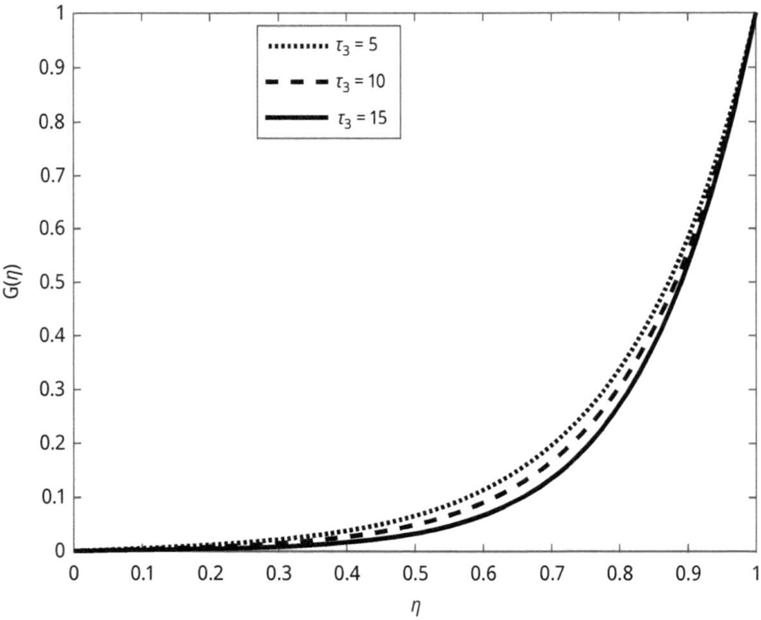

Figure 3.16: TV nature with respect to τ_3.

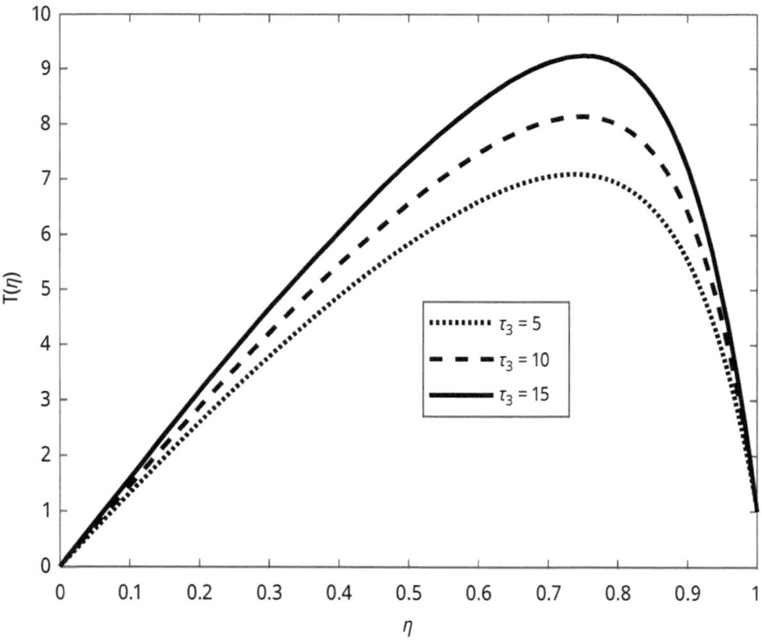

Figure 3.17: Temperature nature with respect to τ_3.

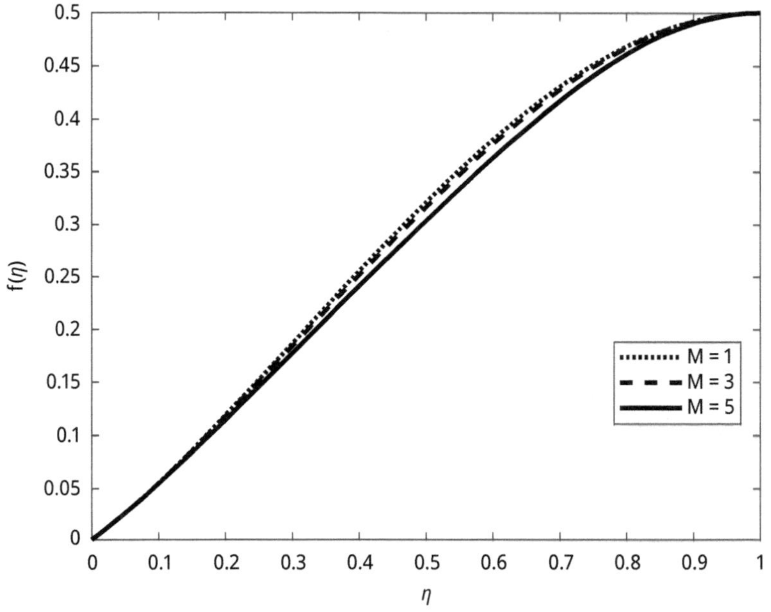

Figure 3.18: AV nature with respect to *M*.

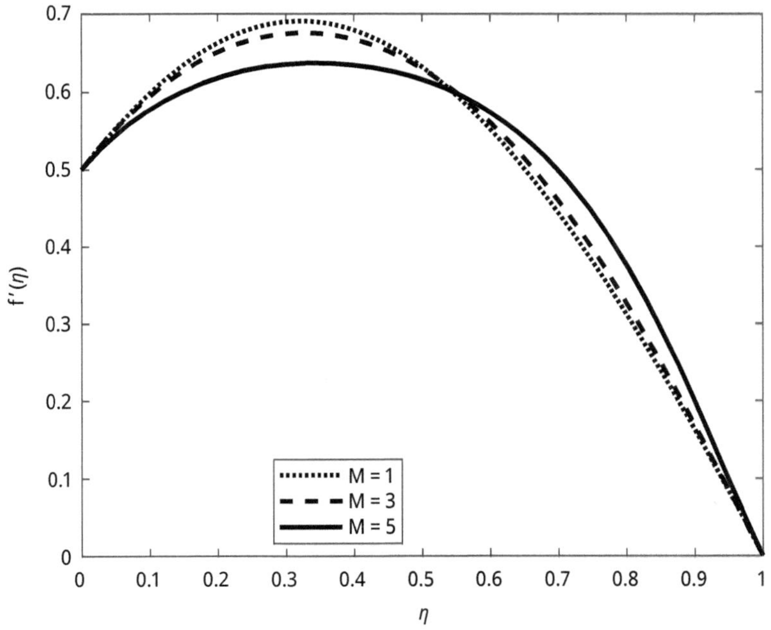

Figure 3.19: RV nature with respect to *M*.

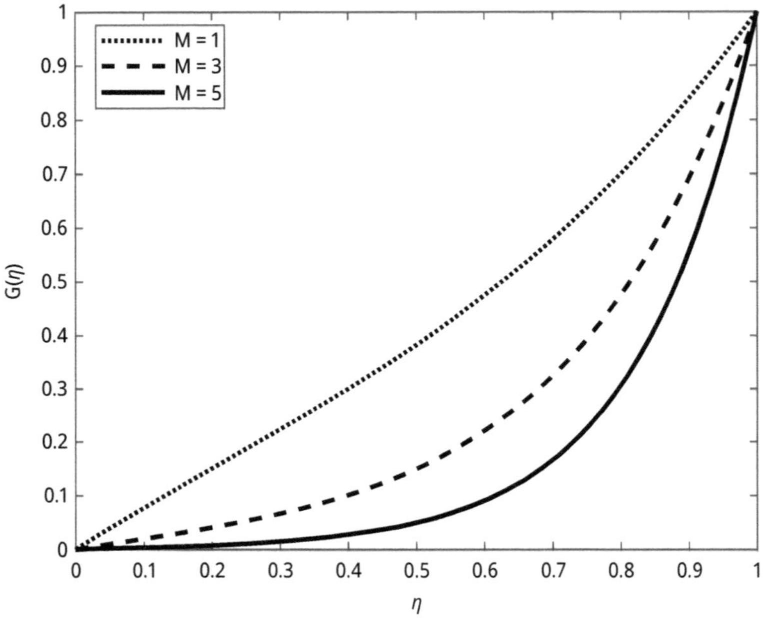

Figure 3.20: TV nature with respect to *M*.

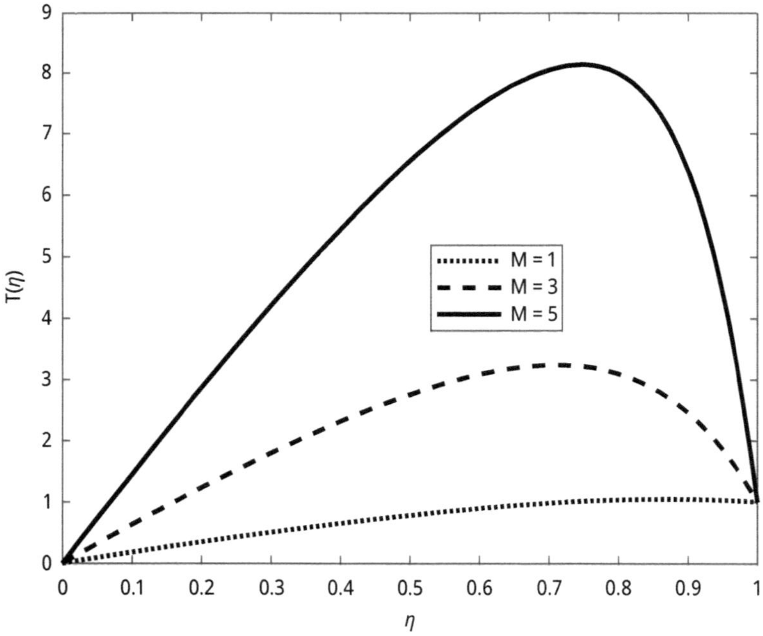

Figure 3.21: Temperature nature with respect to *M*.

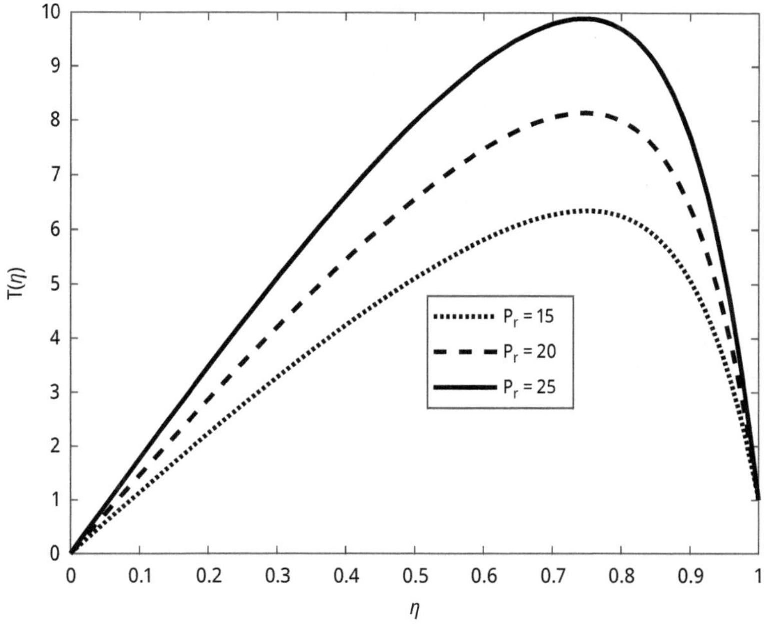

Figure 3.22: Temperature nature with respect to P_r.

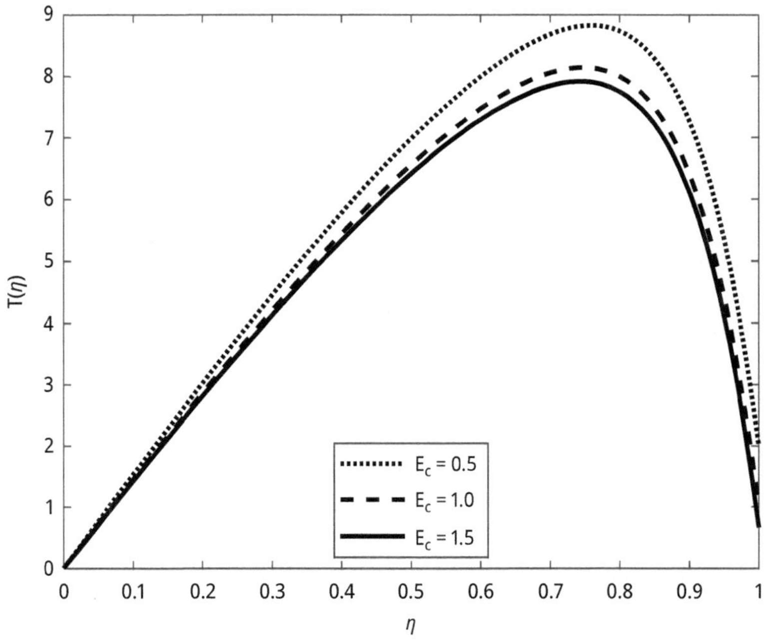

Figure 3.23: Temperature nature with respect to E_c.

similar behavior, but its behavior is opposite near the UD (Figure 3.19). TV falls with more M (Figure 3.20). Figure 3.21 reflects the rising nature of temperature with M.

The impact of the Prandtl and Eckert number on temperature is reflected in Figure 3.22 and Figure 3.23. The temperature rises with rising p_r while falls with the rising E_c.

3.4 Conclusion

The examination of the heat transmission and flow characteristics of non-Newtonian liquids over geometry with physical qualities is successfully studied using DTM via Maple. The representation of non-Newtonian RR flow, continuity, momentum, and concentration are being mathematically formulated in terms of the RR fluid model. The nature of the distribution profiles, which include velocity and temperature, are carefully examined using charts.

References

[1] Von Kármán T. (1921). Uber laminare und turbulente Reibung. Z. Angew. Math. Mech, 1, 233–252.

[2] Batchelor G. K. (1951 Jan 1). Note on a class of solutions of the Navier-Stokes equations representing steady rotationally-symmetric flow. The Quarterly Journal of Mechanics and Applied Mathematics, 4(1), 29–41.

[3] Bödewadt V. U. (1940). Die drehströmung über festem grunde. ZAMM-Journal of Applied Mathematics and Mechanics/Zeitschrift Für Angewandte Mathematik Und Mechanik, 20(5), 241–253.

[4] Jalili B., Ganji A. M., Shateri A., Jalili P., and Ganji D. D. (2023 Sep 1). Thermal analysis of non-Newtonian visco-inelastic fluid MHD flow between rotating disks. Case Studies in Thermal Engineering, 49, 103333.

[5] Zhao T. H., Khan M. I., and Chu Y. M. (2023 Feb). Artificial neural networking (ANN) analysis for heat and entropy generation in flow of non-Newtonian fluid between two rotating disks. Mathematical Methods in the Applied Sciences, 46(3), 3012–3030.

[6] Khan A., Shah R. A., Alam M. K., Rehman S., Shahzad M., Almad S., and Khan M. S. (2021 Jul). Flow dynamics of a time-dependent non-Newtonian and non-isothermal fluid between coaxial squeezing disks. Advances in Mechanical Engineering, 13(7), 16878140211033370.

[7] Gupta R. (2024 Apr). Analytical investigation of flow of a micropolar fluid between disks with vertical magnetic field. International Journal of Applied and Computational Mathematics, 10(2), 91.

[8] Gupta R., and Wakif A. (2023 Dec). Computing neural network to analyze heat and mass transfer in the flow of nanofluid between two disks. Numerical Heat Transfer, Part A: Applications, 8, 1–30.

[9] Gupta R., Selvam J., Vajravelu A., and Nagapan S. (2023 Jan 1). Analysis of a squeezing flow of a Casson nanofluid between two parallel disks in the presence of a variable magnetic field. Symmetry, 15(1), 120.

[10] Gupta R. (2022 Dec). Comparative study of micropolar fluid flow between two disks. Journal of Harbin Institute of Technology, 54(12), 61–76.

[11] Reiner M. (1945 Jul 1). A mathematical theory of dilatancy. American Journal of Mathematics, 67(3), 350–362.

[12] Rivlin R. S. (1948 May 27). The hydrodynamics of non-Newtonian fluids. I. Proceedings of the Royal Society of London. Series A. Mathematical and Physical Sciences, 193(1033), 260–281.

[13] Tabassum M., and Mustafa M. (2018 Aug 1). A numerical treatment for partial slip flow and heat transfer of non-Newtonian Reiner-Rivlin fluid due to rotating disk. International Journal of Heat and Mass Transfer, 123, 979–987.

[14] Puspanathan S., Naganthran K., Hashmi M. M., Hashim I., and Momani S. (2024 Apr 1). Numerical investigation of Reiner–Rivlin fluid flow and heat transfer over a shrinking rotating disk. Chinese Journal of Physics, 88, 198–211.

[15] Behbahan A. S., Alizadeh A. A., Mahmoudi M., Shamsborhan M., Al-Musawi T. J., and Pasha P. (2023 Oct 1). A new Adomian decomposition technique for a thermal analysis forced non-Newtonian magnetic Reiner-Rivlin viscoelastic fluid flow. Alexandria Engineering Journal, 80, 48–57.

[16] Gupta R. Flow of Reiner–Rivlin nanofluid between two rotating disks. In Nanofluids Technology for Thermal Sciences and Engineering, CRC Press, pp. 159–172.

[17] Agarwal R. (2022 Jan 2). An analytical study of non-Newtonian visco-inelastic fluid flowbetween two stretchable rotating disks. Palestine Journal of Mathematics, 11.

[18] Gupta R., Albidah A. B., Noor N. F., and Khan I. (2025 Feb 1). Application of DTM to heat source/sink in squeezing flow of iron oxide polymer nanofluid between electromagnetic surfaces. Case Studies in Thermal Engineering, 66, 105735.

[19] Gupta R., Nandi S., Santra S. S., Saha S., and Vajravelu K. (2024 Oct 12). Numerical simulation of an unsteady ternary hybrid nanofluid along a convectively heated curved stretching sheet: A Tiwari-Das model. Proceedings of the Institution of Mechanical Engineers, Part N: Journal of Nanomaterials, Nanoengineering and Nanosystems, 23977914241286681.

[20] Khan S. A., Razaq A., Alsaedi A., and Hayat T. (2023 Nov 15). Modified thermal and solutal fluxes through convective flow of Reiner-Rivlin material. Energy, 283, 128516.

[21] Sahoo B S. IV. (2019 May 1). Heat transfer due to revolving flow of Reiner-Rivlin fluid over a stretchable surface. Thermal Science and Engineering Progress, 10, 327–336.

[22] Haq F., Hussain A., and Ghazwani H. A. (2024 Nov). Numerical simulation for thermal transport in the chemically reactive flow of bioconvective Reiner-Rivlin nanofluid with magnetic field. Journal of Thermal Analysis and Calorimetry, 149(22), 13117–13128.

[23] Alfvén H. (1942 Oct 3). Existence of electromagnetic-hydrodynamic waves. Nature, 150(3805), 405–406.

[24] Rashidi S., Esfahani J. A., and Maskaniyan M. (2017 Oct 1). Applications of magnetohydrodynamics in biological systems-a review on the numerical studies. Journal of Magnetism and Magnetic Materials, 439, 358–372.

[25] Zhou J. K. (1986). Differential Transformation and Its Applications for Electrical Circuits, vol. 2, Huazhong University Press, Wuhan China (In Chinese) google schola, pp. 413–420.

[26] Gupta R., Khan I., Omer A. S., Aljohani A. F., and Alqahtani A. M. (2024 Oct 1). New insights of heat transfer in pistons and nozzles flow of graphene-transformer oil nanofluid: A differential transform method. Case Studies in Thermal Engineering, 62, 105119.

[27] Gupta R. (2024 Jun 12). Modeling of artificial neural network to analyze heat and mass transfer of ternary hybrid nanofluid between two parallel plates with inclined magnetic field. Numerical Heat Transfer, Part A: Applications, 1–35. https://doi.org/10.1080/10407782.2024.2366445

[28] Gupta R. (2024 May 28). MHD flow of a rotating micropolar fluid between two plates by using DTM. International Journal of Modern Physics C, 35(05), 2450058.

[29] Gupta R., and Agrawal D. (2023 Jun 1). Flow analysis of a micropolar nanofluid between two parallel disks in the presence of a magnetic field. Journal of Nanofluids, 12(5), 1320–1326.

Hemant Yadav

Chapter 4
Heat transfer in forced and natural convection

Abstract: This chapter deals from fundamentals to advanced aspects of transfer of heat in forced and natural convection mode by exploring both theoretical models and practical applications. Forced and natural convection play crucial roles in engineering field where efficient heat transfer is essential, such as in cooling systems, energy production as well as climate control. We begin by examining the governing principles of convection heat transfer, highlighting the differences and interactions between forced and natural convection mechanisms. By a detailed analysis, key parameters such as fluid velocity, temperature gradients, and surface properties that influence heat transfer rates will be explored.

To solve convection problems in complex geometries, computational techniques like finite element and finite volume methods are used, which help to get optimized heat transfer performance in engineering applications. To demonstrate the practical significance of efficient convection heat transfer, different case studies based on industries are also provided, such as in electronic device thermal management, and HVAC systems. A combination of both theoretical insights and practical examples allow researchers and practitioners to enhance their understanding and application of heat transfer in forced and natural convection systems.

Keywords: forced convection, natural convection, finite element method (FEM), computational fluid dynamics (SFD)

4.1 Introduction

Heat transfer is a fundamental concept in engineering, playing a critical role in various applications, from industrial processes to everyday devices [1]. An efficient transfer of heat is essential in numerous fields, including power generation, manufacturing, environmental control, and electronic cooling. As industries strive for better performance and energy efficiency, understanding the nuances of heat transfer mechanisms becomes imperative.

This chapter focuses on the mechanisms of heat transfer through convection, specifically forced and natural convection. Convection, a mode of heat transfer through

Hemant Yadav, Mechanical Engineering Department, Roorkee Institute of Technology, Roorkee,
e-mail: hemantyadav3259@gmail.com

https://doi.org/10.1515/9783111661674-004

fluid motion, is pivotal in systems where thermal management is necessary, such as in cooling systems, energy production, and climate control [2]. The dynamic nature of fluids in motion and their interaction with surfaces leads to complex heat transfer scenarios that are crucial in both industrial applications and natural phenomena.

Forced and natural convection, although governed by similar principles, exhibit distinct characteristics due to the different driving forces behind the fluid motion. Forced convection, induced by external devices such as fans and pumps, is prevalent in engineered systems where precise control over heat transfer rates is required. Conversely, natural convection, driven by buoyancy forces arising from density variations within the fluid due to temperature gradients, often occurs in scenarios where no external mechanisms are available or necessary, such as in atmospheric phenomena or passive cooling systems.

The scope of this chapter encompasses an in-depth exploration of the governing principles of convection heat transfer, emphasizing the critical differences and interactions between forced and natural convection mechanisms. By delving into theoretical underpinnings and practical applications, we aim to provide a comprehensive understanding of convection heat transfer.

In the subsequent sections, we will dissect the fundamental equations that describe fluid flow and heat transfer, analyze key parameters influencing these processes, and introduce advanced computational techniques for solving complex convection problems. Practical applications, illustrated through case studies in industries such as electronics' cooling and HVAC systems, will underscore the real-world relevance of these concepts.

Ultimately, this chapter seeks to bridge the gap between theoretical models and practical implementations, equipping researchers and practitioners with the knowledge to enhance heat transfer efficiency in various engineering systems. By integrating recent advances and future directions, we aim to foster a deeper appreciation of convection heat transfer's role in modern technology and its potential for innovation.

4.2 Governing principles of convection heat transfer

Convection heat transfer involves the movement of heat through a fluid (liquid or gas), caused by the fluid's motion. The two primary modes of convection are forced convection, where the fluid motion is driven by external means (like a fan or pump), and natural convection, which is driven by buoyancy forces due to density differences in the fluid. The fundamental equations governing convection include the Navier-Stokes equations for fluid flow and the energy equation for heat transfer [3].

4.2.1 Forced convection

In forced convection (Figure 4.1), external devices such as fans, blowers, or pumps induce the fluid motion. The heat transfer rate in forced convection depends on several factors, including fluid velocity, surface area, and temperature difference between the fluid and the surface. The key equations include the Navier-Stokes equation for momentum conservation and the energy equation for heat transfer. Forced convection is commonly used in applications like heat exchangers, electronic device cooling, and HVAC systems [4].

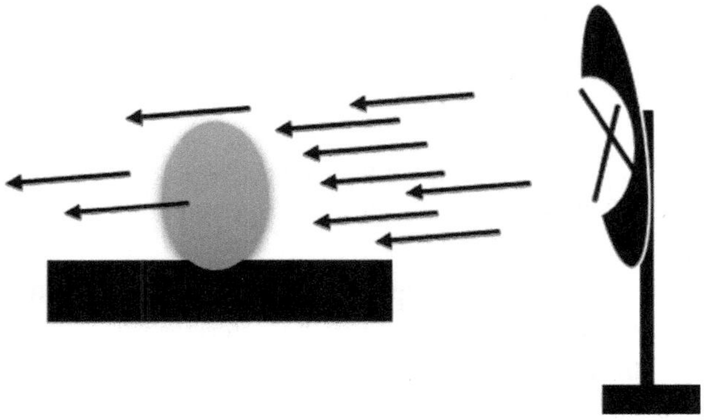

Figure 4.1: Force convection (sample cooled by fan forced air).

4.2.2 Natural convection

Natural convection (Figure 4.2) occurs when fluid motion is generated by buoyancy forces, which arise from density variations due to temperature gradients [5]. This mode of heat transfer is significant in systems where no external mechanical devices are used to move the fluid. The efficiency of natural convection depends on the properties of the fluid, the temperature difference, and the geometry of the system.

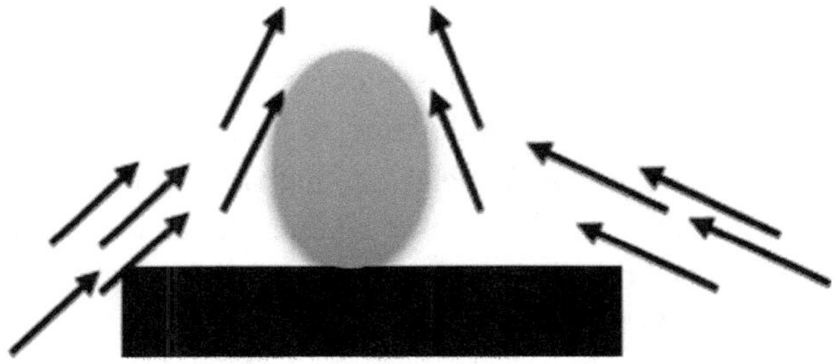

Figure 4.2: Natural convection (sample cooled in air).

4.3 Key parameters influencing heat transfer

Several parameters influence the rate of heat transfer in both forced and natural convection. These include:

- **Reynolds number (Re):** It is a dimensionless number that describes the flow regime (laminar or turbulent). The shift from laminar to turbulent flow is influenced by factors such as surface geometry, roughness, free-stream velocity, surface temperature, and fluid properties. Through extensive experiments in the 1880s, Osborn Reynolds found that the flow regime primarily depends on the balance between inertia and viscous forces in a fluid. This relationship is expressed by the Reynolds number (Re), a dimensionless quantity, defined as follows:

$$R_e = \frac{\rho v L}{\mu} = \frac{\text{Inertia force}}{\text{Viscous force}}$$

  where ρ is the fluid density, v is the velocity, L is the characteristic length, and μ is the dynamic viscosity. A low Reynolds number indicates laminar flow, while a high value suggests turbulence.

  Prandtl number (Pr): Named after Ludwig Prandtl, who introduced the boundary layer theory in 1904 [20], the Prandtl number (Pr) characterizes the relative rates of momentum and heat diffusion in fluids. It varies widely, from below 0.01, for liquid metals, to over 100,000, for heavy oils, with the number for water typically being around 10:

$$\mathbf{Pr} = \frac{\text{Molecular diffusivity of momentum}}{\text{Molecular diffusivity of heat}}$$

- **Grashof number (Gr):** It describes the ratio of buoyancy to viscous forces in natural convection. A higher Grashof number indicates that buoyancy forces dominate, promoting convection, while a lower value suggests that viscous forces suppress fluid motion:

$$\text{Gr} = \frac{g\beta(T_s - T_\infty){L_c}^3}{V^2}$$

  where g is the gravitational acceleration, in m/s^2, β is the coefficient of volume expansion, T_s is the temperature of the surface, in °C, T_∞ is the temperature of the fluid sufficiently far from the surface in °C, L_c is the characteristic length of the geometry in m, and V is the kinematic viscosity of the fluid, in m^2/s.

- **Nusselt number (Nu):**. Nusselt number (Nu) is one group that reduces the number of variables. It represents the ratio of convective to conductive heat transfer in a fluid. It represents the enhancement of heat transfer through a fluid layer, as compared to pure conduction. It is given by

$$\mathrm{Nu} = \frac{hL}{k}$$

where h is the convective heat transfer coefficient, L is the characteristic length, and k is the thermal conductivity of the fluid. A higher Nusselt number signifies stronger convection, while a lower value indicates domination of heat transfer by conduction.

4.4 Review of literature

Yuge [19] experimentally investigated heat transfer between spheres and air flow for Reynolds numbers between 3.5 and 1.44×10^5, and Grashof numbers from 1 to 10^8. Empirical equations for forced, natural, and combined convection were derived and compared with findings from other research. A graphical method, utilizing the experimental data, was employed to forecast the heat transfer performance for combined natural and forced convection scenarios.

Pua et al. [18] hypothesized that downward lighting LEDs require cooling through cylindrical fin heat sinks, which are mounted on top and operate under natural convection. A theoretical model was proposed, and an experimental investigation was conducted to validate its accuracy. Convective heat transfer coefficients were determined for horizontal and vertical bare plates, as well as for vertical plate and cylindrical fin heat sinks under natural and forced convection. Power inputs ranged from 5 to 40 W to maintain temperatures below 100 °C. The study found that surface temperature increased with input power, and natural convection resulted in higher temperatures. The vertical bare plate had lower temperatures than the horizontal one at high heat flux, with minimal differences at low heat input. The plate fin heat sink performed best under natural convection. Experimental heat transfer coefficients were generally higher than those from Nusselt number correlations, with poor agreement for bare plates but satisfactory results for plate and cylindrical fin heat sinks. Actual measurements are recommended for precise thermal management.

Prasad et al. [13] examined mixed-convection heat transfer in a lid-driven cavity with varying depth, combining forced and natural convection effects. Forced convection is generated by a moving lid, while natural convection arises from heating the lower boundary and cooling the upper one. The study spans $\mathrm{Gr}/\mathrm{Re}^2$ ratios from 0.1 to 1000, with Reynolds numbers up to 12,000 and Grashof numbers between 10^3 and 5×10^9. Heat flux measurements and flow visualization reveal that the heat transfer coefficient remains insensitive to $\mathrm{Gr}/\mathrm{Re}^2$ across the tested range. The heat transfer is governed by a relationship resembling pure free convection, indicating the seamless interplay of free- and forced-convection effects. Flow behavior along the boundaries deviates from Blasius flow, influenced by buoyancy-induced destabilization of the boundary layer. These findings provide valuable Nusselt and Stanton number correlations for design applications.

The forced convection heat transfer characteristics of steady, incompressible Newtonian fluid flow over bundles of circular cylinders have been investigated numerically by **Mandhani et al.** [17] using a cell model to approximate inter-cylinder hydrodynamic interactions. The study solves the momentum and energy equations for varying physical and kinematic conditions, focusing on two types of thermal boundary conditions: constant temperature and constant heat flux. Extensive results on temperature fields and the Nusselt number variation across cylinder surfaces are presented for two Prandtl numbers (air and water), with Reynolds numbers ranging from 1 to 500 and void age from 0.4 to 0.99. The surface-averaged Nusselt number increases with decreasing porosity and increasing Prandtl and Reynolds numbers. The results agree with existing data for single cylinders and cylinder arrays, indicating that heat transfer characteristics are mainly influenced by void age, while the geometrical arrangement of the array has a secondary effect. More experimental validation is needed, but cell models prove to be a viable approach for modeling momentum, heat, and mass transfer in multi-cylinder systems.

The low thermal properties of liquids have led to research on nanofluids [16], which are engineered by dispersing nanoparticles (less than 100 nm) in liquids to enhance heat transfer properties and hydrodynamic flow. Both experimental and numerical studies agree that adding nanoparticles increases the heat transfer coefficient, particularly with higher concentrations and smaller particle sizes. The shape of the flow area also affects heat transfer enhancement. Nanofluids show potential for practical applications in fields like transportation, nuclear reactors, and electronics, despite a nominal increase in pressure drop. Future studies should focus on developing generalized correlations for nanofluids, exploring the effects of different tube shapes, and using high-volume concentrations of nanofluids in car radiators for better heat transfer and hydrodynamic performance.

The study investigates forced convective heat transfer from smooth [14], isothermal spheres of varying diameters in air flows at different free-stream velocities. By measuring the steady-state power input to internal heating elements and the temperatures of the air and sphere surfaces, the average heat transfer coefficient was determined, with corrections for thermal radiation and natural convection. The data, covering Reynolds numbers from 7.8×10^3 to 3.3×10^5, extend the range found in the literature. A sudden increase in the Nusselt number was observed at a critical Reynolds number of approximately 2.9×10^5, akin to the "drag crisis" in drag force studies. A correlation between the Nusselt and Reynolds numbers was formulated, accurately describing the data below this critical Reynolds number. The study reports a total uncertainty of about 4.0% for Nusselt number measurements within the specified Reynolds number range.

Baragh et al. [15] showed that heat transfer in porous media is critical for various industrial applications, especially in heat exchanger design and analysis. This study experimentally investigates the forced convection heat transfer of single-phase air flow in a circular cross-sectional channel with different porous media arrangements. The experiments show that porous media enhance heat transfer by creating uniform spaces and high conductivity, leading to increased mean fluid temperature and a reduced temperature difference between the channel wall and the fluid, which in turn raises the heat transfer coefficient. The heat transfer performance ratio and enhancement ratio were defined to evaluate heat transfer alongside pressure drop. Fully filled porous media channels exhibit the best heat transfer enhancement across laminar and turbulent flows. In turbulent flow, channels with annular porous zones near the walls demonstrated superior thermal performance, achieving high heat transfer rates with minimal pressure drop.

Convection heat transfer within cavities, particularly involving nanofluids exhibiting non-Newtonian behavior [10], has garnered significant research interest due to the unique characteristics of these fluids. Unlike Newtonian fluids, non-Newtonian fluids, such as certain nanofluids, do not maintain a linear relationship between shear stress and strain, necessitating the use of models like the power-law and Bingham models to describe their behavior accurately. This review delves into the properties of non-Newtonian nanofluids, summarizing empirical studies and highlighting the utilization of these models in numerical simulations. The analysis focuses on heat transfer within various cavity geometries, emphasizing the distinct properties of nanofluids at different volume fractions. Despite their industrial relevance, non-Newtonian nanofluids have received less attention compared to their Newtonian counterparts. The review underscores the need for more in-depth research into these complex fluids, advocating for the adoption of empirical relationships to better simulate their flow dynamics and enhance the accuracy of heat transfer predictions in engineering applications.

Convection heat transfer in cavities, especially with nanofluids [8, 9, 11], has drawn considerable research attention due to its importance in various industrial applications. Non-Newtonian nanofluids, characterized by a nonlinear relationship between shear stress and strain, deviate from Newton's law of viscosity. Researchers have primarily used the power-law and Bingham models to represent these fluids' behavior, providing numerical solutions to better understand their thermal performance in cavities.

The investigation of non-Newtonian nanofluids in different cavity configurations highlights the crucial role of viscosity in heat transfer efficiency. Studies have summarized relationships for the viscosity of these fluids, emphasizing their unique thermal behaviors. Empirical and numerical findings are often compiled in tables, offering a comprehensive view of current advancements.

In parallel, research on film-wise condensation on enhanced surfaces has shown significant progress. Enhanced structures have been developed to improve natural and forced convection condensation heat transfer coefficients. For natural convection, models focusing on liquid film behavior, semi-empirical data, and numerical analyses have been evaluated, whereas forced convection models have explored gravity and vapor shear dominance.

Findings reveal that three-dimensional fin structures on external surfaces significantly enhance heat transfer compared to two-dimensional fins, though their efficiency decreases with increasing tube rows. Internally enhanced tubes, such as herringbone and pin fin designs, demonstrate superior performance in forced convection.

Despite the progress, further research is needed to optimize fin geometries, develop models for three-dimensional structures, and explore non-Newtonian nanofluids' applications across broader conditions to fully harness their potential.

Solar thermal collectors and heat exchangers have gained significant popularity in research due to their vital role in generating clean thermal energy [12]. These systems are pivotal for human life and industrial applications, with numerous studies exploring the impact of physical factors on their efficiency and performance in various configurations. Current research predominantly evaluates thermal performance in solar receivers and heat exchangers using forced thermal transfer, emphasizing their critical importance in industrial sectors.

This comprehensive review consolidates data from experimental and numerical studies on forced thermal transfer in different geometries of solar collectors and heat exchangers. Both traditional and innovative approaches are examined, focusing on two primary applications: forced convection in solar receivers and heat exchangers. Key parameters like wind speed, solar radiation intensity, and wall types are analyzed to optimize these systems' performance.

The forced convection heat transfer characteristics of steady, incompressible Newtonian fluid flow over bundles of circular cylinders have been investigated numerically using a cell model to approximate inter-cylinder hydrodynamic interactions. The study solves the momentum and energy equations for varying physical and kinematic conditions, focusing on two types of thermal boundary conditions: constant temperature and constant heat flux. Extensive results on temperature fields and the Nusselt number variation across cylinder surfaces are presented for two Prandtl numbers (air and water), with Reynolds numbers ranging from 1 to 500 and voidage from 0.4 to 0.99. The surface-averaged Nusselt number increases with decreasing porosity and increasing Prandtl and Reynolds numbers. The results agree with existing data for single cylinders and cylinder arrays, indicating that heat transfer characteristics are mainly influenced by voidage, while the geometrical arrangement of the array has a secondary effect. More experimental validation is needed, but cell models prove to be a viable approach for modeling momentum, heat, and mass transfer in multi-cylinder systems.

4.5 Computational techniques in convection

Solving convection problems, especially in complex geometries, often requires computational techniques [6, 21]. Finite element and finite volume methods are widely used to simulate and optimize heat transfer in engineering applications. These methods allow for detailed analysis and visualization of heat transfer processes, helping engineers design more efficient systems.

4.6 Practical applications and case studies

To demonstrate the practical significance of convection heat transfer, several case studies are provided:

- **Electronic cooling:** Forced convection is crucial in managing the heat generated by electronic components, ensuring their reliable operation.
- **HVAC systems:** Both forced and natural convection are used in heating, ventilation, and air conditioning to maintain comfortable indoor environments.

Each case study highlights the importance of understanding convection principles in designing effective thermal management solutions.

4.7 Recent advances and future directions

Recent advances in convection heat transfer include the development of new materials with enhanced thermal properties, improved computational algorithms for more accurate simulations, and innovative design techniques for better heat exchanger performance [7]. Future research may focus on integrating renewable energy sources and further optimizing thermal management systems for energy efficiency.

4.8 Conclusion

This chapter has provided a comprehensive overview of heat transfer in forced and natural convection. By understanding the governing principles, key parameters, computational techniques, and practical applications, researchers and engineers can optimize thermal management in various engineering systems. The combination of theoretical insights and practical examples underscores the critical role of convection in modern technology.

References

[1] Bejan A. (2013 Mar 28). Convection Heat Transfer, John Wiley & Sons.

[2] Incropera F. P., DeWitt D. P., Bergman T. L., and Lavine A. S. (1996 Feb 16). Fundamentals of Heat and Mass Transfer, New York: Wiley.

[3] Versteeg H. K. (2007). An Introduction to Computational Fluid Dynamics the Finite Volume Method, 2/E, Pearson Education India.

[4] Kays W. M., and Crawford M. E. (1980). Convective Heat and Mass Transfer. 243.

[5] Churchill S. W., and Chu H. H. (1975 Sep 1). Correlating equations for laminar and turbulent free convection from a horizontal cylinder. International Journal of Heat and Mass Transfer, 18(9), 1049–1053.

[6] Patankar S. (2018 Oct 8). Numerical Heat Transfer and Fluid Flow, CRC press.
[7] Rohsenow W. M., Hartnett, J. P., and Cho Y. I. (1998 May 1). Handbook of Heat Transfer, New York: Mcgraw-Hill.
[8] Khanafer K., and Vafai K. (2017 Oct). A critical synthesis of thermophysical characteristics of nanofluids. In Nanotechnology and Energy, vol. 3, Jenny Stanford Publishing, pp. 279–332.
[9] Mahian O., Kianifar A., Kleinstreuer C., Al-Nimr M. D., Pop I., Sahin A. Z., and Wongwises S. (2013 Oct 1). A review of entropy generation in nanofluid flow. International Journal of Heat and Mass Transfer, 65, 514–532.
[10] Yang L., and Du K. (2020 Jun). A comprehensive review on the natural, forced, and mixed convection of non-Newtonian fluids (nanofluids) inside different cavities. Journal of Thermal Analysis and Calorimetry, 140, 2033–2054.
[11] Ho J. Y., and Leong K. C. (2021 Mar 5). A critical review of film-wise natural and forced convection condensation on enhanced surfaces. Applied Thermal Engineering, 186, 116437.
[12] Rebhi R., Menni Y., Lorenzini G., and Ahmad H. (2022). Forced-convection heat transfer in solar collectors and heat exchangers: A review. Journal of Advanced Research in Applied Sciences and Engineering Technology, 26(3), 1–5.
[13] Prasad A. K., and Koseff J. R. (1996 Oct 1). Combined forced and natural convection heat transfer in a deep lid-driven cavity flow. International Journal of Heat and Fluid Flow, 17(5), 460–467.
[14] Will J. B., Kruyt N. P., and Venner C. H. (2017 Jun 1). An experimental study of forced convective heat transfer from smooth, solid spheres. International Journal of Heat and Mass Transfer, 109, 1059–1067.
[15] Baragh S., Shokouhmand H., Ajarostaghi S. S., and Nikian M. (2018 Dec 1). An experimental investigation on forced convection heat transfer of single-phase flow in a channel with different arrangements of porous media. International Journal of Thermal Sciences, 134, 370–379.
[16] Hussein A. M., Sharma K. V., Bakar R. A., and Kadirgama K. (2014 Jan 1). A review of forced convection heat transfer enhancement and hydrodynamic characteristics of a nanofluid. Renewable and Sustainable Energy Reviews, 29, 734–743.
[17] Mandhani V. K., Chhabra R. P., and Eswaran V. (2002 Feb 1). Forced convection heat transfer in tube banks in cross flow. Chemical Engineering Science, 57(3), 379–391.
[18] Pua S. W., Ong K. S., Lai K. C., and Naghavi M. S. (2019 Mar). Natural and forced convection heat transfer coefficients of various finned heat sinks for miniature electronic systems. Proceedings of the Institution of Mechanical Engineers, Part A: Journal of Power and Energy, 233(2), 249–261.
[19] Yuge T. (1960 Aug 1). Experiments on heat transfer from spheres including combined natural and forced convection. Journal of Heat Transfer (US), 82.
[20] Tani I. (1977 Jan). History of boundary layer theory. Annual Review of Fluid Mechanics, 9(1), 87–111.
[21] Cengel Y. A., and Ghajar A. J. (2020). Heat and Mass Transfer, McGraw-Hill Education.

Sudipta Priyadarshini and Sukanta Nayak

Chapter 5
Numerical study of coupled partial differential equations in heat transfer problems with imprecisely defined parameters

Abstract: The epistemic-type uncertainties are the main focus in connection with the heat transfer problems. The same are arising due to imprecise involved parameters, boundary conditions, experimental errors, and material properties. This chapter includes the numerical approach to investigate heat transfer problems in presence of uncertainties. In addition, to quantify the impreciseness, fuzzy number is used. Using the fuzzy number with the concept of finite element method (FEM), fuzzy finite element method (FFEM) is developed and the same is discussed in detail. Then a case study is illustrated using FFEM to quantify the velocity and temperature profiles. The Rayleigh number (*Ra*) and radiation parameter (*Rd*) are considered as fuzzy. Here, the modelled coupled algebraic equations are simplified through Gauss-Seidel (GS) iterative technique. Further, convergence analysis of the different profiles is discussed and the sensitiveness of the same is analyzed. Finally, using the obtained results the efficiency and usefulness of the FFEM is depicted.

Keywords: fuzzy finite element method (FFEM), coupled partial differential equations, heat transfer problem, convergence

5.1 Introduction

Fluid flow and heat transfer problems have played a significant role with growth of science and technology. In this direction, various applications of the same can be found in [1–3]. The mathematical model of these problems is a challenging task, due to complexity involved in the systems. Generally, the solutions of the contained equations, which are basically a collection of PDEs, describe the fluid movement and energy transfer within the porous medium. Therefore, many numerical strategies are being adopted to address this phenomenon, but the finite element method is given

Sudipta Priyadarshini, Department of Basic Science & Humanities, Vignan's Institute of Information Technology (A), Duvvada, Vishakhapatnam 530049, Andhra Pradesh, India, e-mail: sudipta.grc@gmail.com
Sukanta Nayak, Department of Mathematics, School of Advanced Sciences, VIT-AP University, Inavolu, Besides AP Secretariat, Amaravati, Andhra Pradesh, India, e-mail: sukantgacr@gmail.com

https://doi.org/10.1515/9783111661674-005

more attention because of its adaptability to complicated geometry and boundary conditions.

In this direction, the authors in [4] have investigated the electro kinematic and hydrodynamic flow in a porous medium using finite element method (FEM). Further, [5] describe the solution of a coupled partial differential equation through the FEM procedure. In addition, authors in [6] provided the simplified form of the FEM for heat and mass transfer problems. Then in [7] the authors discussed the conjugate heat transfer in a vertical porous annular cylindrical with the help of FEM. Using the FEM, [8] have investigated the uniform and nonuniform heat transfer in a trapezoidal cavity. Further, in [9] the authors have taken the penalty finite element approach to examine the impact of a square cavity in various thermal boundary conditions.

The abovementioned literature demonstrates that, when the involved variables are given in crisp (exact) form one can use FEM as easier. However, in actual practice, the involved parameters and relevant quantities could be ambiguous. This may occur due to insufficient knowledge about the system, experimental error, imprecise information, etc. Consequently, numerous researchers are succeeding in adopting probabilistic approaches to handle these uncertainties. One of the known methods, viz., Monte Carlo is used to handle the probabilistic type uncertainties. In this manner, [10] has used the Monte Carlo process for the transient radiative transport planner process. But in this approach, one needs a big data set. However, a small data set requires an appropriate method for handling this impreciseness.

In this context, fuzzy set theory [11] has been introduced to manage such uncertainties. But actual fuzzy becomes difficult to handle and as such calculation is tedious. So, the authors have tried many simplified fuzzy approaches. In this regard, the vertex approach was suggested by [12] for computing functions of fuzzy variables. To address uncertain fuzzy variables, [13] employed the fuzzy-weighted average method (FWAM). Further, Yang et al. [14] also updated FWAM and suggested a technique that requires less processing time. Several authors have looked into numerous engineering difficulties using these techniques.

Nicolai et al. [15] have investigated an uncertain heat conduction problem and observed that the obtained solution is better in comparison with the response surface method. Further, Nayak and Chakraverty [16] have analyzed the conjugate heat transfer problem by using the FFEM approach. Additionally, Chakraverty and Nayak [17] have investigated the diffusion problems in a fuzzy environment by taking FFEM techniques; whereas Wang and Zhiping [18] apply a sensitivity-based interval analysis method and a modified version of the intervals known as the interval parameter perturbation method to solve heat conduction. In addition, Nayak [19] used the IFEM to solve the convection-diffusion systems' parametric uncertainty problem. Zhao et al. [20] have investigated the fuzzy observer-based control problem for solving nonlinear coupled hyperbolic PDEs and ODEs.

In this direction, a different numerical strategy is suggested to examine the coupled PDEs in heat transfer problems while accounting for uncertainties. The aim of the chapter is to provide one of the numerical approaches for solving the coupled nonlinear PDEs by using a triangular fuzzy number (TFN). The heat transfer in a square cavity is investigated with a fuzzy environment. Then by using the FFEM technique the involved nonlinear coupled PDEs are transformed into a coupled algebraic system of equations. The solution for it has been carried out by taking the Gauss-Seidel iterative approach in terms of fuzzy. Then the contained uncertain variables are taken as TFN for the study of convergence and sensitive analysis.

5.2 Preliminaries

Here, some of the important definitions related to fuzzy set theory (FST) ([21, 22]) have been discussed.

Definition 5.2.1 ([23, 24]): A fuzzy set $\hat{A}$ is a set of ordered pairs such that $\hat{A} = \{(\mathfrak{x}, \mu_{\hat{A}}(\mathfrak{x})) \mid \forall \mathfrak{x} \in X, \mu_{\hat{A}}(\mathfrak{x}) \in [0,1]\}$ where $\mu_{\hat{A}}(\mathfrak{x})$ denotes the membership function, X indicates the universe of discourse, and $\mu_{\hat{A}}: X \to [0,1]$ is defined as the membership function.

Definition 5.2.2 ([25, 26]): A fuzzy number is a piecewise continuous, convex, and normalized fuzzy set. A triangular fuzzy number $\hat{A} = [a^L, a^M, \ a^R]$is one of the fuzzy numbers in which the membership function is

$$\mu_{\hat{A}}(\mathfrak{x}) = \begin{cases} \mu_L, a^L \le x \le a^M \\ \mu_R, a^M \le x \le a^R \\ 0, \quad \textit{otherwise} \end{cases} \tag{5.1}$$

where $\mu_L = \dfrac{\mathfrak{x} - a^L}{a^M - a^L}$ and $\mu_R = \dfrac{a^R - \mathfrak{x}}{a^R - a^M}$.

Definition 2.3 [23]: Suppose $\hat{\mathfrak{x}} = [\underline{\mathfrak{x}}(\alpha), \overline{\mathfrak{x}}(\alpha)]$ and $\hat{\mathfrak{y}} = [\underline{\mathfrak{y}}(\alpha), \overline{\mathfrak{y}}(\alpha)]$ are defined two fuzzy numbers with the scaler ℓ, then the standard basic operations are defined as follows:

(a) $\hat{\mathfrak{x}} = \hat{\mathfrak{y}}$ if and only if $\underline{\mathfrak{x}}(\alpha) = \underline{\mathfrak{y}}(\alpha)$ and $\overline{\mathfrak{x}}(\alpha) = \overline{\mathfrak{y}}(\alpha)$;

(b) $\hat{\mathfrak{x}} + \mathfrak{y} = [\underline{\mathfrak{x}}(\alpha) + \underline{\mathfrak{y}}(\alpha), \overline{\mathfrak{x}}(\alpha) + \overline{\mathfrak{y}}(\alpha)]$;

(c) $l\hat{\mathfrak{x}} = \begin{cases} [\ell\underline{\mathfrak{x}}(\alpha), \ell\overline{\mathfrak{x}}(\alpha)], l \ge 0 \\ [\ell\overline{\mathfrak{x}}(\alpha), \ell\underline{\mathfrak{x}}(\alpha)], l < 0. \end{cases}$

Here, $\underline{\mathfrak{x}}$, $\underline{\mathfrak{y}}$ and $\overline{\mathfrak{x}}$, $\overline{\mathfrak{y}}$ are denote lower and upper limits of fuzzy numbers.

5.3 Problem analysis

Consider a square cavity, shown in Figure 5.1, with x- and y-directions, respectively [6]. The heat applies on the left-side wall of the cavity. The upper and lower walls of the cavity were maintained adiabatic.

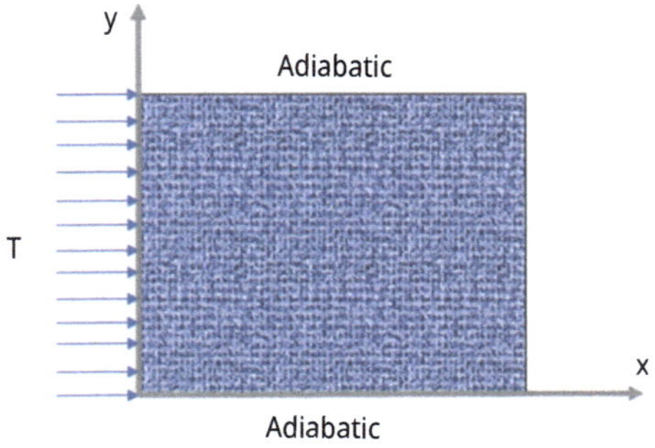

Figure 5.1: Geometry of a square cavity.

The heat transfer in a cavity satisfies the conservation laws. Here, the governing equations are reduced to the nondimensional form written as follows [5], [27]:

Momentum equation:

$$\frac{\partial^2 \bar{\psi}}{\partial \bar{x}^2} + \frac{\partial^2 \bar{\psi}}{\partial \bar{y}^2} = -Ra \frac{\partial \bar{T}}{\partial \bar{x}} \tag{5.2}$$

Energy equation:

$$\frac{\partial \bar{\psi}}{\partial \bar{y}} \frac{\partial \bar{T}}{\partial \bar{x}} - \frac{\partial \bar{\psi}}{\partial \bar{x}} \frac{\partial \bar{T}}{\partial \bar{y}} = \left(\left(1 + \frac{4Rd}{3} \right) \frac{\partial^2 \bar{T}}{\partial \bar{x}^2} + \frac{\partial^2 \bar{T}}{\partial \bar{y}^2} \right) \tag{5.3}$$

where the nondimensional numbers are shown as radiation parameter $(Rd) = \frac{4\sigma T_\infty^3}{\beta_R k}$, Rayleigh number$(Ra) = \frac{g\beta_T \Delta T K l_{ref}}{Va}$, stream function ($\bar{\psi}$), and temperature ($\bar{T}$).

5.4 Method formulation

Generally, solving the coupled system of governing equations in fluid flow and heat transfer problems are complicated, but when it comes under uncertainties it is a more challenging task to handle [28]. The same may occur due to experimental error, insufficient information about the system, boundary conditions, involved variables, material properties, etc. Furthermore, one cannot ignore the involvement of uncertainties in the relevant parameters of the equations. So, for dealing with such kind of situations one can use the non-probabilistic approach such as fuzzy numbers. We have considered the TFN) with the finite element method namely, fuzzy finite element method (FFEM) to handle the uncertain system numerically. In FFEM, we have taken the linear triangular approach to discretize the domain as follows [21, 29]:

$$\begin{aligned}\hat{x} &= \hat{L}_1\hat{x}_1 + \hat{L}_2\hat{x}_2 + \hat{L}_3\hat{x}_3,\\ \hat{y} &= \hat{L}_1\hat{y}_1 + \hat{L}_2\hat{y}_2 + \hat{L}_3\hat{y}_3,\\ \hat{L} &= \hat{L}_1 + \hat{L}_2 + \hat{L}_3.\end{aligned} \tag{5.4}$$

where $\hat{L}$ denotes the shape function in a fuzzy environment and $\hat{}$ denotes the fuzzy. Additionally, eq. (5.4) can be transformed into the following matrix form:

$$\begin{bmatrix} 1 & 1 & 1 \\ \hat{x}_1 & \hat{x}_2 & \hat{x}_3 \\ \hat{y}_1 & \hat{y}_2 & \hat{y}_3 \end{bmatrix} \begin{Bmatrix} \hat{L}_1 \\ \hat{L}_2 \\ \hat{L}_3 \end{Bmatrix} = \begin{Bmatrix} 1 \\ \hat{x} \\ \hat{y} \end{Bmatrix}$$

The fuzzy area of the triangular element $(\hat{A})$ is $\frac{1}{2}\begin{vmatrix} 1 & 1 & 1 \\ \hat{x}_1 & \hat{x}_2 & \hat{x}_3 \\ \hat{y}_1 & \hat{y}_2 & \hat{y}_3 \end{vmatrix}$.
Then, one can find:

$$\begin{gathered}\begin{Bmatrix} \hat{L}_1 \\ \hat{L}_2 \\ \hat{L}_3 \end{Bmatrix} = \begin{bmatrix} 1 & 1 & 1 \\ \hat{x}_1 & \hat{x}_2 & \hat{x}_3 \\ \hat{y}_1 & \hat{y}_2 & \hat{y}_3 \end{bmatrix}^{-1} \begin{Bmatrix} 1 \\ \hat{x} \\ \hat{y} \end{Bmatrix} \begin{Bmatrix} \hat{L}_1 \\ \hat{L}_2 \\ \hat{L}_3 \end{Bmatrix} = \\ \frac{1}{2\hat{A}} \begin{bmatrix} \hat{x}_2\hat{y}_3 - \hat{x}_3\hat{y}_2 & \hat{y}_2 - \hat{y}_3 & \hat{x}_3 - \hat{x}_2 \\ \hat{x}_3\hat{y}_1 - \hat{x}_1\hat{y}_3 & \hat{y}_3 - \hat{y}_1 & \hat{x}_1 - \hat{x}_3 \\ \hat{x}_1\hat{y}_2 - \hat{x}_2\hat{y}_1 & \hat{y}_1 - \hat{y}_2 & \hat{x}_2 - \hat{x}_1 \end{bmatrix} \begin{Bmatrix} 1 \\ \hat{x} \\ \hat{y} \end{Bmatrix} = \frac{1}{2\hat{A}} \begin{bmatrix} \hat{a}_1 & \hat{b}_1 & \hat{c}_1 \\ \hat{a}_2 & \hat{b}_2 & \hat{c}_2 \\ \hat{a}_3 & \hat{b}_3 & \hat{c}_3 \end{bmatrix} \begin{Bmatrix} 1 \\ \hat{x} \\ \hat{y} \end{Bmatrix}.\end{gathered} \tag{5.5}$$

where

$$\begin{gathered}\hat{a}_1 = \hat{x}_2\hat{y}_3 - \hat{x}_3\hat{y}_2,\ \hat{a}_2 = \hat{x}_3\hat{y}_1 - \hat{x}_1\hat{y}_3,\ \hat{a}_3 = \hat{x}_1\hat{y}_2 - \hat{x}_2\hat{y}_1 \\ \hat{b}_1 = \hat{y}_2 - \hat{y}_3,\ \hat{b}_2 = \hat{y}_3 - \hat{y}_1,\ \hat{b}_3 = \hat{y}_1 - \hat{y}_2 \\ \hat{c}_1 = \hat{x}_3 - \hat{x}_2,\ \hat{c}_2 = \hat{x}_1 - \hat{x}_3,\ \hat{c}_3 = \hat{x}_2 - \hat{x}_1\end{gathered} \tag{5.6}$$

The shape functions can be visualized as

$$\begin{Bmatrix} \hat{L}_1 \\ \hat{L}_2 \\ \hat{L}_3 \end{Bmatrix} = \begin{Bmatrix} \dfrac{\hat{a}_1 + \hat{b}_1 x + \hat{c}_1 y}{2\hat{A}} \\ \dfrac{\hat{a}_2 + \hat{b}_2 x + \hat{c}_2 y}{2\hat{A}} \\ \dfrac{\hat{a}_3 + \hat{b}_3 x + \hat{c}_3 y}{2\hat{A}} \end{Bmatrix}$$

The integration formula is given by

$$\iint_{\hat{A}} \hat{L}_1^a \hat{L}_2^b \hat{L}_3^c d\hat{A} = \frac{a!b!c!2\hat{A}}{(a+b+c+2)!} \tag{5.7}$$

Here, the complete diagram of the FFEM approach [19] is apparent in Figure 5.2, which contains three layers. Initially, the parameters are considered as TFN, through the fuzzy input layer. Then, the hidden layer executes the data by using the FEM algorithm to convert the governing equations into algebraic form. Finally, solve the same by using the iterative technique and provide the output in terms of TFN through a fuzzy output layer. In this layer, the output of the system may be distinct in type and nature according to the fuzzy input variables [17, 30, 31].

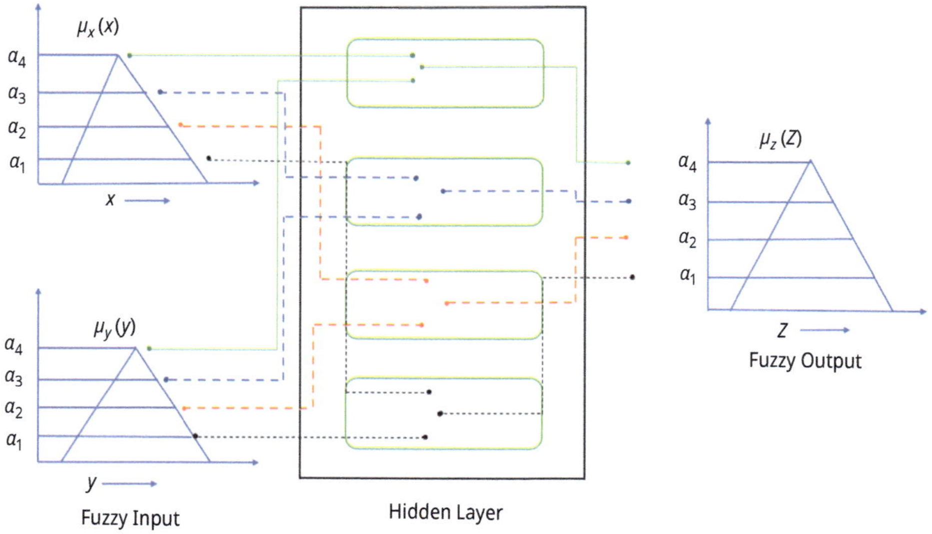

Figure 5.2: The layout of the fuzzy finite element method.

5.5 Solution of fuzzy coupled partial differential equation in heat transfer problem by using FFEM

The governing equation of the system such as momentum and energy equations are presented in eqs. (5.2) and (5.3) are shown coupled in nature. So, to solve the same the Galerkin residual approach is adopted. Accordingly, consider the fuzzy temperature and stream function distribution of the system as follows:

$$\overline{T} = \hat{L}_1\overline{T}_1 + \hat{L}_2\overline{T}_2 + \hat{L}_3\overline{T}_3$$

$$\overline{\psi} = \hat{L}_1\overline{\psi}_1 + \hat{L}_2\overline{\psi}_2 + \hat{L}_3\overline{\psi}_3$$

By using the above approach, eq. (5.2) is written as

$$\hat{R} = -\int_{\hat{A}} [\hat{L}]^T \left[\frac{\partial^2\overline{\psi}}{\partial\overline{x}^2} + \frac{\partial^2\overline{\psi}}{\partial\overline{y}^2} + \hat{R}a\frac{\partial\overline{T}}{\partial\overline{x}}\right] d\hat{A} = 0 \tag{5.8}$$

Here, $\hat{R}$ is defined as the fuzzy residue. Now, integrate each term of eq. (5.8) by applying integration by parts method such as

$$\int_{\hat{A}} [\hat{L}]^T \frac{\partial^2\overline{\psi}}{\partial\overline{x}^2} dA = [\hat{L}]^T \int_{\hat{A}} \frac{\partial^2\overline{\psi}}{\partial\overline{x}^2} d\hat{A} - \int_{\hat{A}} \frac{\partial [\hat{L}]^T}{\partial x} \frac{\partial\overline{\psi}}{\partial\overline{x}} d\hat{A} \tag{5.9}$$

By using Green's theorem, eq. (5.9) is rewritten as

$$\int_{\hat{A}} [\hat{L}]^T \frac{\partial^2\overline{\psi}}{\partial\overline{x}^2} d\hat{A} = -\int_{\hat{A}} \frac{\partial [\hat{L}]^T}{\partial\overline{x}} \frac{\partial\overline{\psi}}{\partial\overline{x}} d\hat{A} \tag{5.10}$$

$$\int_{\hat{A}} [\hat{L}]^T \frac{\partial^2\overline{\psi}}{\partial\overline{x}^2} d\hat{A} = \frac{1}{4\hat{A}^2} \int_{\hat{A}} \begin{bmatrix} \hat{b}_1 \\ \hat{b}_2 \\ \hat{b}_3 \end{bmatrix} \begin{bmatrix} \hat{b}_1 & \hat{b}_2 & \hat{b}_3 \end{bmatrix} \begin{bmatrix} \overline{\psi}_1 \\ \overline{\psi}_2 \\ \overline{\psi}_3 \end{bmatrix} d\hat{A}$$

$$= \frac{1}{4\hat{A}} \begin{bmatrix} \hat{b}_1^2 & \hat{b}_1\hat{b}_2 & \hat{b}_1\hat{b}_3 \\ \hat{b}_1\hat{b}_2 & \hat{b}_2^2 & \hat{b}_2\hat{b}_3 \\ \hat{b}_1\hat{b}_2 & \hat{b}_2\hat{b}_3 & \hat{b}_3^2 \end{bmatrix} \begin{bmatrix} \overline{\psi}_1 \\ \overline{\psi}_2 \\ \overline{\psi}_3 \end{bmatrix}$$

Similarly,

$$\int_{\hat{A}} [\hat{L}]^T \frac{\partial^2\overline{\psi}}{\partial\overline{y}^2} d\hat{A} = \frac{1}{4\hat{A}} \begin{bmatrix} \hat{c}_1^2 & \hat{c}_1\hat{c}_2 & \hat{c}_1\hat{c}_3 \\ \hat{c}_1\hat{c}_2 & \hat{c}_2^2 & \hat{c}_2\hat{c}_3 \\ \hat{c}_1\hat{c}_2 & \hat{c}_2\hat{c}_3 & \hat{c}_3^2 \end{bmatrix} \begin{bmatrix} \overline{\psi}_1 \\ \overline{\psi}_2 \\ \overline{\psi}_3 \end{bmatrix}$$

By using the integral term $\iint_{\hat{A}} \hat{L}_1^a\hat{L}_2^b\hat{L}_3^c d\hat{A} = \frac{a!b!c!2\hat{A}}{(a+b+c+2)!}$, the third term of eq. (5.8) is shown as

$$\int_{\hat{A}} [\hat{L}]^T \widehat{Ra} \frac{\partial \bar{\hat{T}}}{\partial \bar{\hat{x}}} d\hat{A} = -\frac{\widehat{Ra}}{6} \left\{ \begin{bmatrix} \hat{b}_1 \\ \hat{b}_2 \\ \hat{b}_3 \end{bmatrix} \begin{bmatrix} \bar{\hat{T}}_1 & \bar{\hat{T}}_2 & \bar{\hat{T}}_3 \end{bmatrix} \right\} = -\frac{\widehat{Ra}}{6} \begin{Bmatrix} \hat{b}_1\bar{\hat{T}}_1 + \hat{b}_2\bar{\hat{T}}_2 + \hat{b}_3\bar{\hat{T}}_3 \\ \hat{b}_1\bar{\hat{T}}_1 + \hat{b}_2\bar{\hat{T}}_2 + \hat{b}_3\bar{\hat{T}}_3 \\ \hat{b}_1\bar{\hat{T}}_1 + \hat{b}_2\bar{\hat{T}}_2 + \hat{b}_3\bar{\hat{T}}_3 \end{Bmatrix}$$

Finally, the converted algebraic form of the governing momentum eq. (5.2) is presented as

$$\frac{1}{4\hat{A}} \left\{ \begin{bmatrix} \hat{b}_1^2 & \hat{b}_1\hat{b}_2 & \hat{b}_1\hat{b}_3 \\ \hat{b}_1\hat{b}_2 & \hat{b}_2^2 & \hat{b}_2\hat{b}_3 \\ \hat{b}_1\hat{b}_2 & \hat{b}_2\hat{b}_3 & \hat{b}_3^2 \end{bmatrix} + \begin{bmatrix} \hat{c}_1^2 & \hat{c}_1\hat{c}_2 & \hat{c}_1\hat{c}_3 \\ \hat{c}_1\hat{c}_2 & \hat{c}_2^2 & \hat{c}_2\hat{c}_3 \\ \hat{c}_1\hat{c}_2 & \hat{c}_2\hat{c}_3 & \hat{c}_3^2 \end{bmatrix} \right\} \begin{Bmatrix} \bar{\hat{\psi}}_1 \\ \bar{\hat{\psi}}_2 \\ \bar{\hat{\psi}}_3 \end{Bmatrix} = \frac{\widehat{Ra}}{6} \begin{Bmatrix} \hat{b}_1\bar{\hat{T}}_1 + \hat{b}_2\bar{\hat{T}}_2 + \hat{b}_3\bar{\hat{T}}_3 \\ \hat{b}_1\bar{\hat{T}}_1 + \hat{b}_2\bar{\hat{T}}_2 + \hat{b}_3\bar{\hat{T}}_3 \\ \hat{b}_1\bar{\hat{T}}_1 + \hat{b}_2\bar{\hat{T}}_2 + \hat{b}_3\bar{\hat{T}}_3 \end{Bmatrix} \tag{5.11}$$

In a similar manner, the energy equation (5.3) is shown as

$$\hat{R} = -\int_{\hat{A}} [\hat{L}]^T \left[\left[\frac{\partial \bar{\hat{\psi}}}{\partial \bar{\hat{y}}} \frac{\partial \bar{\hat{T}}}{\partial \bar{\hat{x}}} - \frac{\partial \bar{\hat{\psi}}}{\partial \bar{\hat{x}}} \frac{\partial \bar{\hat{T}}}{\partial \bar{\hat{y}}} \right] - \left(\left(1 + \frac{4}{3} \widehat{R_d} \right) \frac{\partial^2 \bar{\hat{T}}}{\partial \bar{\hat{x}}^2} + \frac{\partial^2 \bar{\hat{T}}}{\partial \bar{\hat{y}}^2} \right) \right] d\hat{A} = 0 \tag{5.12}$$

After integrating the above equation, the derived algebraic form of eq. (5.3) is delineated as

$$\frac{1}{12\hat{A}} \begin{Bmatrix} \hat{c}_1\bar{\hat{\psi}}_1 + \hat{c}_2\bar{\hat{\psi}}_2 + \hat{c}_3\bar{\hat{\psi}}_3 \\ \hat{c}_1\bar{\hat{\psi}}_1 + \hat{c}_2\bar{\hat{\psi}}_2 + \hat{c}_3\bar{\hat{\psi}}_3 \\ \hat{c}_1\bar{\hat{\psi}}_1 + \hat{c}_2\bar{\hat{\psi}}_2 + \hat{c}_3\bar{\hat{\psi}}_3 \end{Bmatrix} \begin{bmatrix} \hat{b}_1 & \hat{b}_2 & \hat{b}_3 \end{bmatrix} - \frac{1}{12\hat{A}} \begin{Bmatrix} \hat{b}_1\bar{\hat{\psi}}_1 + \hat{b}_2\bar{\hat{\psi}}_2 + \hat{b}_3\bar{\hat{\psi}}_3 \\ \hat{b}_1\bar{\hat{\psi}}_1 + \hat{b}_2\bar{\hat{\psi}}_2 + \hat{b}_3\bar{\hat{\psi}}_3 \\ \hat{b}_1\bar{\hat{\psi}}_1 + \hat{b}_2\bar{\hat{\psi}}_2 + \hat{b}_3\bar{\hat{\psi}}_3 \end{Bmatrix} \begin{bmatrix} \hat{c}_1 & \hat{c}_2 & \hat{c}_3 \end{bmatrix} + \frac{1}{4\hat{A}} \left[\left(1 + \frac{4}{3} \widehat{R_d} \right) \begin{bmatrix} \hat{b}_1^2 & \hat{b}_1\hat{b}_2 & \hat{b}_1\hat{b}_3 \\ \hat{b}_1\hat{b}_2 & \hat{b}_2^2 & \hat{b}_2\hat{b}_3 \\ \hat{b}_1\hat{b}_2 & \hat{b}_2\hat{b}_3 & \hat{b}_3^2 \end{bmatrix} + \begin{bmatrix} \hat{c}_1^2 & \hat{c}_1\hat{c}_2 & \hat{c}_1\hat{c}_3 \\ \hat{c}_1\hat{c}_2 & \hat{c}_2^2 & \hat{c}_2\hat{c}_3 \\ \hat{c}_1\hat{c}_2 & \hat{c}_2\hat{c}_3 & \hat{c}_3^2 \end{bmatrix} \right] \begin{Bmatrix} \bar{\hat{T}}_1 \\ \bar{\hat{T}}_2 \\ \bar{\hat{T}}_3 \end{Bmatrix} = 0 \tag{5.13}$$

Finally, the compact forms of eqs. (5.11) and (5.13) are shown as follows:

$$[\hat{K}]\left\{\hat{\overline{\psi}}\right\} = \left\{\hat{\overline{f}}\right\} \tag{5.14}$$

$$[\hat{K}]\left\{\hat{\overline{T}}\right\} = \{0\} \tag{5.15}$$

Here, the system element's obtained elemental stiffness matrix has a set of size 3×3. Then, by assembling all those matrices, we get a global stiffness matrix of size $n \times n$ for the entire domain with a force vector is $n \times 1$. The fuzzy iterative process for solving the coupled PDEs is displayed in Figure 5.3.

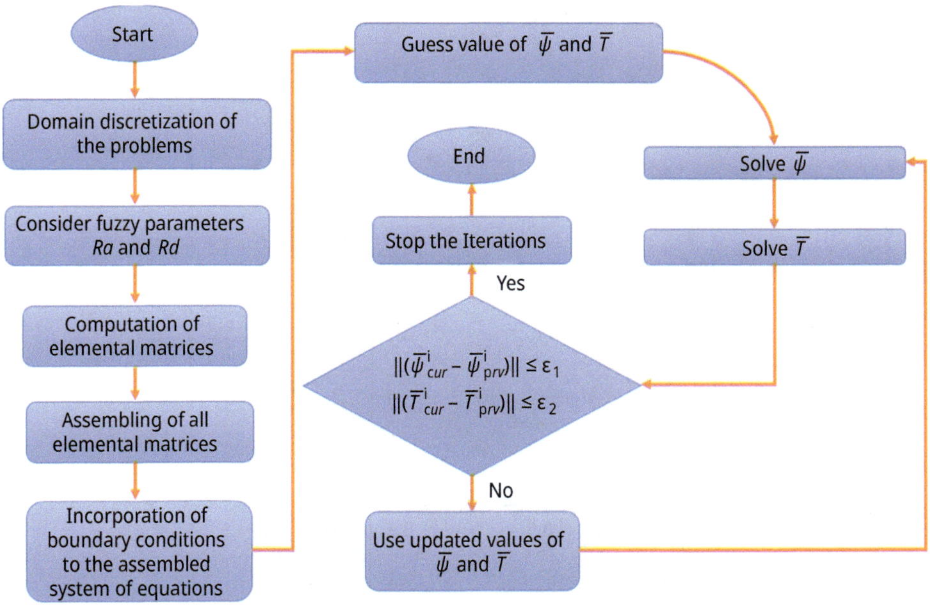

Figure 5.3: Schematic model for the approach to the solution.

Since the equations are coupled in nature, initially take the guess value for the vector $\overline{\psi}$, and $\overline{T}$ and when fed into eq. (5.14) it gives the new value of $\hat{\overline{\psi}}$. Similarly, the updated value is $\hat{\overline{\psi}}$ applied into eq. (5.15) to obtain fresh values of $\hat{\overline{T}}$. In a similar manner, continue this process until the convergence conditions are satisfied. For our research work the tolerance criteria are considered as follows:

$$\left\|\hat{\overline{\psi}}^i_{cur} - \hat{\overline{\psi}}^i_{prv}\right\| \leq \epsilon_1 = 10^{-9} \text{ and } \left\|\hat{\overline{T}}^i_{cur} - \hat{\overline{T}}^i_{prv}\right\| \leq \epsilon_2 = 10^{-9}$$

for $i = 1, 2, \ldots, n$. In this context, cur and prv denote the current and previous values of the iterations. One can see the convergence analysis of the same in [23].

5.6 Case study and discussion

In the present study, we consider a square cavity, and the domain is discretized into two elemental forms. The model diagram of the elements is presented in Figure 5.4. Since the system contains distinct parameters that show the physical characteristics of the heat transfer and fluid flow within the cavity, there may be uncertainties. Here, the involved variables such as *Ra* and *Rd* are considered as fuzzy variables.

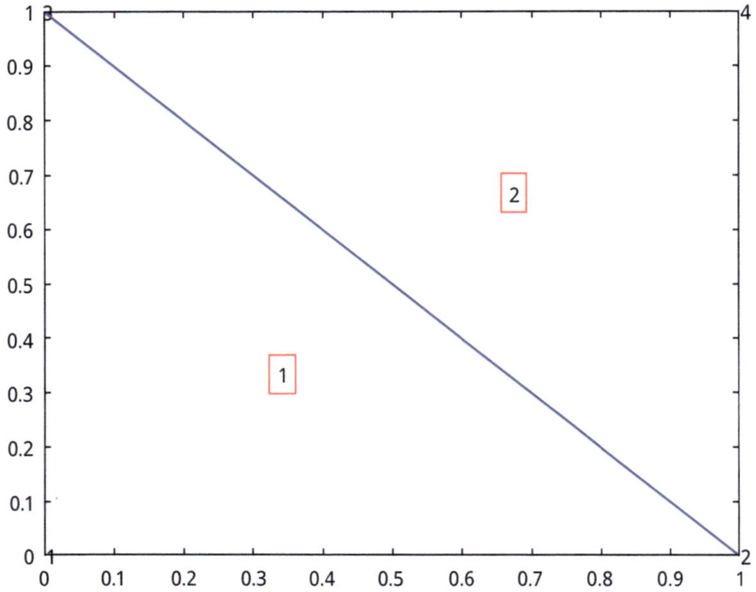

Figure 5.4: Geometry of 2-element discretization.

Table 5.1 describes the crisp (exact) and fuzzy values of the involved uncertain parameters for the current study.

Table 5.1: Crisp and fuzzy values of the parameters.

Parameters	Crisp value	TFN
Ra	1	[0.1,1,0.3]
Rd	1	[0.1,1,0.3]

Here, we discuss the convergence of the uncertain variables $\overline{\hat{\psi}}$ and $\overline{\hat{T}}$ profile with fuzzy parametric values. First, we choose *Ra* as TFN, and the crisp value of *Rd*, then by using abovementioned FFEM, the obtained nodal solution of $\overline{\hat{\psi}}$ and $\overline{\hat{T}}$ are shown in

Figures 5.5–5.9. Similarly, by choosing a single parameter Rd as fuzzy with crisp Ra, the obtained nodal $\tilde{\psi}$ and $\tilde{T}$ are discussed in Figures 5.9 to 5.12. Further, to analyze the sensitivity of the imprecisely defined parameters, the width of the TFN is considered. The convergence of the same is represented in Figures 5.13 and 5.14. Here, we found that, for a single parameter Ra as fuzzy the obtained TFN solution of $\tilde{\psi}$ for node 2 is converged to [0, 0.00000001, 0.00000001], and for node 4 it is converged at [0, 0, 0.00000001]. The graphical analysis of the same is depicted in Figures 5.5 and 5.6. Similarly, the TFN solutions of $\tilde{T}$ for nodes 2 and 4 are shown in Figures 5.7 and 5.8, where the same is converged at [29.99999993, 29.99999999, 29.99999999] and [29.99999999, 29.99999999, 30.00000008].

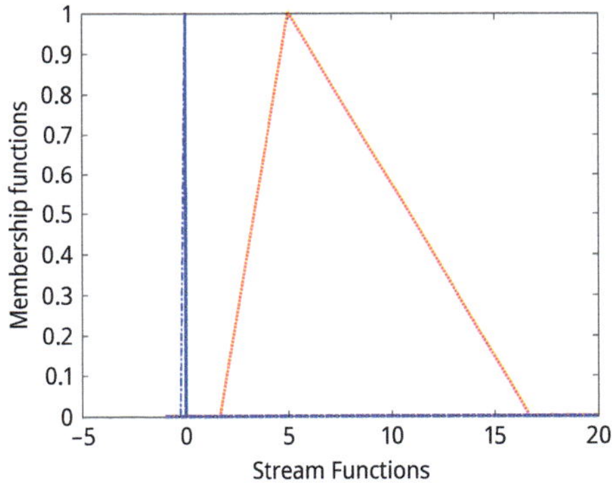

Figure 5.5: TFN solutions of stream function for node 2, when only *Ra* is fuzzy.

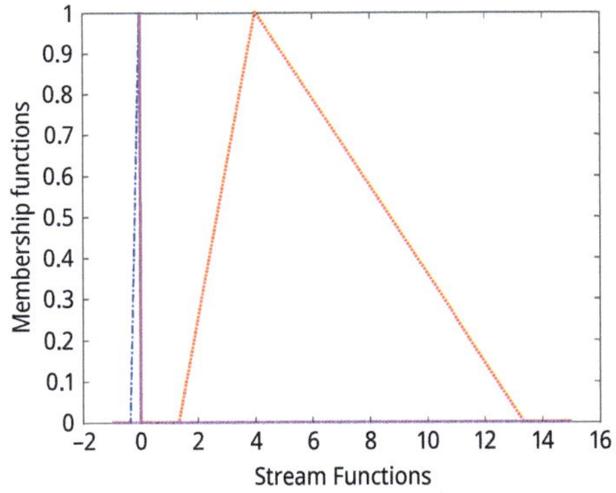

Figure 5.6: TFN solutions of stream function for node 4, when only *Ra* is fuzzy.

Similarly, by taking the single parameter Rd as fuzzy it is found that the TFN solution of $\tilde{\psi}$ for node 3 is converged at [0, 0.00000001, 0.00000004] and the schematic view of the same is presented in Figure 5.9. In this direction, the TFN solution for node 4 is converged to [0, 0, 0.00000003] and the clear picture of the same is presented in Figure 5.10.

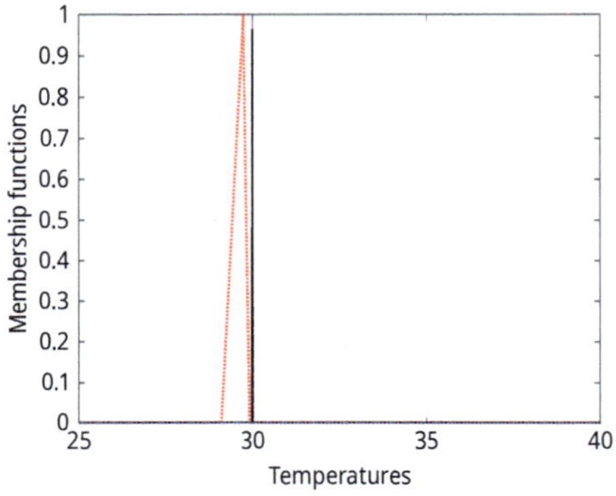

Figure 5.7: TFN solutions of temperature for node 2, when only *Ra* is fuzzy.

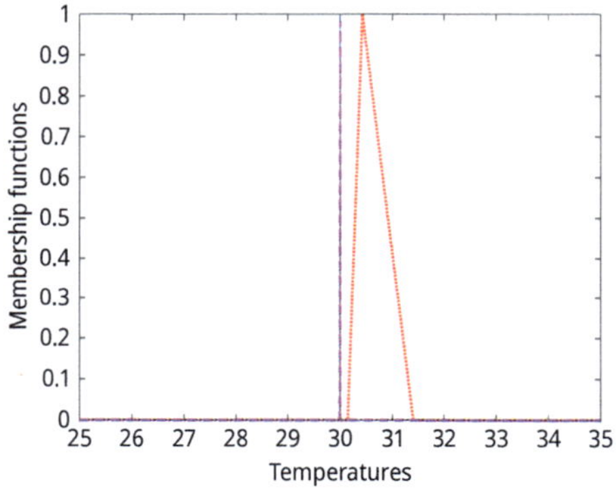

Figure 5.8: TFN solutions of temperature for node 4 when only *Ra* is fuzzy.

In practice, the TFN solution of $\overline{\hat{T}}$ for nodes 2 and 4 are converged to [29.99999999, 29.99999999, 30], [29.99999999, 29.99999999, 30] and the same is shown in Figures 5.11 and 5.12. Here, Figures 5.13 and 5.14 represent the width convergence of the $\overline{\hat{\psi}}$ and $\overline{\hat{T}}$ with respect to different iterations. It is noticed that the *Ra* is more sensitive than *Rd* in terms of $\overline{\hat{\psi}}$ and $\overline{\hat{T}}$, because the width is apparently more. Also, with the variation of *Ra*, the $\overline{\hat{\psi}}$ and $\overline{\hat{T}}$ increase and decrease with respective nodes. Here, we observe that the effect of *Ra* is more, so we can say that the system is unstable when *Ra* contained vagueness.

Similarly, the width of the $\overline{\hat{\psi}}$ and $\overline{\hat{T}}$ are the least when the only parameter *Rd* is fuzzy. That means the imprecise parameter *Rd* is less sensitive as compared to *Ra*, which shows the system is stable when the single parameter *Rd* contained vagueness.

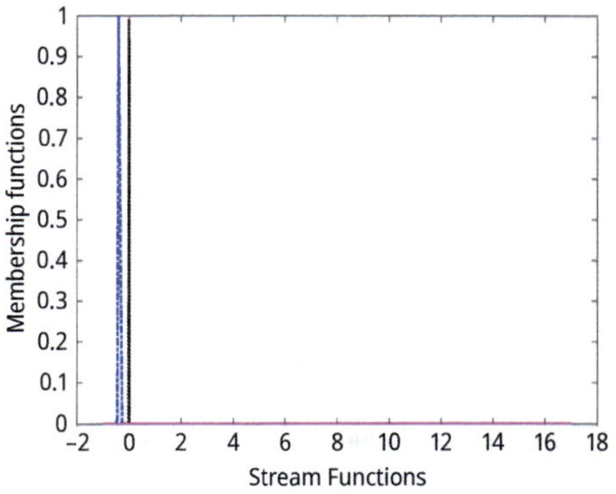

Figure 5.9: TFN solutions of stream function for node 2, when only *Rd* is fuzzy.

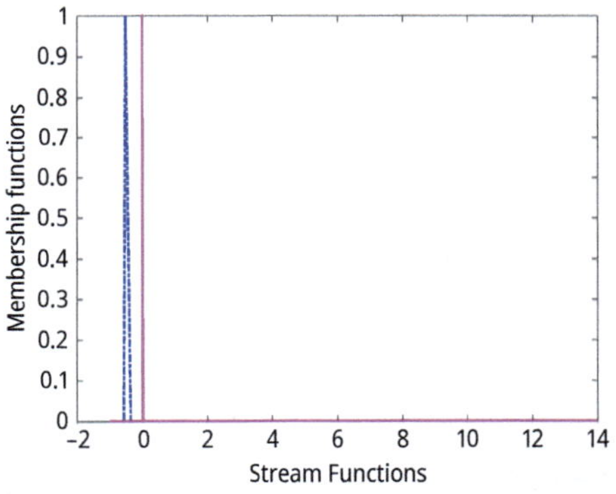

Figure 5.10: TFN solutions of stream function for node 4, when only *Rd* is fuzzy.

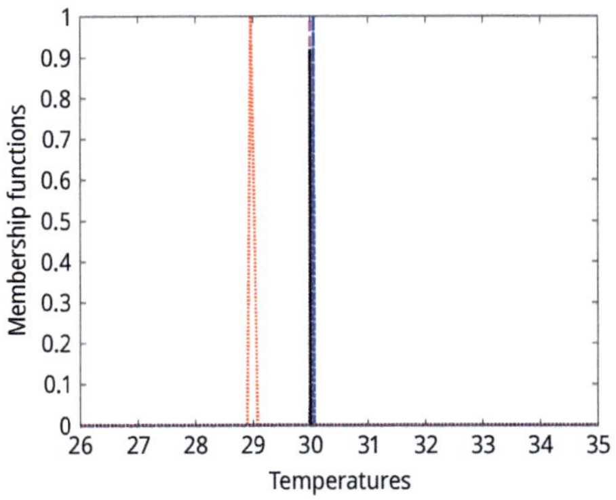

Figure 5.11: TFN solutions of temperature for node 2, when only *Rd* is fuzzy.

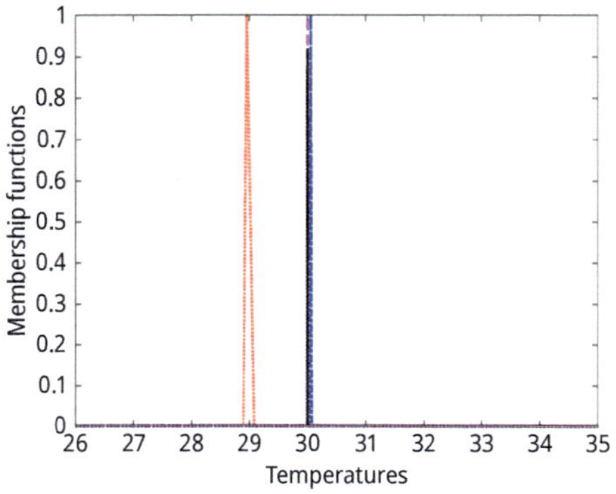

Figure 5.12: TFN solutions of temperature for node 4, when only *Rd* is fuzzy.

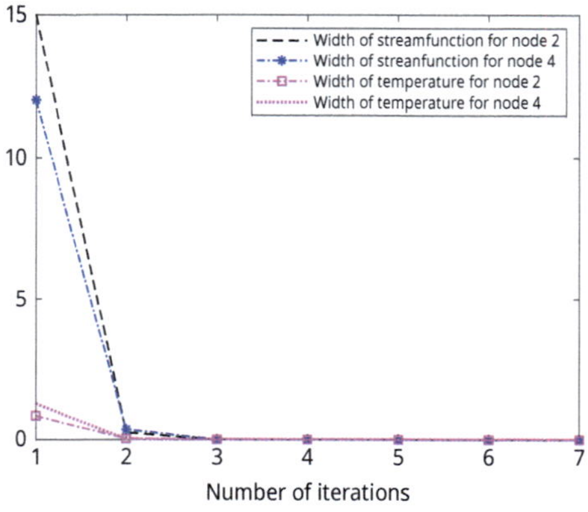

Figure 5.13: The width solution of stream function and temperature while *Ra* is fuzzy.

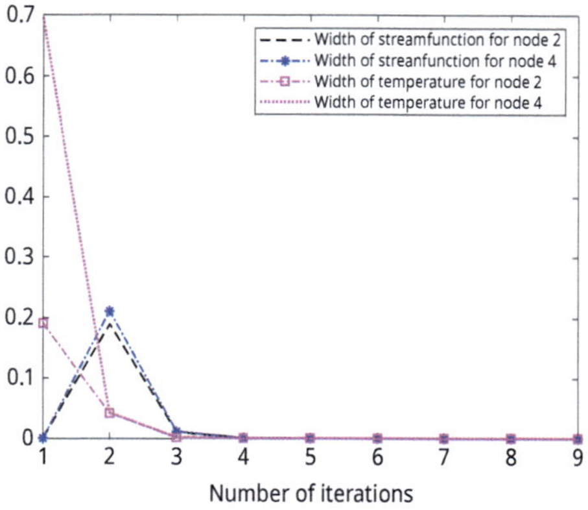

Figure 5.14: The width solution of stream function and temperature while *Rd* is fuzzy.

5.7 Conclusion

In this chapter, the significant role of FFEM is presented for solving coupled PDEs occurred in heat transfer problems. For the investigation, we have considered heat transfer in a square cavity. The governing equations are converted into a set of algebraic equations using FFEM. The solution of the same has been demonstrated by taking the numerical iterative procedure. Then choosing involved uncertain parameters the convergence analysis of stream function and temperature were discussed. As sensitive analysis plays an important role in the control of uncertainty and its consequences, we observed the same. Hence, it is observed that the parameter Rd has less sensitivity when compared to others. In future, the proposed approach can be used to solve other complicated coupled PDEs in different fluid flow and heat transfer problems.

References

[1] Bejan A. (1979). On the boundary layer regime in a vertical enclosure filled with a porous medium. Letters in Heat and Mass Transfer, 6(2), 93–102.

[2] Nield D. A., and Bejan A. (1999). Convection in Porous Media, Springer.

[3] Vafai K. (2009). Handbook of Porous Media, Marcel Dekker.

[4] Lewis R. W., and Garner R. W. (1972). A finite element solution of coupled electrokinetic and hydrodynamic flow in porous media. International Journal for Numerical Methods in Engineering, 5(1), 41–55.

[5] Ahamad N. A., Soudagar M. E. M., Kamangar S., and Irfan B. A. (2017). "Fem formulation of coupled partial differential equations for heat transfer,". IOP Conference Series: Materials Science and Engineering, 225, 012023.

[6] Badruddin I. A., Khan A., Idris M. Y. I., Nik-Ghaali N., N.j S. A., and Al-Rashed A. A. (2017). Simplified finite element algorithm to solve conjugate heat and mass transfer in porous medium. International Journal of Numerical Methods for Heat & Fluid Flow, 27(11), 2481–2507.

[7] N. J S. A., Sarfaraz K., Badruddin A. I., Abdullah A. A. A. A.-R., Quadir G. A., Khaleed H. M. T., and Khan T. M. Y. (2014). Conjugate heat transfer in porous. Journal of Porous Media, 17(12), 1109–1119.

[8] Aparna K., and Seetharamu K. (2017). Investigations on the effect of non-uniform temperature on fluid flow and heat transfer in a trapezoidal cavity filled with porous media. International Journal of Heat and Mass Transfer, 108, 63–78.

[9] Basak T., Pradeep P., Roy S., and Pop I. (2011). Finite element based heatline approach to study mixed convection in a porous square cavity with various wall thermal boundary conditions. International Journal of Heat and Mass Transfer, 54(9–10), 1706–1727.

[10] Wu C. Y. (2009). Monte Carlo simulation of transient radiative transfer in a medium with a variable reflective index. International Journal of Heat and Mass Transfer, 52, 4151–4159.

[11] Zadeh L. A. (1965). Fuzzy sets. Information and Control, 8, 338–353.

[12] Dong W., and Shah H. C. (1987). Vertex method for computing functions of fuzzy variables. Fuzzy Sets and Systems, 24(1), 65–78.

[13] Dong W., and Wong F. (1987). Fuzzy weighted averages and implementation of the extension principle. Fuzzy Sets and Systems, 21(2), 183–199.

[14] Yang H. Q., Yao H., and Jones J. D. (1993). Calculating functions on fuzzy numbers. Fuzzy Sets and System, 55, 273–283.
[15] Nicolai, B., Egea, J., Scheerlinck N., Banga J., and Datta A. (2011). Fuzzy finite element analysis of heat conduction problems with uncertain parameters. Journal of Food Engineering, 103(1), 38–46.
[16] Nayak S., and Chakraverty S. (2013). Non-probabilistic approach to investigate uncertain conjugate heat transfer in an imprecisely defined plate. International Journal of Heat and Mass Transfer, 67, 445–454.
[17] Chakraverty S., and Nayak S. (2013). Fuzzy finite element method in diffusion problems. In Mathematics of Uncertainty Modelling in the Analysis of Engineering and Science, IGI global, pp. 309–328.
[18] Wang C., and Zhiping Q. (2015). Interval analysis of steady-state heat convection–diffusion problem with uncertain-but-bounded parameters. International Journal of Heat and Mass Transfer, 91, 355–362.
[19] Nayak S. (2020). Uncertain quantification of field variables involved in transient convection diffusion problems for imprecisely defined parameters. International Communications in Heat and Mass Transfer, 119, 104894.
[20] Zhao Y., Gao H., and Qiu J. (2019). Fuzzy observer based control for nonlinear coupled hyperbolic PDE-ODE systems. IEEE Transactions on Fuzzy Systems, 27(7), 1332–1346.
[21] Priyadarshini S., and Nayak S. (2022). A numerical approach to study heat and mass transfer in porous medium influenced by uncertain parameters. International Communications in Heat and Mass Transfer, 139, 106411.
[22] Panigrahi P. K., and Nayak S. (2023). Numerical investigation of non-probabilistic systems using inner outer direct search optimization technique. AIMS Mathematics, 8(9), 21329–21358.
[23] Priyadarshini S., and Nayak S. (2023). Effects of imprecisely defined parameters on heat and mass transfer in a vertical annular porous cylinder. International Communications in Heat and Mass Transfer, 149, 107097.
[24] Panigrahi P. K., and Nayak S. (2024). Conjugate gradient with Armijo line search approach to investigate imprecisely defined unconstrained optimisation problem. International Journal of Computational Science and Engineering, 27(4), 458–471.
[25] Priyadarshini S., and Nayak S. (2023). A new hybrid approach to study heat and mass transfer in porous medium influenced by imprecisely defined parameters. Case Studies in Thermal Engineering, 51, 103619.
[26] Panigrahi P. K., and Nayak S. (2024). Numerical approach to solve imprecisely defined systems using inner outer direct search optimization technique. Mathematics and Computers in Simulation, 215, 578–606.
[27] Lewis R. W., and Garner R. W. (1972). A finite element solution of coupled electrokinetic and hydrodynamic flow in porous media. International Journal for Numerical Methods in Engineering, 5(1), 41–55.
[28] Nithiarasu P., Lewis R. W., and Seetharamu K. N. (2016). Fundamentals of the Finite Element Method for Heat and Mass Transfer, 2nd Edition Wiley Series In Computational Mechanics.
[29] Nayak S., and Chakraverty S. (2008). Interval Finite Element Method with MATLAB, Academic Press.
[30] Priyadarshini S., and Nayak S. (2024). Fuzzy finite element solution of uncertain convection-diffusion heat transfer for a rectangular plate. Computational Algorithms and Numerical Dimensions, 3(4), 298–307.
[31] P. L. Durga and S. Nayak, “Modified Levenberg-Marquardt Neural Network Method to Investigate Optimum Height of a Water Tank Problem,” 2024 OITS International Conference on Information Technology (OCIT), Vijayawada, India, 2024, pp. 339–344.

Sudipta Priyadarshini and Sukanta Nayak

Chapter 6
Numerical approach to study the effect of uncertain spectrum of field variables in a porous cavity

Abstract: This chapter investigates the effect of imprecisely defined parameter Rayleigh number (*Ra*) with errors. The uncertainties are caused due to imprecisely defined involved parameters, insufficient knowledge about the system, material properties, etc. As such, the mathematical modeling of a porous cavity becomes uncertain. In practice, the crisp coupled partial differential equations (PDEs) are difficult to solve but in addition to epistemic uncertainties, it becomes more challenging. Therefore, the main objective is to present an alternative method that eases the process as well as the computation with uncertainties. By using the fuzzy finite element method, the involved PDEs are transformed into algebraic equations in fuzzy environment. Further the same has been solved through the Gauss-Seidel iterative method. Finally, the sensitiveness of the variables is analyzed and establishes a convergence. This approach can be utilized efficaciously in various science and engineering problems, viz., fluid flow in porous medium, mechanics, geoscience, structural dynamical problems, etc.

Keywords: fuzzy finite element method (FFEM), triangular fuzzy number (TFN), heat transfer, porous medium

6.1 Introduction

Heat transfer in a porous medium has garnered attention because of its extensive science and engineering applications. These types of occurrences with different sizes, such as nano to the macro can be found in industrial equipment, cavities, and computer chips. Furthermore, Nield and Bejan [1], Pop and Ingham [2], and Vafai [3] have studied applications related to heat transfer in porous medium. Generally, much research is concentrated on relatively easy geometry arrangements such as rectangular or square cavities. Due to popular applications like roofing and the cooling of electronic equipment, the triangular enclosure has been considered for the study.

Sudipta Priyadarshini, Department of Basic Science & Humanities, Vignan's Institute of Information Technology (A), Duvvada, Vishakhapatnam 530049, Andhra Pradesh, India, e-mail: sudipta.grc@gmail.com
Sukanta Nayak, Department of Mathematics, School of Advanced Sciences, VIT-AP University, Inavolu, Besides AP Secretariat, Amaravati, Andhra Pradesh, India, e-mail: sukantgacr@gmail.com

https://doi.org/10.1515/9783111661674-006

To get a better understanding of the above-defined heat transfer problems, mathematical models need to be studied. In general, the mathematical models are constructed using differential equations. Among them, the PDEs can be used to mathematically represent the natural convection in a porous enclosure. Generally, the heat transfer phenomenon in numerous cavities was investigated by using different numerical approaches. Throughout, FEM gets more attention because of its flexibility to use in different boundaries and geometries. The natural convection with rectangular, non-rectangular, and horizontal cavities has been studied with different researchers as ([4–6]). Further, the heat transfer in a vertical annular cylinder by taking the radiation parameter effect was investigated by Badruddin et al. ([7–9]) using the thermal equilibrium mode. Free convection in a vertical porous enclosure has been analyzed by Kumar in [10]. The numerical investigation of the natural convection in a trapezoidal porous cavity has been studied by Baytas et al. [11], where they handle the Darcy and energy equations by adopting implicit FDM approach. In addition, the same author Baytas et al. [12] further investigated free convection in oblique enclosures filled with a porous medium by using the same approach.

Recently, the convection of heat transfer in a triangular permeable porous cavity has been analyzed by Ullah et al. [13]. Further, the experimental study was done by Rasheedin et al. [14] in a triangular prism buried in a porous substance. The hydromagnetic flow and heat transmission of Casson fluid in a partially heated triangular cavity are studied numerically by Khan et al. [15].

The abovementioned literature provides an overview of the convection of heat transfer, where the defined parameters of the systems have been considered as crisp. But in real practice, it may be uncertain. This makes it more challenging to solve. In this regard, to deal with the uncertainties, one may use either probabilistic or non-probabilistic approaches. But the probabilistic approach is authentic when the data set is large. Several authors have employed probabilistic techniques to handle the same; one can find this in [16, 17]. Therefore, in case of a small data set this approach may fail to produce efficient results.

In this direction, in 1965 Zadeh [18] has introduced the fuzzy set and number to overcome the uncertainties in the system. In this regard, Nayak and Chakraverty [19] have investigated the numerical solution of uncertain neutron diffusion equation for imprecisely defined homogeneous triangular bare reactor, while the interval parameter perturbation method (IPPM) has been used in [20] by the author Wang et al. to analyze the sensitivity of heat convection-diffusion problems. Further, Chakraverty and Nayak [21] have investigated the uncertain heat conduction problems by using the non-probabilistic approach. Hazarika et al. [22] have studied numerically the MHD flow in a fuzzy environment with an induced magnetic field.

It has been noticed that the implementation of uncertainties with different regular and non-regular geometries makes the problem difficult. Current investigation has considered the heat transfer in a porous cavity with uncertain environment. The numerical study of the same has been performed by adopting the fuzzy finite element method (FFEM). Initially, the imprecisely defined governing equations are converted into uncertain algebraic equations, then by using the Gauss-Seidel iterative method the equations are solved and the convergence analysis is performed. The investigation is carried out by taking Rayleigh number (*Ra*) with 5% and 8% error of uncertainties and the sensitivity analysis of the parameters is discussed.

6.2 Preliminaries

Definition 6.2.1 ([23]): Let X be a group of objects denoted by x, then a fuzzy set $\hat{A}$ in X is defined as a set of ordered pairs such as:

$$\hat{A} = \left\{ \left(x, \mu_{\hat{A}}(x)\right) | \forall x \in X, \mu_{\hat{A}}(x) \in [0,1] \right\} \tag{6.1}$$

where $\mu_{\hat{A}}(x)$ is called the membership function of x.

Definition 6.2.2 ([23, 24]): A fuzzy set $\hat{A}$ is said to be a fuzzy number, while $\hat{A}$ is normal, which means the height $h(\hat{A}) = 1$, $\hat{A}$ is convex and the membership function $\mu_{\hat{A}}(x)$ is at least piecewise continuous.

Definition 6.2.3 ([23, 24]): A fuzzy number $\hat{A} = \left[a^L, a^M, a^R \right]$ is a triangular fuzzy number (TFN) present in Figure 6.1 with the membership function $\mu_{\hat{A}}(x)$ and defined as

$$\mu_{\hat{A}}(x) = \begin{cases} \mu_L, & a^L \le x \le a^M \\ \mu_R, & a^M \le x \le a^R \\ 0, & \text{otherwise} \end{cases} \tag{6.2}$$

where $\mu_L = \dfrac{x - a^L}{a^M - a^L}$ and $\mu_R = \dfrac{a^R - x}{a^R - a^M}$.

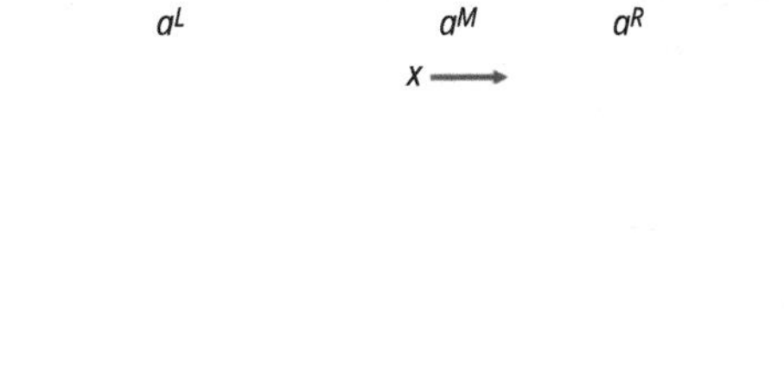

Figure 6.1: Triangular fuzzy number.

Definition 6.2.4 ([23]): Suppose that $\hat{x} = [\underline{x}(\alpha), \overline{x}(\alpha)]$ and $\hat{y} = [\underline{y}(\alpha), \overline{y}(\alpha)]$ are two fuzzy numbers; then using the above definitions, fuzzy arithmetic operations can be written as:

Addition

$$\hat{x} + \hat{y} = [\underline{x}(\alpha) + \underline{y}(\alpha), \overline{x}(\alpha) + \overline{y}(\alpha)]$$

Subtraction

$$\hat{x} - \hat{y} = [\underline{x}(\alpha) - \overline{y}(\alpha), \overline{x}(\alpha) - \underline{y}(\alpha)]$$

Multiplication

$$\hat{x} \times \hat{y} = \left[\min\left\{\underline{x}(\alpha) \times \underline{y}(\alpha), \underline{x}(\alpha) \times \overline{y}(\alpha), \overline{x}(\alpha) \times \underline{y}(\alpha), \overline{x}(\alpha) \times \overline{y}(\alpha)\right\},\right.$$
$$\left.\max\left\{\underline{x}(\alpha) \times \underline{y}(\alpha), \underline{x}(\alpha) \times \overline{y}(\alpha), \overline{x}(\alpha) \times \underline{y}(\alpha), \overline{x}(\alpha) \times \overline{y}(\alpha)\right\}\right].$$

Division

$$\frac{\hat{x}}{\hat{y}} = \left[\min\left\{\frac{\underline{x}(\alpha)}{\overline{y}(\alpha)}, \frac{\underline{x}(\alpha)}{\underline{y}(\alpha)}, \frac{\overline{x}(\alpha)}{\overline{y}(\alpha)}, \frac{\overline{x}(\alpha)}{\underline{y}(\alpha)}\right\}, \max\left\{\frac{\underline{x}(\alpha)}{\overline{y}(\alpha)}, \frac{\underline{x}(\alpha)}{\underline{y}(\alpha)}, \frac{\overline{x}(\alpha)}{\overline{y}(\alpha)}, \frac{\overline{x}(\alpha)}{\underline{y}(\alpha)}\right\}\right],$$

where

$$\underline{y}(\alpha) \neq 0$$

and

$$\overline{y}(\alpha) \neq 0$$

6.3 Formulation of fuzzy finite element method (FFEM)

Generally, solving the coupled system of governing equations in heat transfer problems is complicated, but when it comes with uncertainties it is more challenging to solve. Further, participation of impreciseness in the corresponding parameters of the equations cannot be ignored. To deal with situations such as experimental error, vague information about the involved parameters, boundary conditions, etc., fuzzy numbers may be helpful. So, here we have considered the TFN with finite element method namely, fuzzy finite element method (FFEM) [25, 26] to handle the uncertain system numerically. In FFEM, we have taken the linear triangular approach to discretize the domain as follows:

$$\begin{aligned}
\hat{x} &= \hat{\mathcal{M}}_1\hat{x}_1 + \hat{\mathcal{M}}_2\hat{x}_2 + \hat{\mathcal{M}}_3\hat{x}_3 \\
\hat{y} &= \hat{\mathcal{M}}_1\hat{y}_1 + \hat{\mathcal{M}}_2\hat{y}_2 + \hat{\mathcal{M}}_3\hat{y}_3 \\
\hat{\mathcal{M}} &= \hat{\mathcal{M}}_1 + \hat{\mathcal{M}}_2 + \hat{\mathcal{M}}_3
\end{aligned} \tag{6.3}$$

where $\hat{\mathcal{M}}$ denotes the shape function in fuzzy environment and $\hat{}$ denote the fuzzy. Additionally, eq. (6.3) can be transformed into the following matrix form:

$$\begin{bmatrix} 1 & 1 & 1 \\ \hat{x}_1 & \hat{x}_2 & \hat{x}_3 \\ \hat{y}_1 & \hat{y}_2 & \hat{y}_3 \end{bmatrix} \begin{Bmatrix} \hat{\mathcal{M}}_1 \\ \hat{\mathcal{M}}_2 \\ \hat{\mathcal{M}}_3 \end{Bmatrix} = \begin{Bmatrix} 1 \\ \hat{x} \\ \hat{y} \end{Bmatrix}$$

The fuzzy area of the triangular element $(\hat{A})$ is $\frac{1}{2}\begin{vmatrix} 1 & 1 & 1 \\ \hat{x}_1 & \hat{x}_2 & \hat{x}_3 \\ \hat{y}_1 & \hat{y}_2 & \hat{y}_3 \end{vmatrix}$.

Then, one can find:

$$\begin{Bmatrix} \hat{\mathcal{M}}_1 \\ \hat{\mathcal{M}}_2 \\ \hat{\mathcal{M}}_3 \end{Bmatrix} = \begin{bmatrix} 1 & 1 & 1 \\ \hat{x}_1 & \hat{x}_2 & \hat{x}_3 \\ \hat{y}_1 & \hat{y}_2 & \hat{y}_3 \end{bmatrix}^{-1} \begin{Bmatrix} 1 \\ \hat{x} \\ \hat{y} \end{Bmatrix} \Rightarrow \begin{Bmatrix} \hat{\mathcal{M}}_1 \\ \hat{\mathcal{M}}_2 \\ \hat{\mathcal{M}}_3 \end{Bmatrix} =$$
$$\frac{1}{2\hat{A}} \begin{bmatrix} \hat{x}_2\hat{y}_3 - \hat{x}_3\hat{y}_2 & \hat{y}_2 - \hat{y}_3 & \hat{x}_3 - \hat{x}_2 \\ \hat{x}_3\hat{y}_1 - \hat{x}_1\hat{y}_3 & \hat{y}_3 - \hat{y}_1 & \hat{x}_1 - \hat{x}_3 \\ \hat{x}_1\hat{y}_2 - \hat{x}_2\hat{y}_1 & \hat{y}_1 - \hat{y}_2 & \hat{x}_2 - \hat{x}_1 \end{bmatrix} \begin{Bmatrix} 1 \\ \hat{x} \\ \hat{y} \end{Bmatrix} = \frac{1}{2\hat{A}} \begin{bmatrix} \hat{a}_1 & \hat{b}_1 & \hat{c}_1 \\ \hat{a}_2 & \hat{b}_2 & \hat{c}_2 \\ \hat{a}_3 & \hat{b}_3 & \hat{c}_3 \end{bmatrix} \begin{Bmatrix} 1 \\ \hat{x} \\ \hat{y} \end{Bmatrix} \tag{6.4}$$

where

$$\begin{aligned}
&\hat{a}_1 = \hat{x}_2\hat{y}_3 - \hat{x}_3\hat{y}_2,\ \hat{a}_2 = \hat{x}_3\hat{y}_1 - \hat{x}_1\hat{y}_3,\ \hat{a}_3 = \hat{x}_1\hat{y}_2 - \hat{x}_2\hat{y}_1 \\
&\hat{b}_1 = \hat{y}_2 - \hat{y}_3,\ \hat{b}_2 = \hat{y}_3 - \hat{y}_1,\ \hat{b}_3 = \hat{y}_1 - \hat{y}_2 \\
&\hat{c}_1 = \hat{x}_3 - \hat{x}_2,\ \hat{c}_2 = \hat{x}_1 - \hat{x}_3,\ \hat{c}_3 = \hat{x}_2 - \hat{x}_1
\end{aligned} \tag{6.5}$$

Here, the complete diagram of the FFEM approach is apparent in Figure 6.2. The method involves three layers, input, hidden and output layer. Initially, we have taken the uncertain parameters in terms as a TFN with different α combinations. The hidden layer proposes FEM techniques to convert the governing equations into algebraic form and solve the same by using the iterative technique. At last, in output layer one can find the fuzzy output. In this layer, the output of the system may be different in type and nature according to the fuzzy input parameters.

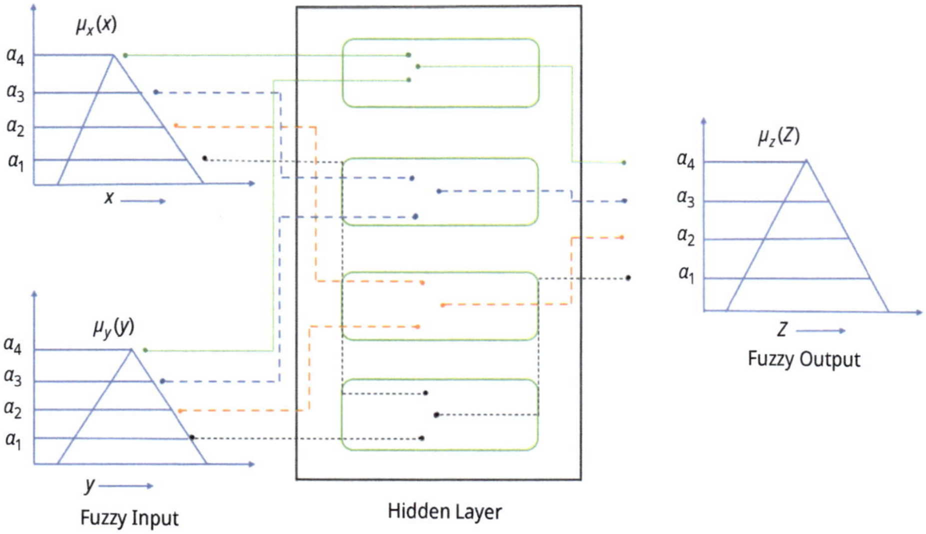

Figure 6.2: The layout of fuzzy finite element method (FFEM).

6.4 Fuzzy coupled partial differential equation in a porous cavity

The heat transfer in a triangular porous cavity is governed by two nonlinear coupled partial differential equations. Here, the corresponding nondimensional momentum and energy equation of the system in an uncertain circumstance is defined as follows:

Momentum equation:

$$\frac{\partial^2 \bar{\psi}}{\partial \bar{x}^2} + \frac{\partial^2 \bar{\psi}}{\partial \bar{y}^2} = -Ra\frac{\partial \bar{T}}{\partial \bar{x}} \tag{6.6}$$

Energy equation:

$$\frac{\partial \bar{\psi}}{\partial \bar{y}}\frac{\partial \bar{T}}{\partial \bar{x}} - \frac{\partial \bar{\psi}}{\partial \bar{x}}\frac{\partial \bar{T}}{\partial \bar{y}} = \frac{\partial^2 \bar{T}}{\partial \bar{x}^2} + \frac{\partial^2 \bar{T}}{\partial \bar{y}^2} \tag{6.7}$$

where *Ra* is defined as the Rayleigh number and it is a physical nondimensional parameter that controls the phenomenon. Due to the uncertainties involved in the model the obtained governing equations can be defined imprecisely. Let us consider the temperature and stream function in fuzzy environment. Here, $\widehat{}$ denotes the fuzzy non-dimensional form:

$$\begin{aligned}\overline{\hat{T}} &= \hat{\mathcal{M}}_1\overline{\hat{T}}_1 + \hat{\mathcal{M}}_2\overline{\hat{T}}_2 + \hat{\mathcal{M}}_3\overline{\hat{T}}_3 \\ \overline{\hat{\psi}} &= \hat{\mathcal{M}}_1\overline{\hat{\psi}}_1 + \hat{\mathcal{M}}_2\overline{\hat{\psi}}_2 + \hat{\mathcal{M}}_3\overline{\hat{\psi}}_3.\end{aligned} \tag{6.8}$$

Now, by using the Galerkin residual technique in uncertain environment eq. (6.6) is integrated by using integration by parts method as

$$R = -\int_{\hat{A}} \left[\hat{\mathcal{M}}\right]^T \left[\frac{\partial^2\overline{\hat{\psi}}}{\partial\overline{\hat{x}}^2} + \frac{\partial^2\overline{\hat{\psi}}}{\partial\overline{\hat{y}}^2} + \widehat{Ra}\frac{\partial\overline{\hat{T}}}{\partial\overline{\hat{x}}}\right] d\hat{A} = 0 \tag{6.9}$$

In the same manner, eq. (6.7) can be rewritten as

$$R = -\int_{\hat{A}} \left[\hat{\mathcal{M}}\right]^T \left[\left[\frac{\partial\overline{\hat{\psi}}}{\partial\overline{\hat{y}}}\frac{\partial\overline{\hat{T}}}{\partial\overline{\hat{x}}} - \frac{\partial\overline{\hat{\psi}}}{\partial\overline{\hat{x}}}\frac{\partial\overline{\hat{T}}}{\partial\overline{\hat{y}}}\right] - \left(\frac{\partial^2\overline{\hat{T}}}{\partial\overline{\hat{x}}^2} + \frac{\partial^2\overline{\hat{T}}}{\partial\overline{\hat{y}}^2}\right)\right] d\hat{A} = 0 \tag{6.10}$$

After solving eqs. (6.9) and (6.10), the secure algebraic equations are presented below:

$$\begin{aligned}&\frac{1}{4\hat{A}}\left\{\begin{bmatrix}\hat{b}_1^2 & \hat{b}_1\hat{b}_2 & \hat{b}_1\hat{b}_3 \\ \hat{b}_1\hat{b}_2 & \hat{b}_2^2 & \hat{b}_2\hat{b}_3 \\ \hat{b}_1\hat{b}_3 & \hat{b}_2\hat{b}_3 & \hat{b}_3^2\end{bmatrix} + \begin{bmatrix}\hat{c}_1^2 & \hat{c}_1\hat{c}_2 & \hat{c}_1\hat{c}_3 \\ \hat{c}_1\hat{c}_2 & \hat{c}_2^2 & \hat{c}_2\hat{c}_3 \\ \hat{c}_1\hat{c}_3 & \hat{c}_2\hat{c}_3 & \hat{c}_3^2\end{bmatrix}\right\}\left\{\begin{matrix}\overline{\hat{\psi}}_1 \\ \overline{\hat{\psi}}_2 \\ \overline{\hat{\psi}}_3\end{matrix}\right\} = \\ &\frac{\widehat{Ra}}{6}\left\{\begin{matrix}\hat{b}_1\overline{\hat{T}}_1 + \hat{b}_2\overline{\hat{T}}_2 + \hat{b}_3\overline{\hat{T}}_3 \\ \hat{b}_1\overline{\hat{T}}_1 + \hat{b}_2\overline{\hat{T}}_2 + \hat{b}_3\overline{\hat{T}}_3 \\ \hat{b}_1\overline{\hat{T}}_1 + \hat{b}_2\overline{\hat{T}}_2 + \hat{b}_3\overline{\hat{T}}_3\end{matrix}\right\}\end{aligned} \tag{6.11}$$

In a similar manner, eq. (6.10) can be shown as

$$\begin{aligned}&\frac{1}{12\hat{A}}\left\{\begin{matrix}\hat{c}_1\overline{\hat{\psi}}_1 + \hat{c}_2\overline{\hat{\psi}}_2 + \hat{c}_3\overline{\hat{\psi}}_3 \\ \hat{c}_1\overline{\hat{\psi}}_1 + \hat{c}_2\overline{\hat{\psi}}_2 + \hat{c}_3\overline{\hat{\psi}}_3 \\ \hat{c}_1\overline{\hat{\psi}}_1 + \hat{c}_2\overline{\hat{\psi}}_2 + \hat{c}_3\overline{\hat{\psi}}_3\end{matrix}\right\}\left[\hat{b}_1 \quad \hat{b}_2 \quad \hat{b}_3\right] \\ &-\frac{1}{12\hat{A}}\left\{\begin{matrix}\hat{b}_1\overline{\hat{\psi}}_1 + \hat{b}_2\overline{\hat{\psi}}_2 + \hat{b}_3\overline{\hat{\psi}}_3 \\ \hat{b}_1\overline{\hat{\psi}}_1 + \hat{b}_2\overline{\hat{\psi}}_2 + \hat{b}_3\overline{\hat{\psi}}_3 \\ \hat{b}_1\overline{\hat{\psi}}_1 + \hat{b}_2\overline{\hat{\psi}}_2 + \hat{b}_3\overline{\hat{\psi}}_3\end{matrix}\right\}\left[\hat{c}_1 \quad \hat{c}_2 \quad \hat{c}_3\right]\end{aligned}$$

$$+\frac{1}{4\hat{A}}\left[\begin{bmatrix}\hat{b}_1^2 & \hat{b}_1\hat{b}_2 & \hat{b}_1\hat{b}_3\\ \hat{b}_1\hat{b}_2 & \hat{b}_2^2 & \hat{b}_2\hat{b}_3\\ \hat{b}_1\hat{b}_3 & \hat{b}_2\hat{b}_3 & \hat{b}_3^2\end{bmatrix}+\begin{bmatrix}\hat{c}_1^2 & \hat{c}_1\hat{c}_2 & \hat{c}_1\hat{c}_3\\ \hat{c}_1\hat{c}_2 & \hat{c}_2^2 & \hat{c}_2\hat{c}_3\\ \hat{c}_1\hat{c}_3 & \hat{c}_2\hat{c}_3 & \hat{c}_3^2\end{bmatrix}\right]\begin{Bmatrix}\overline{\hat{T}}_1\\ \overline{\hat{T}}_2\\ \overline{\hat{T}}_3\end{Bmatrix}=0 \tag{6.12}$$

The obtained compact form of the above obtained equations is expressed in eq. (6.13), where the global stiffness matrix denotes $[\hat{K}]$, and $\{\hat{f}\}$ is defined as the force vector of the system:

$$\begin{aligned}[\hat{K}]\left\{\overline{\hat{\psi}}\right\}&=\left\{\hat{f}\right\}\\ [\hat{K}]\left\{\overline{\hat{T}}\right\}&=\{0\}\end{aligned} \tag{6.13}$$

Due to the coupled behavior of eq. (6.13) the solution to it may be derived by taking a fuzzy iterative approach. Then by applying the boundary conditions the system of algebraic equations is solved through the Gauss-Seidel iteration approach with fuzziness. One can find the details of the same in [23, 27–31]. The current investigation has contained the convergence as follows. Here, p defines the present and prv defines the previous number of iterations:

$$\left|\left|\left(\overline{\hat{\psi}}_p^i-\overline{\hat{\psi}}_{prv}^i\right)\right|\right|\le\epsilon_1=10^{-9},\ \text{and}\ \left|\left|\left(\overline{\hat{T}}_p^i-\overline{\hat{T}}_{prv}^i\right)\right|\right|\le\epsilon_1=10^{-9} \tag{6.14}$$

for $i=1,\ 2,\ \ldots,n$.

6.5 Case study

In current investigation, the triangular porous cavity [30] has been considered and the discretization of the same is presented in Figure 6.3. In practice, the design of building roofs and attics, as well as the cooling of electrical gadgets like TVs and PCs, are among the main uses for the same. Here, one can find the uncertainties due to involved parameters, boundary conditions, etc. The current results have been discussed in terms of stream function and temperature with uncertain percentage error, viz., 5% and 8%.

The aim of this investigation is to effectively implement the FFEM for the given scenario, and examine the sensitivity of the imprecisely defined parameters. Here the crisp and fuzzy values of the defined uncertain parameter are represented in Table 6.1.

The next section describes the quantitative characterization and estimation of uncertainties of the imprecisely defined involved variables using various Ra configurations.

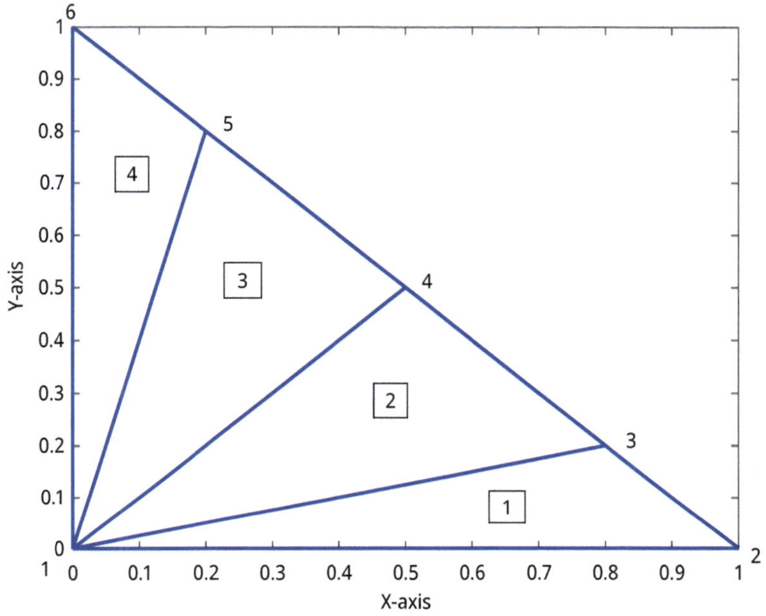

Figure 6.3: Model diagram for the 4-element discretization of a triangular porous cavity.

Table 6.1: Imprecisely defined decision values.

Parameter	**Crisp value with uncertain error**
Ra	1 ± 5%
	1 ± 8%

6.5.1 Case 1 (at 5% uncertain percentage error)

In case 1, we have picked up *Ra* as the fuzzy variable with a 5% uncertain percentage error. Next, by smoothly implementing the Gauss-Seidel (GS) FFEM procedure, the obtained results are analyzed in Figures such as Figures 6.4 to 6.7. One can see the variation of TFN solution in case of stream function ($\hat{\psi}$) and temperature ($\bar{\hat{T}}$) with different nodes in Figures 6.4 and 6.5, where the $\bar{\hat{\psi}}$ for node 3, 4, 5 and 6 are converged to [0.999972040.99997810.99998305], [0.999892030.99991540.99993454], [0.999779050.999826880.99986604], and [0.999715040.999776730.99982724]. The average width of $\bar{\hat{\psi}}$ is 6.31775E-05. Figure 6.6 shows the nodal width of the $\hat{\psi}$ with respect to different alpha values, such as that for $\alpha = 0$ and $\alpha = 1$.

Similarly, the TFN solutions of $\bar{\hat{T}}$ for nodes 3 and 4 are converged to [30.0000060730.00000784 30.00001001] and [30.0000111830.0000144530.00001846], whereas, the width

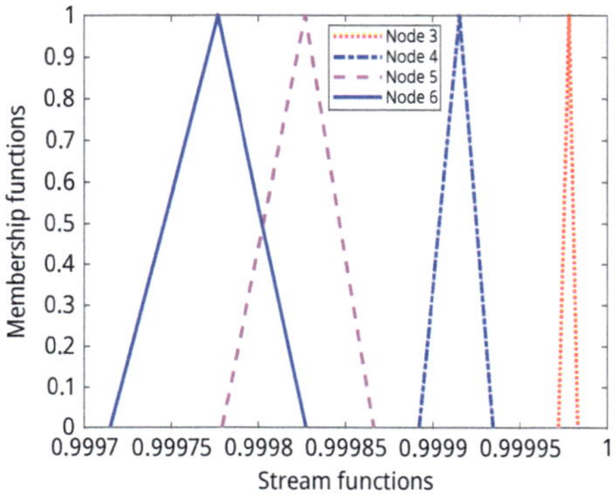

Figure 6.4: Nodal TFN solution of stream function in case 1.

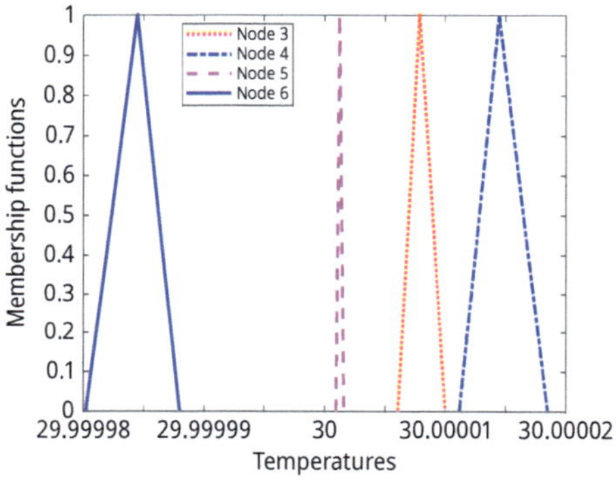

Figure 6.5: Nodal TFN solution of temperature in case 1.

wise convergence of the $\hat{\overline{T}}$ with different alpha values such as $\alpha = 0$ and $\alpha = 1$ is presented in Figure 6.7. The average width of the same is 4.9075E-06. In practice, the convergence value with respect to nodes 5 and 6 are [30.0000009530.0000012330.00000159], and [29.9999802129.9999844829.99998798].

From case 1 study, we observe that each nodal solution has contained fewer uncertainties in case of temperature as compared with stream function.

6.5.2 Case 2 (at 8% uncertain percentage error)

In this case, the imprecisely defined parameter Ra has been considered with 8% uncertain percentage error. Figures 6.8 to 6.11 have presented the solution graph of the system after implementing the GS FFEM approach. Here, Figure 6.8 shows the nodal TFN solution of the $\hat{\overline{\psi}}$. The obtained TFN solution of $\hat{\overline{\psi}}$ for nodes 3 and 4 is converged to [0.999967810.99997810.99998556] and [0.999875690.99991540.99994424]. In addition, the nodal width convergence of the same is presented in Figure 6.10 by taking different alpha values such as and $\alpha = 0$ and $\alpha = 1$. The average width of the $\hat{\overline{\psi}}$ is

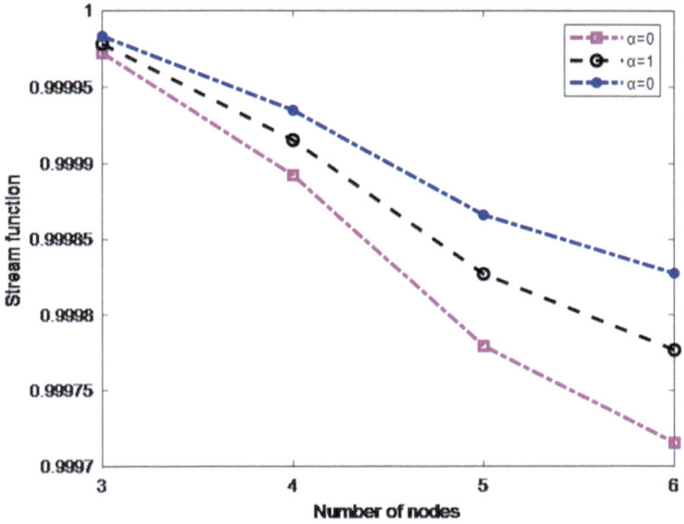

Figure 6.6: Width solution of stream function in case 1.

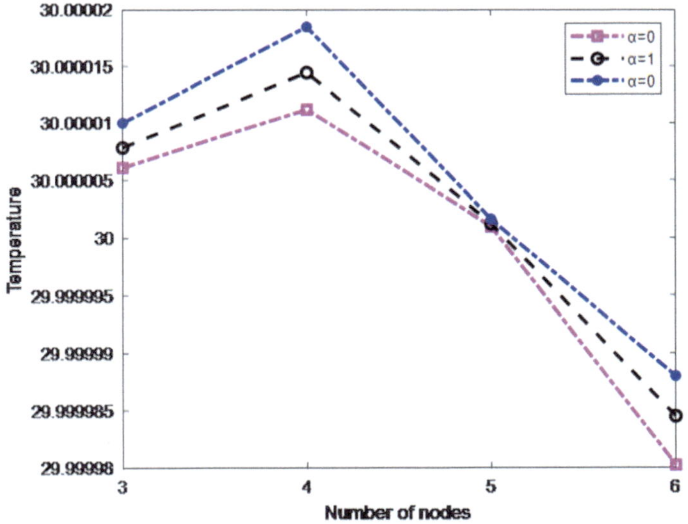

Figure 6.7: Width solution of temperature in case 1.

0.000101873. The TFN solution of the same with respect to nodes 5 and 6 is converged to [0.999745620.999826880. 9998859] and [0.999671940.99977673 0.99985285].

In a similar manner, the nodal TFN and width wise solutions with respect to $\alpha = 0$ and $\alpha = 1$ of $\bar{\tilde{T}}$ are shown in Figures 6.9 and 6.11. The TFN solutions of the system with respect to nodes 3, 4, 5, and 6 are converged to [30.0000051730.0000078430.00001153], [30.00000953 30.0000144530.00002124], [30.00000081 30.0000012330.00000182], and [29.

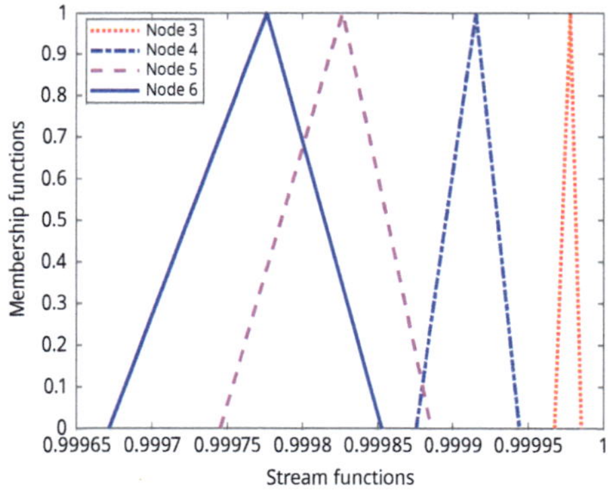

Figure 6.8: Nodal TFN solution of stream function in case 2.

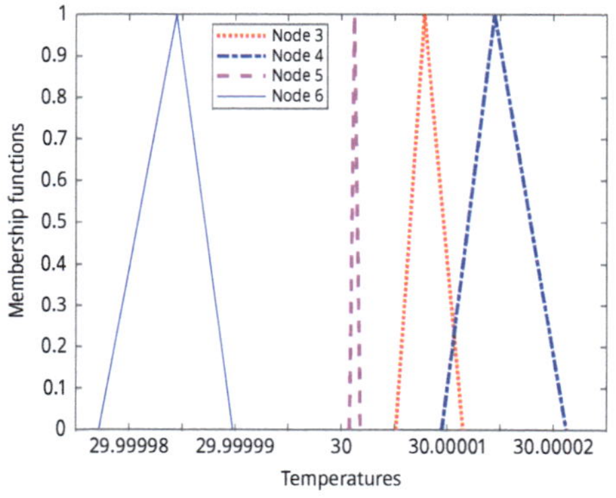

Figure 6.9: Nodal TFN solution of temperature in case 2.

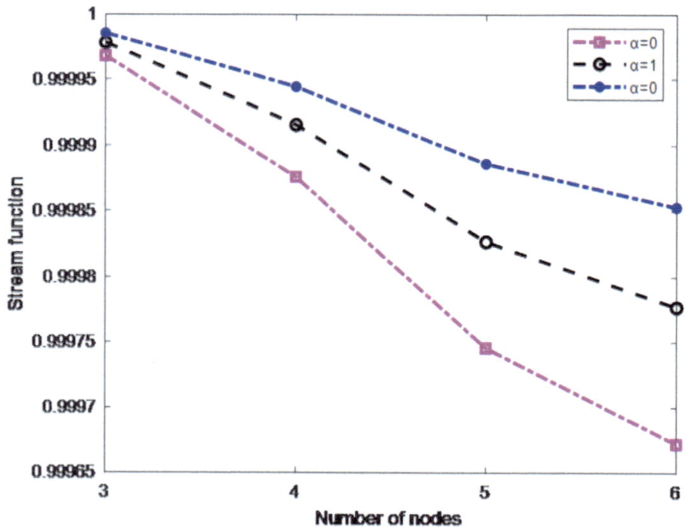

Figure 6.10: Width of stream function in case 2.

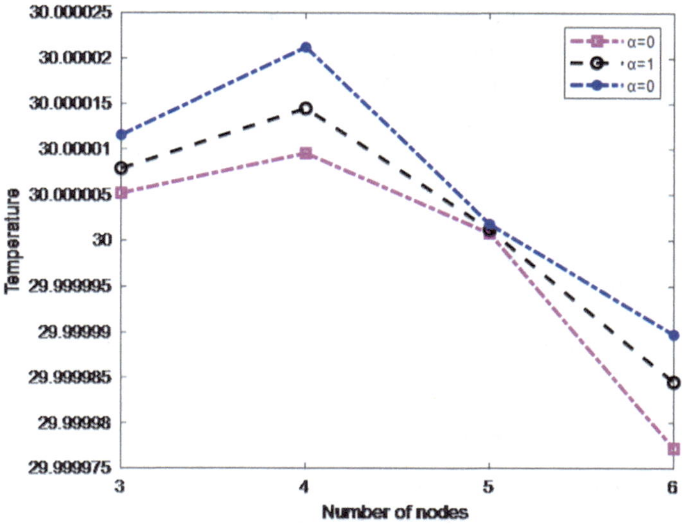

Figure 6.11: Width of temperature in case 2.

999977229.9999844829.99998976]. The average width of $\overline{\hat{T}}$ is 7.91E-06. From this case we observed that the uncertain contained nodal solution is less in the case of $\overline{\hat{T}}$ by comparing with $\overline{\hat{\psi}}$.

When the three cases are compared, it is found that the uncertain parameter *Ra* is more sensitive when the uncertain error is 8% in case of stream function. Similarly, the parameter *Ra* is less sensitive when the uncertain error is 5% in terms of temperature. That means the stability of the system can be analyzed by observing the sensitivity of the uncertain parameters. Also, it is observed that the system is stable at a 5% uncertain error.

6.6 Conclusion

The main purpose of this work is to provide a numerical model to investigate the triangular cavity problem with imprecisely defined variables. Initially, the mathematical modeling of the problem with an uncertain environment has been studied. In addition, the uncertain governing equations are solved by using the proposed FFEM approach. Then for solving the uncertain coupled system of algebraic equations the Guess-Seidel iterative techniques with TFN is considered. The sensitiveness of the parameters is studied with respect to error term. Finally, the stability of the system is analyzed based on the sensitivity of uncertain parameters.

As a future work, one can adopt this proposed numerical approach to solve various engineering and science problems in irregular geometry; it can also be used to analyze the sensitiveness of the involved parameters.

References

[1] Nield D. A., and Bejan A. (1999). Convection In Porous Media, Springer.

[2] Pop I., and Ingham D. B. (2001). Convective Heat Transfer, Elsevier.

[3] Vafai K. (2009). Handbook of Porous Media, Marcel Dekker.

[4] Bejan A. (1979). On the boundary layer regime in a vertical enclosure filled with a porous medium. Letters in Heat and Mass Transfer, 6(2), 93–102.

[5] Kumar B. R., and Kumar B. (2004). Parallel computation of natural convection XE "natural convection" in trapezoidal porous enclosures. Mathematics and Computers in Simulation, 65(2), 221–229.

[6] Murthy P. V. S. N., Kumar B. V. R., and Singh P. (1997). Natural convection heat transfer from a horizontal wavy surface in a porous enclosure. Numerical Heat Transfer, Part A: Applications, 31(2), 207–221.

[7] Badruddin I. A., Zainal Z., Narayana P. A., and Seetharamu K. (2006). Heat transfer by radiation and natural convection through a vertical annulus embedded in porous medium. International Communications in Heat and Mass Transfer, 33, 500–507.

[8] Badruddin I. A., Zainal Z., Narayana P. A., and Seetharamu K. (2006). Thermal non-equilibrium modeling of heat transfer through vertical annulus embedded with porous medium. International Journal of Heat and Mass Transfer, 49, 4955–4965.

[9] Badruddin I. A., Zainal Z., Khan Z. A., and Mallick Z. (2007). Effect of viscous dissipation and radiation on natural convection in a porous medium embedded within vertical annulus. International Journal of Thermal Sciences, 46, 221–227.

[10] Kumar B. V. R. (2000). A study of free convection induced by a vertical wavy surface with heat flux in a porous enclosure. Numerical Heat Transfer, Part A: Applications, 37(5), 493–510.

[11] Baytaş A., and Pop I. (2001). Natural convection in a trapezoidal enclosure filled with a porous medium. International Journal of Engineering Science, 39(2), 125–134.

[12] Baytas A., and Pop I. (1999). Free convection in oblique enclosures filled with a porous medium. International Journal of Heat and Mass Transfer, 42(6), 1047–1057.

[13] Ullah M. Z., Hussain S. T., Ul Haq R., Alzahrani A. K., and Mallawi F. (2022). Thermal energy performance due to convection process of nanofluid in a porous medium due to split lid motion in a right triangular enclosure. Journal of Computational Design and Engineering, 9(3), 890–906.

[14] Rasheed S. A., and Hasan A. J. (2022). Effect of orientation on the natural convection XE "natural convection" heat transfer from a heated triangular prism embedded in porous media. Case Studies in Thermal Engineering, 35, 102134.

[15] Khan Z. H., Usman M., Khan W. A., Hamid M., and Ul Haq R. (2022). Thermal treatment inside a partially heated triangular cavity filled with casson fluid with an inner cylindrical obstacle via FEM approach. The European Physical Journal Special Topics, 231(13), 2683–2694.

[16] Wu C. Y. (2009). Monte Carlo simulation of transient radiative transfer in a medium with a variable reflective index. International Journal of Heat and Mass Transfer, 52, 4151–4159.

[17] Kovtanyuk, A. E., Botkin N. D., and Hoffmann K. H. (2012). Numerical simulations of a coupled radiative-conductive heat transfer model using a modified Monte Carlo method. International Journal of Heat and Mass Transfer, 55, 649–654.

[18] Zadeh L. (1965). Information and control. Fuzzy Set, 8, 338–353.

[19] Nayak S., and Chakraverty S. (2015). Numerical solution of uncertain neutron diffusion equation for imprecisely defined homogeneous triangular bare reactor. Sadhana, 40(7), 2095–2109.
[20] Wang C., and Zhiping Q. (2015). Interval analysis of steady-state heat convection–diffusion problem with uncertain-but-bounded parameters. International Journal of Heat and Mass Transfer, 91, 355–362.
[21] Chakraverty S., and Nayak S. (2012). Fuzzy finite element method for solving uncertain heat conduction problems. Coupled Systems Mechanics, 1(4), 345–360.
[22] Hazarika G. C., Dutta P., and Borah J. (2021). Numerical analysis of an MHD flow in fuzzy environment in presence of induced magnetic field. Heat Transfer, Wiley, 1–11.
[23] S. Priyadarshini and S. Nayak, "A numerical approach to study heat and mass transfer in porous medium influenced by uncertain parameters," International Communications in Heat and Mass Transfer, 139, 106411, 2022.
[24] Panigrahi P. K., and Nayak S. (2023). Numerical investigation of non-probabilistic systems using inner outer direct search optimization technique. AIMS Mathematics, 8(9), 21329–21358.
[25] Nayak S., and Chakraverty S. (2008). Interval Finite Element Method with MATLAB, Academic Press.
[26] Nithiarasu P., Lewis R. W., and Seetharamu K. N. (2016). Fundamentals of the Finite Element Method for Heat and Mass Transfer, 2nd ed. Wiley Series In Computational Mechanics.
[27] Panigrahi P. K., and Nayak S. (2024). Numerical approach to solve imprecisely defined systems using inner outer direct search optimization technique. Mathematics and Computers in Simulation, 215, 578–606.
[28] Priyadarshini S., and Nayak S. (2023). Effects of imprecisely defined parameters on heat and mass transfer in a vertical annular porous cylinder. International Communications in Heat and Mass Transfer, 149, 107097.
[29] Priyadarshini S., and Nayak S. (2023). A new hybrid approach to study heat and mass transfer in porous medium influenced by imprecisely defined parameters. Case Studies in Thermal Engineering, 51, 103619.
[30] Varol, Y., Oztop H. F., and Varol A. (2006). Free convection in porous media filled right-angle triangular enclosures. International Communications in Heat and Mass Transfer, 33, 1190–1197.
[31] Priyadarshini S., and Nayak S. (2024). Fuzzy finite element solution of uncertain convection-diffusion heat transfer for a rectangular plate. Computational Algorithms and Numerical Dimensions, 3(4), 298–307.
[32] P. L. Durga and S. Nayak, "Modified Levenberg-Marquardt Neural Network Method to Investigate Optimum Height of a Water Tank Problem," 2024 OITS International Conference on Information Technology (OCIT), Vijayawada, India, 2024, 339–344,

Alok Dhaundiyal and Laszlo Toth

Chapter 7
Investigation of the thermal fluid system using direct numerical simulation

Abstract: The chapter focuses on predicting the state properties of the thermodynamic system through computational fluid dynamics (CFD). A laboratory reactor of 1.28 kW_{th} was developed and considered whilst examining the processed loose biomass. The transient change in the thermo-fluid properties of gas in the core element of the unit was evaluated. The wave number and time step considered during the analysis were 18.47 mm^{-1} and 0.21, respectively. The system's hydrostatic pressure was omitted in the analysis. The flow is incompressible and subsonic. The standard deviation between the solution sets and the experimental data for velocity, pressure, and temperature were 0.02 m/s, 0.019 mbar, and 0.94 K, respectively. The maximum predicted static gas pressure through the direct numerical simulation was 227.60 mbar. Similarly, the gas temperature did not exceed 383 K across the core element.

Keywords: biomass, reactor, microscale, fluid dynamics, simulation

7.1 Introduction

Computational fluid dynamics (CFD) is one of the versatile methodologies for predicting fluid flow, heat and mass transfer, and related applications, which can be incorporated through the numerical solution of the governing mathematical equations related to the conservation of mass, momentum, energy, and species mass. Through this scheme, the various fluid and gas dynamics applications can be solved by implementing different inbuilt algorithms. The modus operandi of the CFD relies on forecasting the fluid properties by approximating the solution of the governing partial differential equations (PDEs). Among the numerical modeling techniques, the CFD has owned the position, which primarily emphasizes the finite volume method (FVM) for discretizing fluid-related problems, such as FLUENT [1, 2]. Owing to its multitasking attribute, it almost covers the whole gamut of fluid-related problems, such as the wind-induced motions around the modern super-tall building in computational wind engineering. In the primal phase, CFD was applied to examine the regular shape, which involves studying

Alok Dhaundiyal, Centre for Energy Research, Konkoly utca. 29-33, 1121 Budapest, Hungary, e-mail: alok.ekcer@gmail.com
Laszlo Toth, Institute of Technology, Hungarian University of Agriculture and Life Sciences, 2100 Godollo, Hungary

https://doi.org/10.1515/9783111661674-007

flow fields around the bluff bodies. The investigation was mainly on the 2D cross sections that were related to authenticating the location of stagnation point, pressure, and velocity distribution, and the traits at the proximity of the surface boundary and far regions [3, 4]. The minor flow-related problem to the cross-flow turbines through a physical model is a costly and cumbersome job, therefore it becomes essential to invoke a complementary technique to experimental investigate flow patterns for different characteristic curves of water turbine; therefore the computational simulation can be an alternative measure to visualize the flow field around the runner [5, 6]. Another beneficial aspect of the CFD is to optimize the performance of the energy storage system via simulation of fluid flow within the system, which can further be analyzed by knowing the heat characteristic of the substrate and rectifying any critical problem that occurs during the working stage [7]. However, the application of the CFD is not limited to hydraulic machines, wind tunnel experiments and energy storage systems but can also be implemented to evaluate energy generation reactors.

A tangentially fired coal boiler of power generation of 500 MW was examined for different settings of ammonia injection with varying configurations of boiler burners. The objective was to offset carbon generation during operation through the co-firing technique of burning coal with 20% ammonia. The CFD numerical technique was employed to determine the effect of different ammonia injection strategies to counter NO_x emission during its operation. It was concluded that a single-burner injection relatively curtailed the carbon emission to 0.30% as compared to a setting employing five burners [8]. In another case, prediction of the ignition behavior of 40 MW of tangentially fired coal boiler was performed using the CFD technique. Coal devolatilization and char combustion were modeled by involving the first-order harmonic approximation in the radiative transfer equation. For simulation purposes, the standard k-ε model and the Lagrangian tracking model with a random walk model were adopted [9]. The different perspective was discussed by incorporating the CFD application to study the structural properties of catalyst in a catalytic fast pyrolysis of biomass based on multiphase flow. It was noticed that the CFD had successfully shown the rapid cracking of biomass and was able to analyze the mass transfer process [10]. The solar pyrolysis of beech wood pellets was studied using the CFD platform ANSYS FLUENT 14.0. The objective of the work is to examine the biomass degradation of the pellets and to validate the solution with the experimental findings. The biomass pyrolysis had undergone degradation from 600 °C to 2,000 °C at two heating rates of 10 $°C{\cdot}s^{-1}$ and 50 $°C{\cdot}s^{-1}$. Based on the 2D unsteady modeling of biomass pellet under radiatively induced fast pyrolysis, it was noticed the heating rate dropped significantly as moving towards the center. Consequently, it led to a nonuniform charring profile inside the particle [11]. It is also possible to model the pyrolysis reactors by incorporating the two-fluid model (TFM) and CFD with the discrete element method (DEM) to investigate the behavior of feedstock inside the pyrolysis reactor. Primarily, TFM assumed that the particulate feedstock and the carrier gas phase are two interpenetrating and coexisting continua that may allow exchange of mass, momentum, and heat. As these processes are driven by physical and chemical phenomena, the particulate phase can be represented with the particle number or mass density, the mean particle velocity and specific enthalpy, which can be easily represented through the average values over all particles in a control volume of the system and can be related to statistics via particle property distribution. The evolution of property distribution with changes in coordinates and time can be represented by the population balance equation (PBE) [12, 13]. A bench-scale fluidized bed reactor was examined by using the Eulerian multiphase CFD model. Biomass particles and the bed material (sand) were assumed to be part of particulate phases and modeled considering the kinetic theory of granular flow. The quadrature method of the moment was applied to solve the PBE. A 3-parameter shrinkage model was considered to calculate the particle shrinking rate as well as record the variation in the apparent density of the particle. It was reported that the density of particles decreased continuously for all three cases (devolatization, char formation, and variation in size) owing to the continuous mass loss and pore formation [14]. Another task considered the Eulerian approach to model the multiphase flow in fluidized beds by coupling the kinetic theory of granular flows. It was claimed that the convection process dominated the heating up of the sawdust particles, whereas the ratio of catalyst mass to tar flow rate could affect the final product yield distribution in space-time [15]. A granular multiphase flow model was applied to examine the pyrolysis kinetics of a continuous flow drop tube reactor . Based on weight, 45% of biofuel, 25% of gas, 10% water, and 20% of biochar was forecasted through the Eulerian technique at a reactor wall temperature of 773 K. A mammoth share of biofuel was claimed to be extracted through cellulose [16]. A two-fluid technique was considered to examine the mass and heat transfer process inside a laboratory-scale externally heated fixed-bed coal reactor. It was stated that the evaporation of water content would determine the coal pyrolysis process. Due to this, the voltage of the coal bed along the reactor wall is relatively high as compared to the inner region of the bed [17]. To study a fluidized bed, a transient multi-fluid model (MFM) based on the Eulerian approach was adopted to predict hydrodynamics, and temperature and chemical species distribution. It was reported that the reaction rate was considerably high at the beginning of the reaction but reduced with the reduction of algae biomass, which eventually, transformed the bed regime to a bubbling phase in the pyrolysis reactor [18]. In another work, a CFD model based on the Lagrangian and Euler methods was applied to determine the effects of droplet behavior related to its collision and breakage and solvent evaporation on the droplet size, flow field, and collection efficiency during the formation of ZnO particles in ultrasonic atomization pyrolysis reactor. The results claimed that droplet collision and breakage would enhance the growth of droplet size and widen the droplet size distribution, resulting in offering resistance during its transportation [19]. The k-ε model was employed to simulate the cold state of fluid in a fluidized bed reactor. It was reported that the reduction in the gas velocity below the minimum fluidization velocity caused the formation of discontinuous bubbles in the reactor [20]. The MFM or TFM was employed to describe the behavior of the fluidized bed reactors and to model the shrinkage effect in biomass. It was reported that the shrinkage model with no change in the particle size supported segregation and tended to provide a coarse gradient between the dense and the freeboard zones of the fluidized bed. It was also noticed that the biomass mass fraction roughly dropped by half in the both dense and freeboard zones. Conversely, the char and metaplastic mass fraction was enhanced by 50% in the dense zone and gradually increased in the freeboard [21]. It was noticed in most studies that they primarily dealt with turbulent models, where the properties are anisotropic for the given integral size. They are relatively fast and require less computation effort but do not cover the smallest dissipative scales.

In this chapter, the objective was to emphasize the direct numerical simulation that involves solving Navier-Stokes equations without considering any turbulent model and to cover the whole range of spatial and temporal scales [22]. To implement this scheme, a laboratory-sized fixed bed reactor was used to predict the fluid behavior of generated gas.

7.2 Materials and methods

This section is based on the designed unit and the simulation technique applied whilst investigating the flow behavior of the producer gas along the core element and around the pellet.

7.2.1 Geometrical configuration of the pyrolysis unit

The reactor's 3D model for the simulation process was created on AutoCAD (Autodesk, 2021). The inner core of the system is made of a 1.45-mm thick welded carbon steel plate. The height of the pyrolysis unit was 430 mm, whereas the rock wool insulation between the inner core and the outer body was 30 mm. Similarly, the circumferential insulation around the core element had a thickness of 41.5 mm. The material used to fabricate the outer body was galvanized steel. The inner and outer radii of the unit were 42.60 mm and 82.50 mm, respectively. The measurement ports were provided along the length of the unit to measure the temperature of the fixed bed and the producer gas. Two outlet pipes of different diameters were retrofitted with the 4.9-mm steel cover plate of 93.70 mm diameter to measure the dynamic pressure (current-based transducer, Huba Control) and to analyze the gas composition. A lid of 71.60 mm in length and 118.40 in diameter was also provided for the additional heat insulation between the main body and the ambient air. A grate of 79.55 mm diameter was retrofitted within the lower bottom of the reactor to hold the feedstock. A 2-mm thick cast iron plate was welded to separate the heating chamber from the fixed bed.

A concentric ceramic-made heating component of 153 mm height was used to hold the nichrome wire. The inner and outer radii of the concentric casing were 23.50 mm and 36.50 mm, respectively. An opening of 26.80 mm in diameter was carved out 30 mm above the base for the electrical supply to the heating unit. The 3D model of the laboratory-type reactor is illustrated in Figure 7.1.

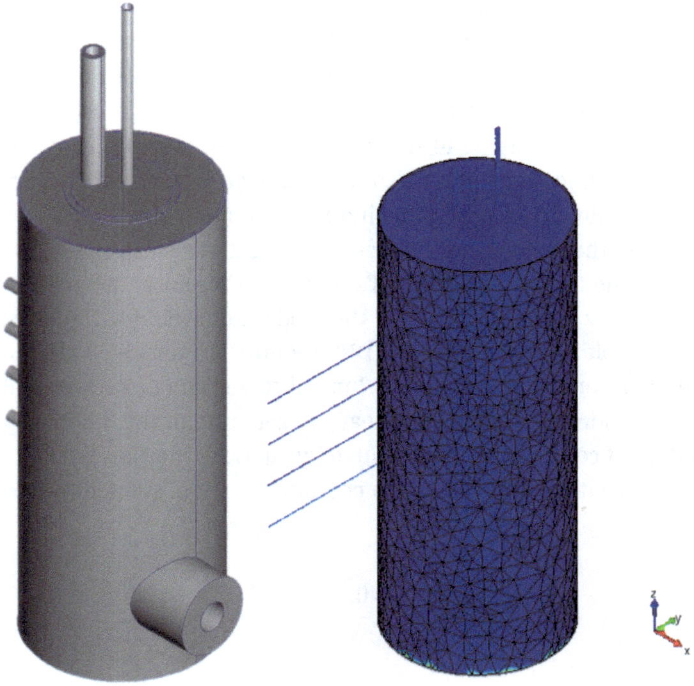

Figure 7.1: The 3D model of the bench-scale reactor.

7.2.2 Direct numerical simulation

Since practically implementing the turbulent velocity field is a cumbersome task, an alternative is needed to generate computer-based simulations to determine the characteristics of turbulent flow. This technique is referred to as CFD, which relies on numerical algorithms and the interface to interact with the end user. Nonetheless, it does not warrant that the CFD can cover the whole gamut of the fluid-related problems as the complexity of turbulence sometimes requires validation through the experimental data set. At present, it encompasses data validation and numerical coding to examine the flow characteristic during thermal decomposition of the feedstock. For simulation purposes, Autodesk CFD 2018 (Autodesk, 2018) was used.

Direct numerical simulation (DNS) solves the Navier-Strokes equations to estimate the 3D and time-dependent solution of the instantaneous fluid as a function of space and time. However, in most cases of flow problems, it is difficult to obtain a solution owing to the complexity of the transport equations. Generally, the DNS is applied to a selective geometry, such as flow in a channel or over a flat plate and additionally, it requires a considerable amount of computational effort. Whilst applying CFD to any application, the objectives related to numerical accuracy, solution stability, convergence, specification of boundary conditions, and optimal use of available computing resources are the primary concern. The equations relevant to the flow field must be solved using grids with a mesh domain fine enough to resolve most of the individual eddies, which can vary from 30 to 100 µm in a low-viscosity fluid such as alcohol. Furthermore, small time steps must also be employed to monitor the unsteady behavior of turbulent flow. The reason behind the mammoth size of memory allocation is the reduction in the scale of the smallest eddies and the steep rise in the fluctuation frequency with the increase in Reynold's number. Therefore, the applicability of the direct numerical solution is limited to the moderate level of turbulence. Even for relatively low Reynolds numbers, typical DNS requires around 5–20 million nodes in the flow field, therefore its application is limited to verifying experimental results and for investigating ideal flow that is not easy to generate in the laboratory [23–26]. The basic equation of continuity is represented by eq. (7.1). The Navier-Stokes equations for momentum variation along Cartesian coordinates are presented by eqs. (7.2)–(7.4):

$$\frac{\partial v_x}{\partial x} + \frac{\partial v_y}{\partial y} + \frac{\partial v_z}{\partial z} = 0 \tag{7.1}$$

x-Component

$$\rho\left(\frac{\partial v_x}{\partial t} + v_x\frac{\partial v_x}{\partial x} + v_y\frac{\partial v_x}{\partial y} + v_z\frac{\partial v_x}{\partial z}\right) = -\frac{\partial P}{\partial x} + \mu\left(\frac{\partial^2 v_x}{\partial x^2} + \frac{\partial^2 v_x}{\partial y^2} + \frac{\partial^2 v_x}{\partial z^2}\right) + \rho g_x \tag{7.2}$$

y-Component

$$\rho\left(\frac{\partial v_y}{\partial t} + v_x\frac{\partial v_y}{\partial x} + v_y\frac{\partial v_y}{\partial y} + v_z\frac{\partial v_y}{\partial z}\right) = -\frac{\partial P}{\partial y} + \mu\left(\frac{\partial^2 v_y}{\partial x^2} + \frac{\partial^2 v_y}{\partial y^2} + \frac{\partial^2 v_y}{\partial z^2}\right) + \rho g_y \tag{7.3}$$

z-Component

$$\rho\left(\frac{\partial v_z}{\partial t} + v_x\frac{\partial v_z}{\partial x} + v_y\frac{\partial v_z}{\partial y} + v_z\frac{\partial v_z}{\partial z}\right) = -\frac{\partial P}{\partial z} + \mu\left(\frac{\partial^2 v_z}{\partial x^2} + \frac{\partial^2 v_z}{\partial y^2} + \frac{\partial^2 v_z}{\partial z^2}\right) + \rho g_z \tag{7.4}$$

Here, ρ, P, and μ are the density, pressure, and dynamic viscosity of the generated gas, respectively. Similarly, the components of instantaneous velocity and gravity along the x-, y-, and z-axis are represented by v_x, v_y, and v_z, and g_x, g_y, and g_z, respectively.

In the real flow problem, the kinetic energy is transferred from larger eddies to smaller ones, whereas the same energy is changed into internal energy through dissipation and the transit between the scales is referred to as the cascade process. The momentum transportation and kinetic energy dissipation rate estimate the smallest turbulent length scale. This scale is known as the Kolmogorov length scale, ζ [27].

The dissipative Kolmogorov microscale can be written as 1/4th power of the ratio of cubic order of diffusivity of momentum (ν) to the rate of kinetic energy dissipation (ε) due to eddies formation, i.e.:

$$\zeta = \left(\frac{\nu^3}{\varepsilon}\right)^{\frac{1}{4}}$$

or it can be represented in terms of Reynold's number (*Re*):

$$\frac{\zeta}{L} = \frac{1}{Re^{\frac{3}{4}}} \tag{7.5}$$

Here, L is the integral length scale.

To satisfy the above condition or to resolve the microscale, the number of points (n) and the incremental step (h) along the mesh direction must satisfy the following inequality:

$$nh > L \text{ and } h \leq \zeta$$

By combining the above inequality with the asymptotic relation of kinetic energy dissipation rate $\left(\varepsilon \approx \frac{v_{\text{rms}}^3}{L}\right)$ with root mean square velocity and integral length scale, the number of mesh points (n^3) required for a given problem can be estimated by

$$n^3 = Re^{\frac{9}{4}} \tag{7.6}$$

Here, the Reynold's number is $\frac{v_{rms}L}{\nu}$.

To maintain accuracy or in order not to lead to a nonphysical solution, the Courant number (C) must be less than one for estimating the time step requirement of a transient simulation for a given mesh requirement:

$$C = \frac{v_{\text{rms}}\Delta t}{h} < 1 \tag{7.7}$$

Here, Δt is the time step of the simulation that can be solved through an explicit method: $\Delta t \sim Re^{(-1/2)}$ [28].

7.3 Results and discussion

This section was examined based on the thermo-fluid properties of the generated gas during the pyrolysis of pine pellets.

7.3.1 Pressure distribution across the core element of the unit

The stagnant pressure distribution along with the absolute pressure and static pressure due to the flows of generated gas inside the pyrolysis unit are illustrated in Figure 7.2(a–c).

The maximum stagnant pressure (total) was marked near the vicinity of the heating chamber, which varied from 120 to 130 kPa. Conversely, it was reduced by 16.66% to 19.23% as the gas moved out of the reactor chamber. Similarly, the stagnant pressure across the fixed bed was relatively low for the layer of pine pellets in proximity to the heating chamber. It was in the range of 100–115 kPa, which increased by 8.69% for the pellets located near the wall of the core elements and the upper region of the bed. Surprisingly, the variation in the absolute pressure of the gas was rather much wider than that of the total pressure of the gas. The absolute pressure varied from 95 to 100 kPa at the outlet of the reactor, whereas it was enhanced by 10–10.52% in the region between the upper layer of the fixed bed and the outlet of the reactor. As it moved near the fixed bed, it rose from 120 to 125 kPa with a drastic perturbation of 95–110 kPa near the grate of the reactor. A similar variation was noticed in the absolute pressure distribution in the lower stack of pine pellets. The maximum absolute pressure was recorded for the pellets located along the wall of the core part of the reactor and the upper stack, which was 13.63–26.31% higher than the lower section of the fixed bed. The static pressure variation in the producer gas as shown in Figure 7.2 (c) had a negative growth of 9 kPa near the outlet of the reactor, which increased to 3–10 kPa for the region between the fixed bed and the cover plate. The maximum variation in the static pressure was noticed for the region between the grate and the fixed bed. The static pressure across the upper layer of the fixed bed varied from 21 kPa to 24 kPa, which was curtailed to 3–9 kPa for the pellets located near the grate. The average static pressure generated due to the change in the state of generated gas for the given time was 57.66 kPa. Statistically, the generated solution for the pressure variation across the core element and the fixed bed was positively skewed about its mean value. The density function of pressure variation was noticed to exhibit the trait of the raised cosine distribution about its average value.

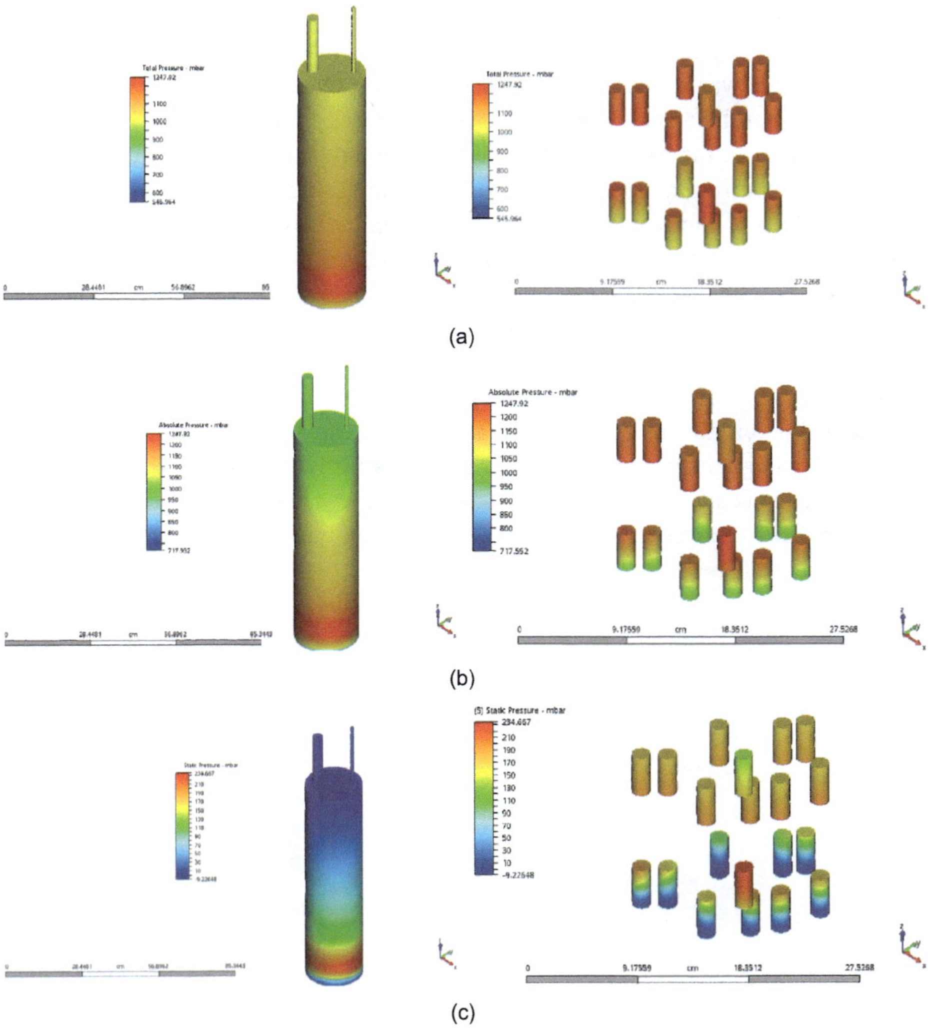

Figure 7.2: The transient change in pressure distribution of the producer gas inside the core component: (a) total pressure, (b) absolute pressure, and (c) static pressure.

7.3.2 Transient change in temperature of generated gas

The temperature distribution along the length of the core component and the fixed bed is shown in Figure 7.3(a and b). The average gas temperature obtained through the DNS was 383 K with a standard deviation of 19.58 K. Upon comparing the solution derived through the DNS with experimental data, it shrunk to 0.94 K. The density function of temperature with temporal variation in gas temperature was negatively skewed about its mean value. The kurtosis obtained for the transient simulation of the temperature of the generated gas was noticed to exhibit the characteristics of normal and cosine distribution functions. The maximum recorded temperature across the core element was 414.44 K. In contrast, the lowest temperature noticed during the transient temperature change was 352.44 K. Likewise, the fixed pine pellet bed was also examined and the gas temperature across the top pellet layer varied from 1,000 to 1,100 K. In contrast, a significant perturbation in gas temperature was seen at the bottom surface of the bed, where the limit extended from 500 to 1,200 K. With a change in the bed length, the temperature was noticed to become steady with time. Conversely, once the local velocity of generated gas was reduced to zero, the stagnant temperature dropped by 61.69% across the upper layer of the fixed bed as depicted in Figure 7.3(a). This happened due to the approximate fall in the polytropic index of the gas by 22% as it moves across the fixed bed and at the same time, it would have a reverse impact on the ratio of stagnant to static pressure ratio as mentioned in Figure 7.2. The stagnant pressure will increase drastically with time. When the flow along the carbon steel cylinder surface was considered, the total temperature was enhanced by 2.12%. However, it was curtailed by the margin of 17.16% at the top cover plate and near the heating chamber.

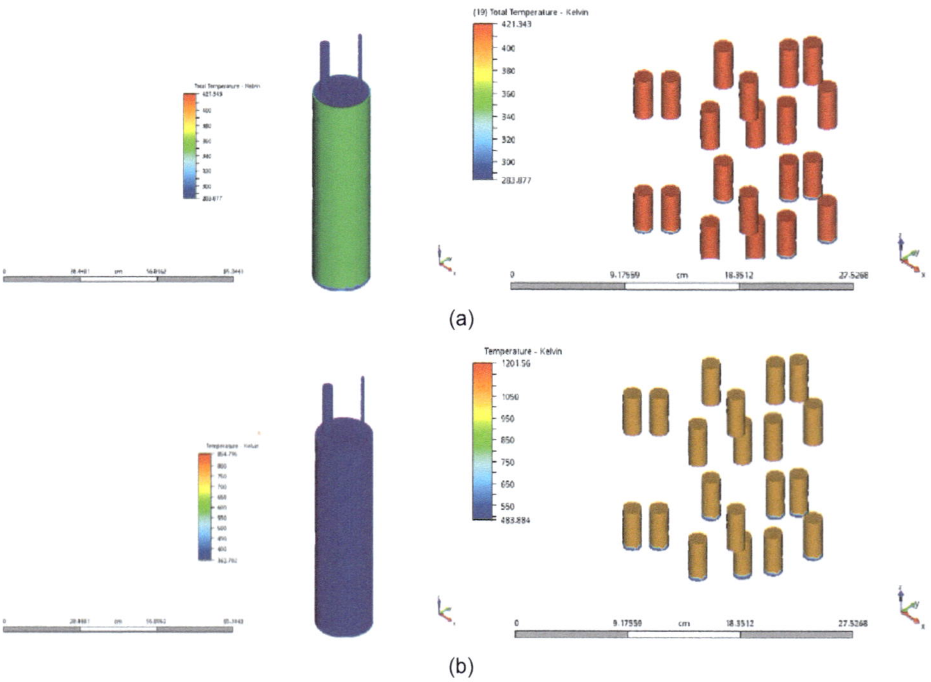

Figure 7.3: Time-bound variation in temperature of the producer gas inside the core component: (a) total temperature and (b) temperature.

7.3.3 Shear stress and the flow characteristic of the generated gas

The shear stress (τ) exerted on the surface of the pellets and carbon steel plate owing to the velocity gradient at the constant viscosity of the generated gas is presented graphically in Figure 7.4(a–d). The maximum τ generated due to the flow of gas along the *YZ*-, *XY*-, and *ZX*-planes of the cylindrical shell were 0.17 kPa, 0.047 kPa, and 0.13 kPa, respectively. Similarly, the corresponding minimum τ along the *YZ*-, *XY*-, and *ZX*-planes of the shell were −0.20 kPa, −0.070 kPa, and −0.159 kPa. The value of τ along the *YZ*-plane varied from −0.020 to 0.020 kPa for the region between the outlet of the reactor and the fixed bed, which implied the reverse flow of gas along the *YZ* section of the core element or shell with the sharp increase in the pressure gradient. The maximum τ in the range of 0.095–0.1 kPa was developed alone in the *YZ* section of the upper layer of pine pellets, whereas it reversed its direction near the top and bottom boundaries of the same plane. Conversely, the value of τ developed along the surface of the bottom region of the fixed bed varied from 0 to −0.01 kPa. With the assumption of a constant viscosity of the producer gas, it could be concluded the reverse flow of gas at the vicinity of the solid wall of the pine pellets along *YZ*-plane. Similarly, the negative value of τ along *YZ*-plane was also observed near the grate of the reactor. The zone of the shell (*YZ*) bounded by the fixed bed experienced the maximum τ of 0.04–0.12 kPa. Unlike the *YZ*-plane, the *XY*-plane of the shell and the fixed bed were noticed to have maximum τ around the grate surface and along the radial directions of the lower layer of the pine pellets, whereas the reversal of τ direction was seen among the pine pellets located on the top surface of the bed. A rise of 0.40 kPa in τ was seen along the *XY*-plane of the gas outlet.

In the case of the *ZX*-plane, τ around the grate surface varied from −0.159 to 0.135 kPa. As it moved to the fixed bed, the range of τ shrank from −0.159 to −0.040 kPa. The value of τ above the fixed bed changed from −20 to 20 kPa along the *ZX*-plane. Likewise, for the fixed bed, τ *was* noticed to vary from −0.010 to 0.1 kPa. The resultant of τ tensor depicted the overall variation along the length of the core elements as well as the fixed bed and it was seen that τ would lie in the domain of 0–0.06 kPa between the fixed bed and the gas outlet. The shear stress would attain its maximum value of 0.144 kPa around the grate and lower bed of the pine pellets. Along the length of the fixed bed, τ was in the range of 0.06–0.12 kPa. The distribution of τ would vary from 0 to 0.06 kPa for the upper bed of the pine pellets, whereas the maximum τ variation of 0.06–0.144 kPa was observed in the lower part of the bed. It can be concluded based on the resultant value of τ that the gas would encounter maximum retardation whilst flowing around the bluff surface of the pine pellet located near the grate surface (the surface that holds the fixed bed), whereas it would decrease with the increase in the length of the bed.

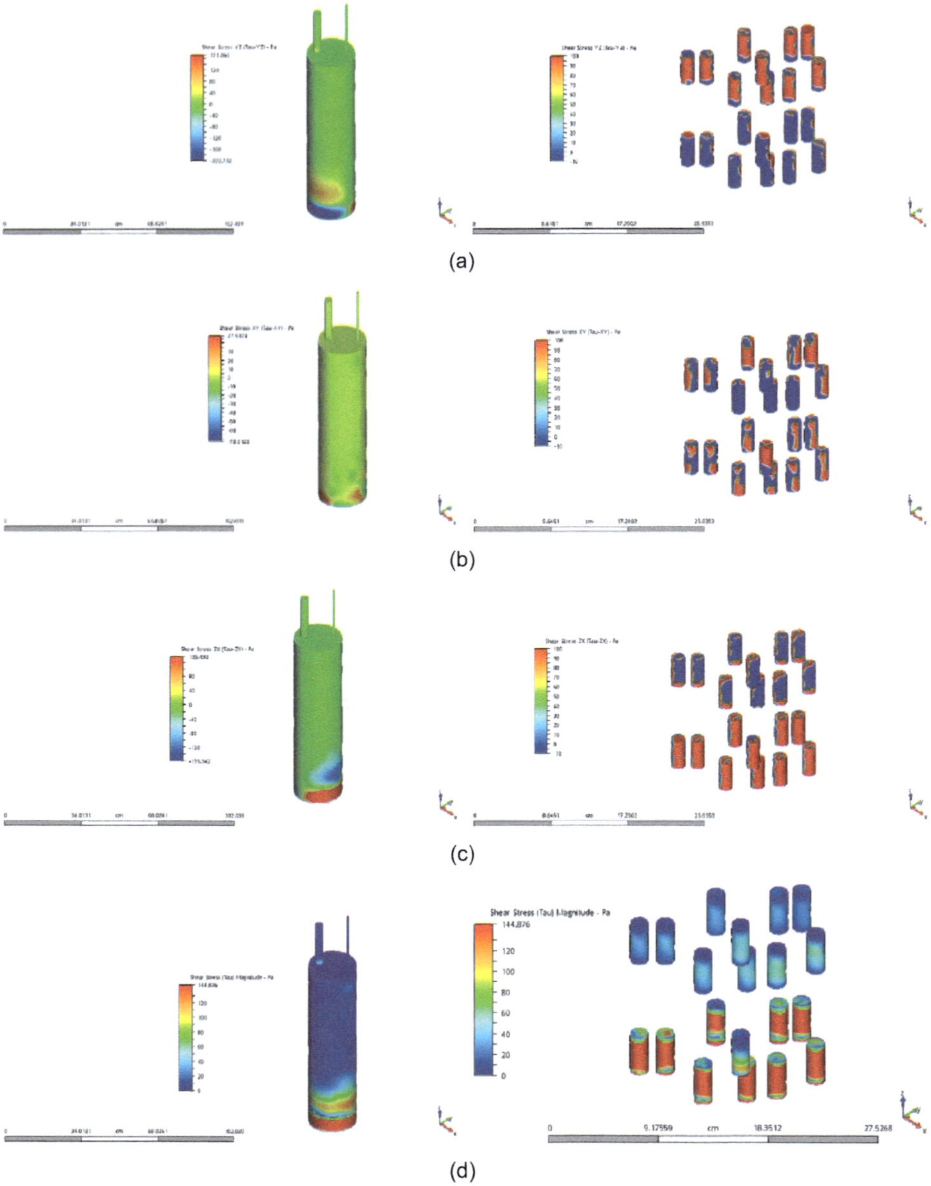

Figure 7.4: The shear stress distribution across the feedstock and core element surface: (a) *YZ*-plane, (b) *XY*-plane, (c) *ZX*-plane, and (d) resultant shear stress.

The velocity distribution along the length of the core element and the fixed bed is illustrated in Figure 7.5. The average change in the gas velocity inside the chamber was 0.15 m/s with a standard deviation of 0.04 m/s. The distribution of velocity about its mean value was estimated to be positively skewed with a skewness of 1.95. Similarly, the

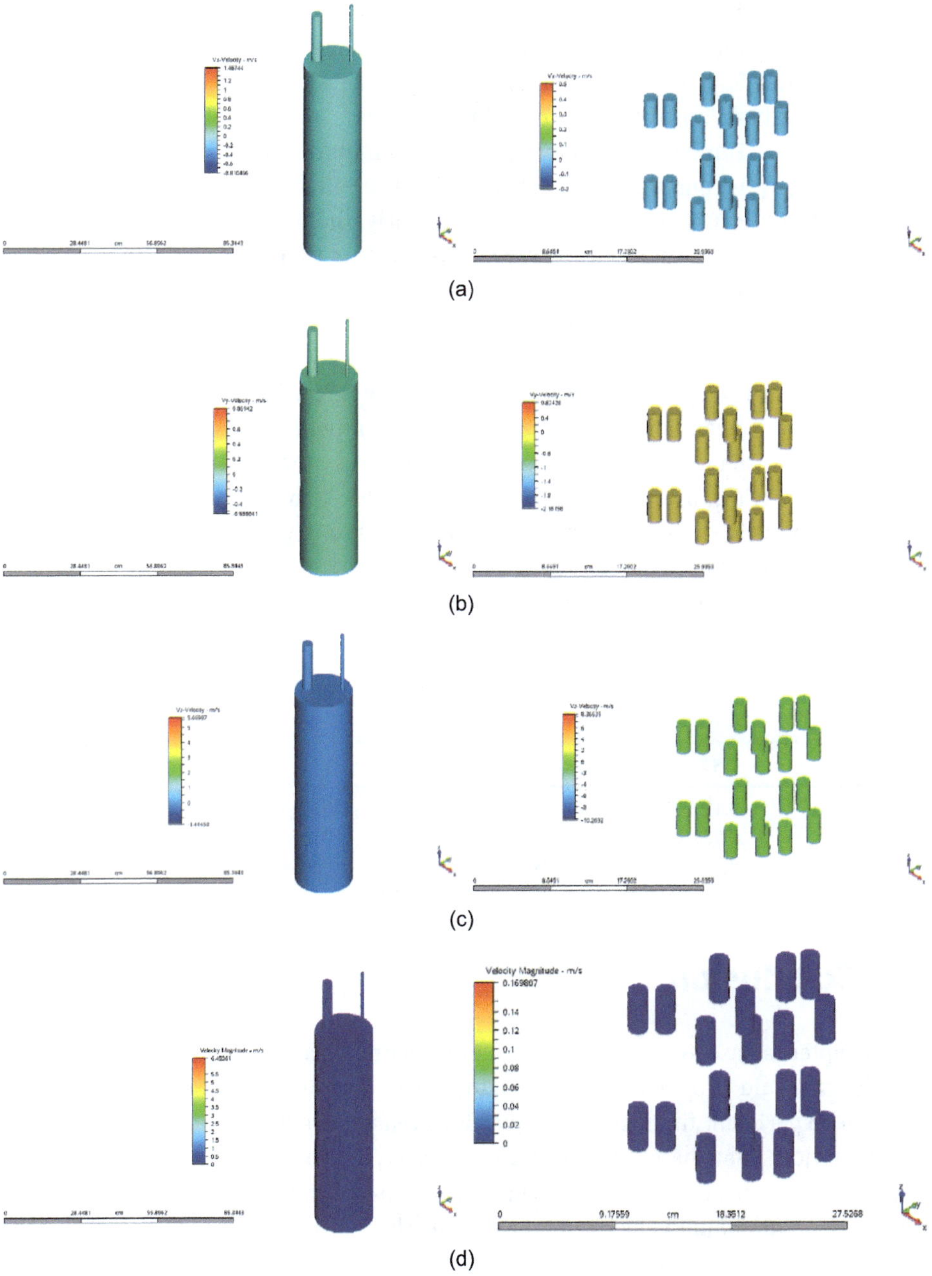

Figure 7.5: The velocity distribution across the length of the core element and feedstock: (a) *x*-direction, (b) *y*-direction, (c) *z*-direction, and (d) resultant local velocity.

density function of the velocity about its mean value exhibits the trait of the binomial distribution. The local velocity was noticed to vary steadily with spatial coordinates.

The reverse flow in the gas velocity was seen along all the directions. The estimated average velocities along the *x*-, *y*-, and *z*-directions were, respectively, 0.26 m/s, −0.16 m/s, and 3.08 m/s. Similarly, the maximum and minimum changes in the local velocity components, v_x, v_y, and v_z, were, correspondingly, 0.043 m/s, 0.006 m/s, and 0.151 m/s, and −0.011 m/s, −0.024 m/s, and 0.001 m/s. Similarly, the distribution functions associated with v_x and v_z were positively skewed about their average value. Conversely, the solution obtained for the local velocity, v_y was seen to exhibit negative skewness. The distribution functions of the local velocities obtained along the *x*-, *y*-, and *z*-directions belong to the super-Gaussian class. The resultant local velocity was noticed to vary from 0 to 0.5 m/s along the length of the core element. Likewise, it falls in the domain of 0–0.01 m/s while examining the distribution pattern around the blunt body of pine pellet. For comparison purposes, the experimental test was conducted to validate the simulation results, which are tabulated in Table 7.1. So, the standard deviation (χ) between the predicted value of velocity and the experimental one was 0.02 m/s. Similarly, the estimated values of χ for the state properties, pressure, and temperature, of the thermodynamic system (gas) were, respectively, 0.019 kPa and 0.94 K.

Table 7.1: Comparison between the experimental and DNS data sets along with χ.

Parameters	DNS	Experimental value	χ
Velocity	0.15 m/s	0.10 m/s	0.02 m/s
Pressure	0.570 kPa	0.550 kPa	0.019 kPa
Temperature	383 K	384.94 K	0.94 K

7.4 Conclusion

This chapter is based on the application of direct numerical simulation in measuring the thermo-fluid properties of the producer gas. The derived results were aligned with the experiment findings for the pyrolysis reactor. Upon comparing the solutions, it was noticed that the tendency of boundary layer separation would be relatively high along the surface of the pine pellets as compared to the cylindrical carbon steel surface. The minimum velocity obtained along the *z*-direction was 0.12 m/s. The polytropic index of gas dropped by 21% during its flow around the bluff body of pine pellets. The stagnant temperature was dropped by 61.69%, which consequently increased the stagnant pressure to 124.79 kPa. The shear stress was significantly high around the grate surface of the core element, but with the increase in the length of the bed, it dropped by 58.58%. The velocity distribution across the lower bed was relatively steady as compared to the freeboard surface.

References

[1] Ansys. (2017). Academic research fluent. Release, 18, (Accessed: 1 October 2024).

[2] Blocken B. (2011). Computational wind engineering: Theory and applications. Environmental Wind Engineering and Design of Wind Energy Structures, 55–93. https://doi.org/10.1007/978-3-7091-0953-3_3.

[3] Murakami S. (1997). Current status and future trends in computational wind engineering. Journal of Wind Engineering and Industrial Aerodynamics, 67–68, 3–34. https://doi.org/10.1016/s0167-6105(97)00230-4.

[4] Rodi W. (1993). On the simulation of turbulent flow past bluff bodies. Computational Wind Engineering, 1, 3–19. https://doi.org/10.1016/b978-0-444-81688-7.50005-x.

[5] Gourdain N., Sicot F., Duchaine F., and Gicquel L. (2014). Large eddy simulation of flows in industrial compressors: A path from 2015 to 2035. Philosophical Transactions of the Royal Society A: Mathematical, Physical and Engineering Sciences, 372(2022), 20130323. https://doi.org/10.1098/rsta.2013.0323.

[6] Investigation of the effects of number of blades on the performance of cross-flow turbine using STAR CCM+. (2017). Polytechnic Journal, https://doi.org/10.25156/ptj.2017.7.4.45.

[7] Yaïci W., Ghorab M., Entchev E., and Hayden S. (2013). Three-dimensional unsteady CFD simulations of a thermal storage tank performance for optimum design. Applied Thermal Engineering, 60(1–2), 152–163. https://doi.org/10.1016/j.applthermaleng.2013.07.001.

[8] Zeng Y., Kweon J., Kim G.-M., and Jeon C.-H. (2024). Carbon-free power generation strategy in South Korea: CFD simulation for ammonia injection strategies through boiler burner configurations in tangentially fired boiler. Energy, 309, 133076. https://doi.org/10.1016/j.energy.2024.133076.

[9] Asotani T., Yamashita T., Tominaga H., Uesugi Y., Itaya Y., and Mori S. (2008). Prediction of ignition behaviour in a tangentially fired pulverized coal boiler using CFD. Fuel, 87(4–5), 482–490. https://doi.org/10.1016/j.fuel.2007.04.018.

[10] Qiu Y., Xu Q., Pang S., and Ma X. (2021). CFD modelling of biomass catalytic fast pyrolysis in bubbling fluidized reactor: Effects of catalyst parameters. https://doi.org/10.22541/au.163256457.75515964/v1.

[11] Soria J., Zeng K., Asensio D., Gauthier D., Flamant G., and Mazza G. (2017). Comprehensive CFD modelling of solar fast pyrolysis of beech wood pellets. Fuel Processing Technology, 158, 226–237. https://doi.org/10.1016/j.fuproc.2017.01.006.

[12] Attanayake D. D., Sewerin F., Kulkarni S., Dernbecher A., Dieguez-Alonso A., and Van Wachem B. (2023). Review of modelling of pyrolysis processes with CFD-DEM. Flow, Turbulence and Combustion, 111(2), 355–408. https://doi.org/10.1007/s10494-023-00436-z.

[13] Rigopoulos S. (2010). Population balance modelling of polydispersed particles in reactive flows. Progress in Energy and Combustion Science, 36(4), 412–443. https://doi.org/10.1016/j.pecs.2009.12.001.

[14] Liu B., Papadikis K., Gu S., Fidalgo B., Longhurst P., Li Z., and Kolios A. (2017). CFD modelling of particle shrinkage in a fluidized bed for biomass fast pyrolysis with quadrature method of moment. Fuel Processing Technology, 164, 51–68. https://doi.org/10.1016/j.fuproc.2017.04.012.

[15] Dong N., Gu S., Armstrong L.-M., Papadikis K., and Luo K. (2012). CFD modelling of thermo-chemical process in catalytic pyrolysis of sawdust in bubbling fluidized beds. In Proceeding of Proceedings of CHT-12. ICHMT International Symposium on Advances in Computational Heat Transfer, pp. 299–315. https://doi.org/10.1615/ichmt.2012.cht-12.180

[16] Tobo Y., Lotfi A., Virla L. D., and Mahinpey N. (2023). Fast pyrolysis multiphase CFD-kinetics model in a drop tube reactor. Fuel, 340, 127524. https://doi.org/10.1016/j.fuel.2023.127524.

[17] Qian Y., Yu Y., Xu G., and Liu X. (2018). CFD modeling of coal pyrolysis in externally heated fixed-bed reactor. Fuel, 233, 685–694. https://doi.org/10.1016/j.fuel.2018.06.100.

[18] Azizi S., and Mowla D. (2015). CFD modeling of algae flash pyrolysis in the batch fluidized bed reactor including heat carrier particles. International Journal of Chemical Reactor Engineering, 14(1), 463–480. https://doi.org/10.1515/ijcre-2014-0185.
[19] Wang J., Wu J., Yuan S., and Yan W.-C. (2021). CFD simulation of ultrasonic atomization pyrolysis reactor: The influence of droplet behaviors and solvent evaporation. International Journal of Chemical Reactor Engineering, 19(2), 167–178. https://doi.org/10.1515/ijcre-2020-0229.
[20] Li R., Jing L. J., and He M. M. (2011). Research on CFD of fluidization of biomass waste fast pyrolysis reactor. Advanced Materials Research, 201–203, 708–713. https://doi.org/10.4028/www.scientific.net/amr.201-203.708.
[21] Thoharudin T. (2024). CFD simulation of biomass pyrolysis in a fluidized bed reactor: Effect of particle shrinkage model. Formosa Journal of Science and Technology, 3(5), 899–916. https://doi.org/10.55927/fjst.v3i5.9435.
[22] Orszag S. A. (1970). Analytical theories of turbulence. Journal of Fluid Mechanics, 41(2), 363–386. https://doi.org/10.1017/s0022112070000642.
[23] Doran P. M. (2013). Fluid Flow. Bioprocess Engineering Principles, 201–254. https://doi.org/10.1016/b978-0-12-220851-5.00007-1.
[24] Larsson J., and Lele S. K. (2009). Direct numerical simulation of canonical shock/turbulence interaction. Physics of Fluids, 21(12). https://doi.org/10.1063/1.3275856.
[25] Shi F., Gao Z., Jiang C., and Lee C. (2019). Investigation on noise from shock/isotropic turbulence interaction using direct numerical simulation. In 25th AIAA/CEAS Aeroacoustics Conference. https://doi.org/10.2514/6.2019-2657
[26] Shi F., Gao Z., Jiang C., and Lee C. (2019). Investigation on noise from shock/isotropic turbulence interaction using direct numerical simulation. In 25th AIAA/CEAS Aeroacoustics Conference. https://doi.org/10.2514/6.2019-2657
[27] Kolmogorov A. N. (1968). Local structure of turbulence in an incompressible viscous fluid at very high Reynolds numbers. Soviet Physics Uspekhi, 10(6), 734–746. https://doi.org/10.1070/pu1968v010n06abeh003710.
[28] Kim J., Moin P., and Moser R. (1987). Turbulence statistics in fully developed channel flow at low Reynolds number. Journal of Fluid Mechanics, 177, 133–166. https://doi.org/10.1017/s0022112087000892.

Satyvir Singh

Chapter 8
Dynamics of shock-accelerated V-shaped gas interface

Abstract: This study investigates the dynamic evolution of a shock-accelerated V-shaped gas interface, which is composed of nitrogen and sulfur hexafluide gases. The main focus is highlighted on the interplay of shock-accelerated instabilities, interface deformation, and vorticity generation. A compressible two-component Euler equation in two-dimensional is solved using a high-order modal discontinuous Galerkin method. For numerical simulations, two-distinct Mach numbers (M_s = 1.22, 1.42) are considered. The results show that the Mach numbers have a significant effect on the flow field, interface deformation, complex wave patterns, inward jet growth, and vorticity creation in the shock-accelerated V-shaped interface. As the Mach number increases, the V-shaped interface deforms differently, and the distance between the triple points and the Mach stem also varies. Greater rolled-up vortex chains are produced by higher Mach values than by smaller ones. A thorough analysis of the Mach number impact identifies the factors that propel the creation of vorticity during the encounter phase. Finally, the time evolution of enstrophy and kinetic energy are investigated.

Keywords: Richtmyer-Meshkov Instability, shock wave, V-shaped interface, numerical simulations, vorticity

8.1 Introduction

The Richtmyer-Meshkov (RM) instability is a fundamental phenomenon in fluid dynamics that occurs when a shock wave interacts with a density gradient, such as at the interface between two different gases [1, 2]. In this phenomenon, the misalignment of pressure and density gradients across the interface generates baroclinic vorticity, causing the initial perturbations to amplify over time. Unlike the Rayleigh-Taylor (RT) instability, where growth is driven by continuous acceleration, RM instability is triggered by impulsive acceleration due to the passage of the shock [3]. The amplitude of the interface perturbations grows linearly at early stages and transitions into nonlinear and turbulent regimes

Acknowledgment: The author acknowledges funding through the German Research Foundation within the research unit DFG-FOR5409.

Satyvir Singh, Institute for Applied and Computational Mathematics, RWTH Aachen University, Germany, e-mail: singh@acom.rwth-aachen.de

https://doi.org/10.1515/9783111661674-008

as the flow evolves. Shock-accelerated gas interfaces are a classic setting for observing and studying RM instability, which has wide-ranging implications for astrophysical phenomena, inertial confinement fusion, and supersonic combustion [4].

Over the past few decades, there has been a lot of interest in the RM instability on a range of complicated shock-accelerated gas interfaces, such as cylindrical, spherical, elliptical, polygonal, single, and multi-modal. Numerous theoretical, experimental, and computational studies have been conducted to examine the emergence of RM instability on such nonhomogeneous interfaces. A comprehensive experimental study of shock-accelerated cylindrical/spherical gas interactions was carried out by Haas and Sturtevant [5]. The flow physics of RM instability at a spherical gas interface was experimentally studied by Ranjan et al. [6] in a divergent shock-refraction configuration. Quirk and Karni [7] replicated the experimental findings of Haas and Sturtevant [5] and provided a thorough computational analysis of the interaction of a weak shock wave with an isolated cylindrical gas interface. Singh et al. [8] investigated numerically the impact of bulk viscosity on the flow morphology of a shock-accelerated cylindrical light interface in diatomic and polyatomic gases. Later, Singh et al. [9] studied the effects of Atwood numbers on the flow physics of shock-driven elliptical gas interface based on the computational simulations. More focus has recently been placed on analyzing the evolution of RM instability induced by the interaction of shock waves with polygonal interfaces [10–13]. These polygonal interfaces create more intricate flow fields through their transmitted shock waves, reflections, reciprocal shock collisions, and regular and irregular refractions.

Due to its ease of theoretical treatment and ability to form complex interfaces, previous work has mostly concentrated on the linear and nonlinear evolutions of RM instability at single-mode interfaces [14–16]. Using large-eddy simulations, Thornber et al. [17] investigated how different 3D multi-mode beginning circumstances affected the growth of RM instability. The RM instability growth of single-shock-accelerated mixing was studied by Lombardini et al. [18] using large-eddy simulations. They proposed that a gas mixing layer eventually evolves to a fully developed turbulent flow at a late time, with a Kolmogorov-like inertial subrange governed by a power law, for high enough incident Mach numbers. To quantify the effect of initial conditions on the evolution of the RM instability, Mohaghar et al. [19] carried out an experimental study. They examined the impact of incident shock strength on the mixing transition, in particular.

Among the various configurations studied, the dynamics of a shock-accelerated V-shaped gas interface presents unique challenges and insights. This topic involves the interaction of a shock wave with a geometrically structured interface, leading to complex flow features, such as vortices, interface deformation, and mixing instabilities. The V-shaped interface introduces additional complexities compared to planar or cylindrical interfaces. Its sharp angles and geometric asymmetry cause localized intensification of shock effects, leading to highly nonuniform velocity and pressure distributions. These characteristics make the dynamics of shock-accelerated V-shaped interfaces a fertile ground for understanding instability mechanisms such as the RM instability and subsequent turbulent mixing. Luo et al. [20] conducted an experimental investigation of a shock-accelerated V-shaped air/SF_6 gaseous interface, with the primary findings being the impact of the initial interface amplitude across the wavelength on the mixing breadth. The RM instability development on a shock-accelerated V-shaped air/helium gaseous interface was then experimentally investigated by Zhai et al. [21] to investigate the effects of initial conditions, specifically the initial amplitude, on the flow characteristics. These experiments also validate computational models, ensuring their applicability to real-world scenarios.

In this chapter, we present a novel investigation into the complex behavior of gas interfaces subjected to shock wave interactions, focusing on the unique dynamics induced by a V-shaped configuration. Using advanced numerical simulations based on the high-fidelity discontinuous Galerkin method, the research delves into the deformation, instability growth, and mixing processes of the interface under shock acceleration. The simulations capture the intricate interplay of vorticity generation, baroclinic effects, and shock-interface interactions, providing detailed insights into the temporal and spatial evolution of the interface. This work not only enhances understanding of shock-driven flows and interface dynamics but also has significant implications for high-speed propulsion, inertial confinement fusion, and astrophysical applications.

The remaining parts of this chapter are organized as follows: Section 8.2 provides an overview of the initial setup and governing equations. Section 8.3 introduces the computational strategies, its validation, and grid refinement study. Section 8.4 presents the numerical results and corresponding discussion for the shock-accelerated V-shaped gas interface. Finally, Section 8.5 summarizes key findings and outlook for further research on shock-interface interactions in complex geometries.

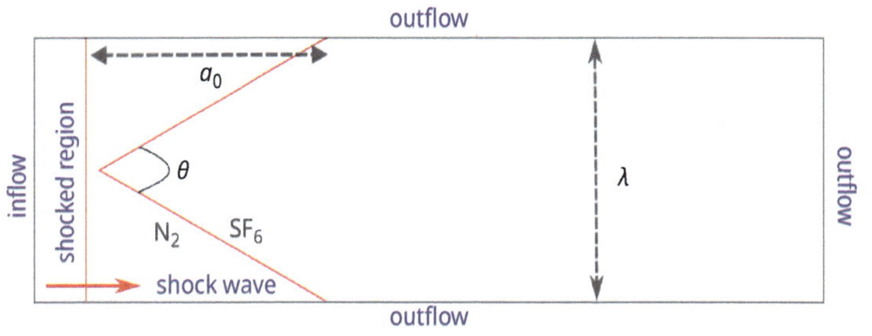

Figure 8.1: Diagram of the initial setup for a shock-accelerated V-shaped gas interface.

8.2 Initial setup and governing equations

8.2.1 Initial setup

Figure 8.1 presents the computational setup for the problem involving a shock-accelerated V-shaped gas interface. The simulation domain is a rectangular region of dimensions 300 × 100 mm^2. A V-shaped interface separating nitrogen (N_2) and sulfur hexafluoride (SF_6) gases is positioned downstream of the incident shock (IS) wave. The IS wave propagates from left to right along the x-direction and starts 20 mm from the left domain boundary, traveling through N_2 gas. The interface is situated 5 mm behind the IS wave. The geometry of the V-shaped interface is characterized by its vertex angle $\theta = 60°$, initial amplitude a_0, and wavelength $\lambda = L$. These parameters are related by the equation $a_0/\lambda = 1/4\tan(\theta/2)$. The location of the interface is determined by $z = L/8 + L/2\ \tan(\theta/2)$.The boundaries of the domain are defined as follows: the left boundary serves as an inflow boundary, while the top, bottom, and right boundaries are designated as outflow boundaries. The initial pressure and temperature in the region surrounding the interface are $P_0 = 101{,}325$ Pa and $T_0 = 273$ K, respectively.

8.3 Governing equations

Building on the previous study on shock-accelerated gas interface, the current study additionally simulates two-component gas flows using a two-dimensional system of compressible Euler equations [11]. While the impacts of surface tension, reactions, viscosity, ionization, and interactions between several inhomogeneities are thought to be negligible and so disregarded, the mechanisms of shocked acceleration, vorticity creation, and mixing are investigated separately in this computational scenario. Interestingly, the discretization and chosen shock limits act as a dissipation source to limit the range of scales. The derived equations' conservative form is provided as

$$\begin{aligned}
\frac{\partial \rho}{\partial t} + \frac{\partial(\rho u)}{\partial x} + \frac{\partial(\rho v)}{\partial y} &= 0,\\
\frac{\partial(\rho u)}{\partial t} + \frac{\partial\left(\rho u^2 + p\right)}{\partial x} + \frac{\partial(\rho uv)}{\partial y} &= 0,\\
\frac{\partial(\rho v)}{\partial t} + \frac{\partial(\rho uv)}{\partial x} + \frac{\partial\left(\rho v^2 + p\right)}{\partial y} &= 0,\\
\frac{\partial(\rho E)}{\partial t} + \frac{\partial((\rho E + p)u)}{\partial x} + \frac{\partial((\rho E + p)v)}{\partial y} &= 0,\\
\frac{\partial(\rho \phi)}{\partial t} + \frac{\partial(\rho u \phi)}{\partial x} + \frac{\partial(\rho v \phi)}{\partial y} &= 0,
\end{aligned} \tag{8.1}$$

where ρ is the density; u and v are the velocity components in the x- and y- directions, respectively; E is the energy, ϕ is the mass fraction; and p is the pressure, which is evaluated from the mathematical expression as follows [21]:

$$p = \gamma_{\max}\left(\rho E - \frac{1}{2}\left(u^2 + v^2\right)\right) \tag{8.2}$$

In this framework, $\gamma_{\max}$ denotes the specific heat ratio of the gas mixture. The relationship among pressure (p), density (ρ), temperature (T), and the mixture-specific gas constant (R) is defined by ideal gas law: $p = \rho RT$. It is assumed that both gas components are in thermal equilibrium and can be modeled as calorically perfect gases, as discussed in [21]. These gases are characterized by their specific heats at constant pressure (C_{p1}, C_{p2}), specific heats at constant volume (C_{v1}, C_{v2}), and specific heat ratios (γ_1, γ_2), The specific heat ratio of the overall mixture is determined based on the thermodynamic properties of the individual components and their respective contributions to the mixture:

$$\gamma_{\max} = \frac{C_{p1}\phi + C_{p2}(1-\phi)}{C_{v1}\phi + C_{v2}(1-\phi)} \tag{8.3}$$

where bubble and ambient gas are indicated by subscripts 1 and 2, respectively.

8.4 Computational strategies, validation, and grid refinement study

8.4.1 Computational strategies

In this study, we solve a two-dimensional system of the two-component compressible Euler equations using an internal modal Discontinuous Galerkin (DG) solver [11, 22]. The computational domain is partitioned into nonoverlapping rectangular elements, within which scaled Legendre polynomial basis functions are employed. Both the flux and volume integrals are evaluated using the Gauss-Legendre quadrature rule for enhanced accuracy. To compute numerical fluxes across element interfaces in two-component flows, the HLLC (Harten-Lax-van Leer-Contact) scheme is applied [23]. The solution within each finite element is approximated using a third-order accurate expansion in scaled Legendre polynomials. For time advancement, we utilize an explicitly strong stability-preserving Runge–Kutta method of third-order accuracy. Additionally, to mitigate spurious oscillations and maintain stability in the solution, we incorporate a high-order moment limiter as proposed by Krivodonova [22].

8.4.2 Validity study

Figure 8.2 illustrates a comparison of the present computational results with experimental findings from Luo et al. [20], for a shock-accelerated V-shaped air/SF_6 inter-

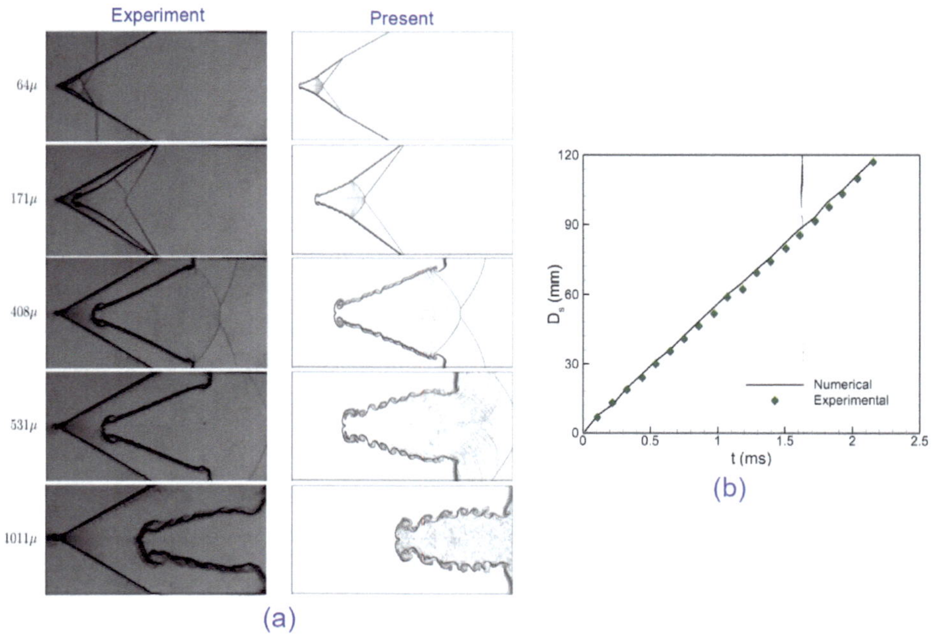

Figure 8.2: Comparison of (a) shadowgraph images and (b) time variation of the upstream interface displacement between the results obtained from the experiment [20] and the present solver in a shock-accelerated V-shaped gas interface.

face. A shock Mach number of $M_S = 1.2$ and a vertex angle of $\theta = 60°$ were used in the experimental investigation. Figure 8.2(a) displays shadowgraph images comparing experimental results (left column) with numerical simulations (right column) at various times. The results demonstrate how the propagating shock wave distorts the V-shaped interface, producing complex wave patterns and deformations. The numerical simulation effectively replicates the alterations observed in the experiment, capturing the interface dynamics and the evolving structures. Figure 8.2(b) presents the time evolution of the upstream interface displacement (D_S), comparing the numerical and experimental data. The simulation results (solid line) show excellent agreement with the experimental measurements (data points), accurately capturing the displacement trend over time. This validates the ability of the present computational method to reproduce the dynamics of the shock-accelerated V-shaped interface.

8.4.3 Grid refinement study

In the following simulations, normalized time is used to display flow morphology snapshots. The characteristic time $t_0 = \lambda / W_i$ is used to standardize the actual computa-

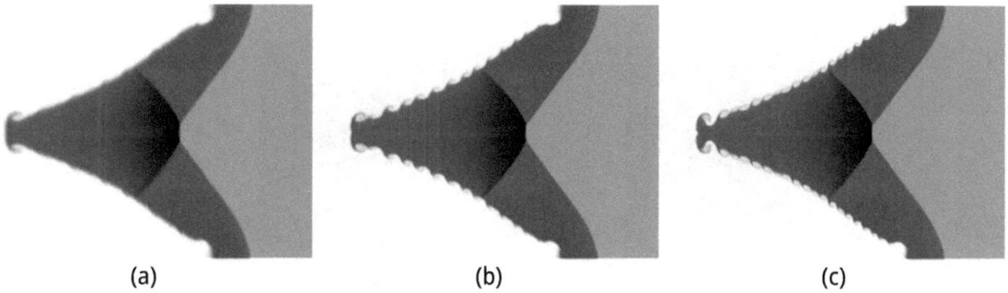

Figure 8.3: Grid refinement study: density contours with (a) 900 × 300, (b) 1,200 × 400, and (c) 1,500 × 500 grid points in the shock-accelerated V-shaped gas interface at $\tau = 15$.

tional time. In this case, $\tau = t/t_0 = tW_i/\lambda$, where W_i is the IS wave velocity and λ is the V-shaped interface's initial wavelength.

Figure 8.3 presents the results of a grid refinement study for the shock-accelerated V-shaped gas interface, showing density contours for three grid resolutions: 900 × 300, 1,200 × 400, and 1,500 × 500 grid points. As the grid resolution increases, the simulations capture more detailed features of the interface deformation, including sharper density gradients and small-scale instabilities caused by the shock acceleration. The finest grid resolution (1,500 × 500) provides the most detailed representation of the interface dynamics, demonstrating the importance of high-resolution grids in accurately resolving complex flow features. This study emphasizes the necessity of grid refinement to achieve numerical convergence and reliable simulations for shock-accelerated gas interfaces.

8.5 Results and discussion

Figure 8.4 illustrates the time evolution of density contours for the shock-accelerated V-shaped gas interface at two Mach numbers: $M_s = 1.22$, and $M_s = 1.62$. The evolution is illustrated over time (τ) as the interface deforms and exhibits various instabilities. This process is primarily driven by the interaction of shock waves with the V-shaped interface, resulting in complex physical phenomena that evolve as the shock propagates and interacts with the interface. The IS wave moves in the direction of the interface in the early phases ($\tau = 2, 5$). A portion of the shock gets reflected back as a reflected shock (RS) when the shock wave interacts with the density gradient across the interface after reaching the V-shaped discontinuity. Another portion of the shock simultaneously transmits as a direct transmitted shock (DTS) into the downstream gas. The contact deforms as a result of the pressure gradients and shear forces created by these shock interactions. These interactions cause the upper vortex ring (UVR) to emerge early on, which signals the beginning of vorticity formation at the interface. With increasing time ($\tau = 10, 20$), the interface deformation is more noticeable. Kelvin-

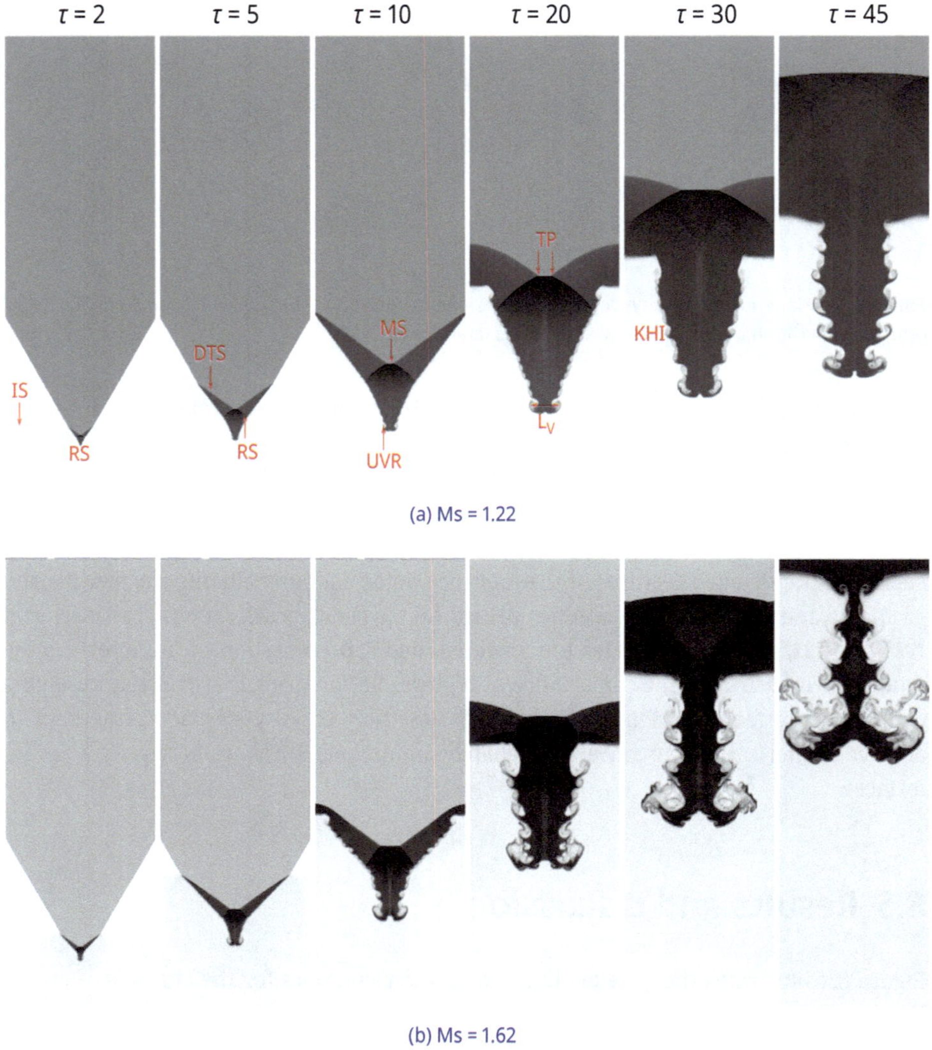

Figure 8.4: Flow field evolution in the shock-accelerated V-shaped gas interface: (a) M_s = 1.22, and (b) M_s = 1.62 IS denotes incident shock; RS denotes the reflected shock; DTS means the direct transmitted shock; UVR denotes the upper vortex ring; KHI denotes the Kelvin–Helmholtz instability; Lv is vertex spacing; MS is the Mach stem; and TP is triple point.

Helmholtz instability (KHI), a phenomenon brought on by variations in velocity between the two interacting fluids, develops as a result of the combined effect of pressure gradients and velocity shear. Along the interface, vortex structures with periodic properties start to appear. Vertex spacing (L_V), the distance between these features, is visible at this point. Furthermore, when it travels along the interface, the triple point (TP), which denotes the point where the incident shock, reflected shock, and contact surface meet, affects the flow dynamics and intensifies the instability. Because of the increased energy supplied by the stronger shock, these instabilities evolve more quickly and show sharper, more exaggerated features for the higher Mach number situation ($M_s = 1.62$). Later on ($\tau = 30, 45$), the contact experiences a great deal of mixing and deformation. When the KHI is dominant, complex vortex structures are created, which improves gas mixing. At this point, there is more interface distortion and the flow seems chaotic. In contrast to the lower Mach number case ($M_s = 1.22$), the higher Mach number example ($M_s = 1.62$) exhibits more complex vortex patterns, more violent mixing, and greater interface deformation. Later on, the flow field becomes more turbulent and chaotic due to the instabilities being amplified by the increased energy associated with a higher Mach number. The distinctions between the two scenarios demonstrate how Mach number has a major influence on how shock-driven instabilities develop. Stronger shocks cause more intense mixing, more noticeable vortex formations, and faster instability growth for $M_s = 1.62$. On the other hand, fewer chaotic flow shapes and slower instability growth are seen at lower Mach numbers ($M_s = 1.22$). This comparison emphasizes how shock strength affects how the gas interface behaves and how instabilities change over time.

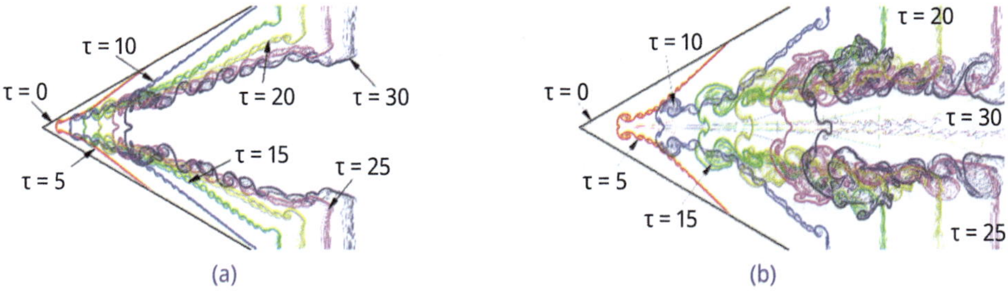

Figure 8.5: Interface deformation history in shock-accelerated V-shaped gas interface: (a) $M_s = 1.22$ and (b) $M_s = 1.62$.

Figure 8.5 illustrates the interface deformation history in a shock-accelerated V-shaped gas interface for two Mach numbers $M_s = 1.22$ and 1.62 over time. For $M_s = 1.22$, the interface deformation progresses gradually, with small disturbances appearing by $\tau = 5$, followed by the formation of periodic vortex structures at $\tau = 10$, and moderate instability growth through $\tau = 30$. The deformation is relatively controlled, with limited mixing and smaller, periodic vortices. In contrast, for $M_s = 1.62$, the higher Mach number causes more rapid and pronounced deformation, with larger disturbances forming earlier at $\tau = 5$ and more chaotic, irregular vortex structures dominating by $\tau = 30$. The KHI grows faster, resulting in enhanced mixing and overlapping vortex patterns. This comparison highlights the influence of Mach number, where a stronger shock $M_s = 1.62$ amplifies instability growth, vortex complexity, and mixing intensity, while a weaker shock $M_s = 1.22$ leads to slower, more periodic deformation.

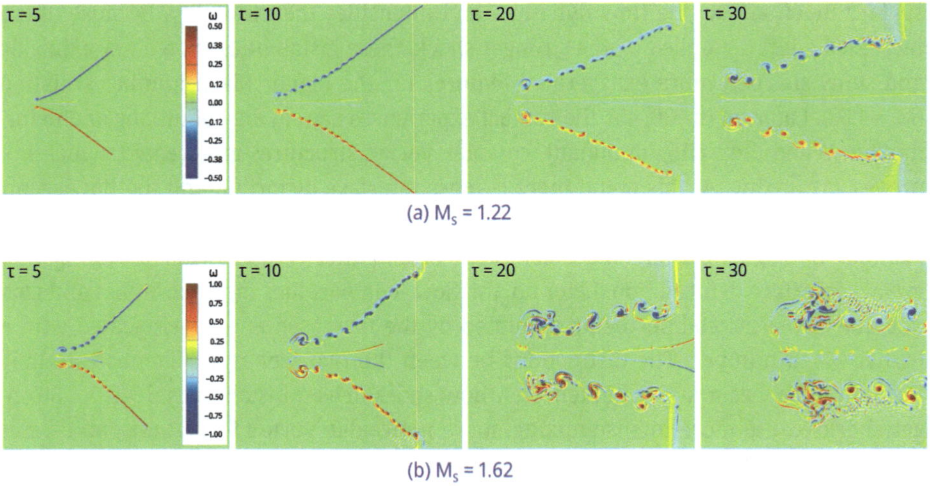

Figure 8.6: Vorticity generation in the shock-accelerated V-shaped gas interface: (a) $M_s = 1.22$ and (b) $M_s = 1.62$.

Figure 8.6 shows the generation of vorticity over time for two Mach numbers ($M_s = 1.22$ and $M_s = 1.62$) in a shock-accelerated V-shaped gas interface. This figure highlights key physical phenomena, such as baroclinic torque and KHI, which drive the evolution of vorticity. Baroclinic torque arises from the misalignment of pressure and density gradients during the shock-interface interaction, initiating vorticity generation. The KHI, driven by shear forces at the interface, amplifies the vorticity and generates vortex structures that grow in size and complexity over time. Positive and negative values in contrasting colors show how vorticity fields evolve, highlighting the emergence of flow instabilities, vortex structures, and mixing processes that are impacted by the Mach number. The development of vorticity is slower when $M_s = 1.22$, as shown in Figure 8.6(a). At $\tau = 5$, vorticity formation is initiated by the baroclinic torque generated by the interaction between the incoming shock and the density gradient at the interface. KHI begins when tiny and recurring vortices appear along the interface at $\tau = 10$. These vortices continue to be symmetrically dispersed and rather weak. The vortex patterns become more pronounced and frequent as time increases to $\tau = 20$, suggesting a substantial growth in instability. With regulated mixing between the two gases and little overlap, the vorticity field displays a regular pattern of symmetric vortices by $\tau = 30$, which is indicative of the weaker shock's reduced energy input. On the other hand, the higher Mach number causes the vorticity to evolve more quickly and energetically for $M_s = 1.62$, as illustrated in Figure 8.6(b). Compared to the lower Mach number example, baroclinic torque produces more pronounced initial vorticity along the interface at $\tau = 5$, with more distinct zones of positive and negative vorticity. The interface displays larger and more irregular vortex formations by $\tau = 10$, which suggests that KHI is growing quickly. The vortex cores are more concentrated and the vorticity gradients are more pronounced. The vorticity field becomes extremely erratic at $\tau = 20$, with larger and more chaotic vortex shapes caused by nonlinear processes and increased instability growth. By $\tau = 30$, overlapping and interacting vortices dominate the interface, resulting in chaotic mixing. More energy is injected into the system by the stronger shock at $M_S = 1.62$, which causes more noticeable vorticity and intricate flow dynamics.

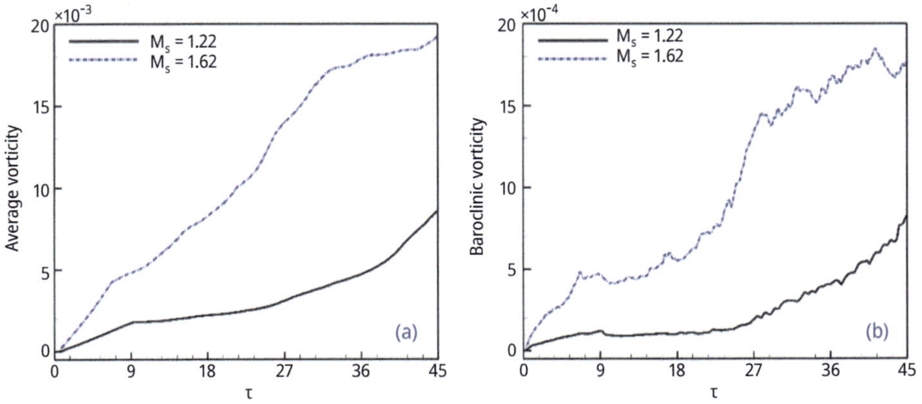

Figure 8.7: Spatially integrated fields of (a) average vorticity and (b) baroclinic vortcity in the shock-accelerated V-shaped gas interface with two different Mach numbers Ms = 1.22 and 1.62.

Two important spatially integrated fields – average vorticity and baroclinic vorticity production – are examined in order to gain a better understanding of vorticity generation phenomena:

$$Average\ vorticity = \frac{\int |\omega| dxdy}{\int dxdy}, \tag{8.4}$$

$$Baroclinic\ vorticity = \frac{\int \left| \frac{1}{\rho^2} \nabla \rho \times \nabla p \right| dxdy}{\int dxdy}. \tag{8.5}$$

where ω is the vorticity.

Figure 8.7 shows the temporal evolution of spatially integrated fields for two Mach numbers ($M_S = 1.22$ and $M_S = 1.62$) in a shock-accelerated V-shaped gas interface for (a) average vorticity and (b) baroclinic vorticity production. According to Figure 8.7 (a), the average vorticity, which gauges the flow's rotational strength, grows noticeably faster and has consistently higher values than $M_S = 1.22$. This discrepancy is a result of the higher Mach number's larger energy intake, which intensifies vorticity formation and instability. A similar measure of vorticity generation brought on by pressure and density gradient misalignment is the baroclinic vorticity production, which is displayed in Figure 8.7(b). With sharper spikes and variations, baroclinic production is significantly larger for $M_s = 1.62$, suggesting more dynamic flow behavior and stronger baroclinic effects. Due to a weaker shock and less instability development, both fields grow more slowly for $M_s = 1.22$. Baroclinic effects are crucial in determining rotational flow dynamics, as seen by the steady relationship between average vorticity and improved baroclinic vorticity production. Overall, Figure 8.8 emphasizes the effect of Mach number, showing that, in contrast to the smoother and more controllable development seen for $M_s = 1.22$, bigger shocks at $M_s = 1.62$ result in faster growth, higher magnitudes, and more chaotic vorticity dynamics.

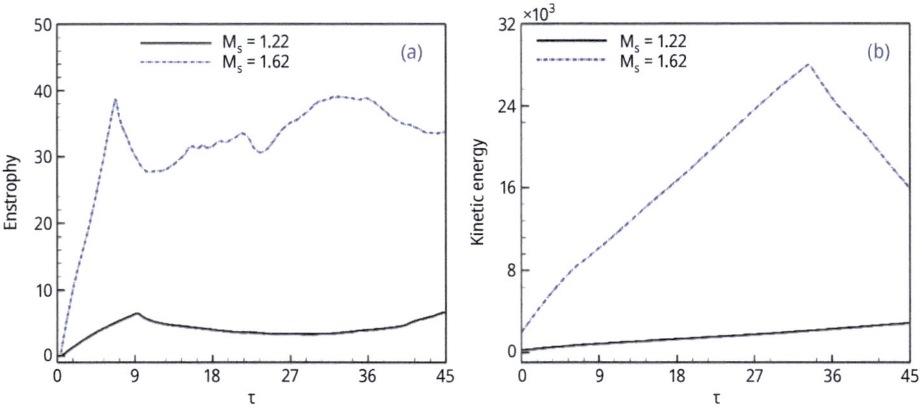

Figure 8.8: Spatially integrated fields of (a) enstrophy and (b) kinetic energy in the shock-accelerated V-shaped gas interface with two different Mach numbers $M_s = 1.22$ and 1.62.

Finally, we investigate the dynamics of the shock-accelerated V-shaped gas interface by analyzing two more quantities – enstrophy and kinetic energy. The spatially integrated field of these quantities are given as

$$Enstrophy = \frac{1}{2}\iint \omega^2 dxdy. \tag{8.6}$$

$$Kinetic\ energy = \frac{1}{2}\iint \mathbf{u}^2 dxdy. \tag{8.7}$$

Figure 8.8 shows the evolution of spatially integrated fields for enstrophy and kinetic energy in a shock-accelerated V-shaped gas interface for two Mach numbers ($M_s = 1.22$ and $M_s = 1.62$). Figure 8.8(a) illustrates enstrophy, which is proportional to the square of the vorticity and gauges the strength of rotational flow. Because of the stronger shock-induced instabilities, enstrophy shows a fast initial spike for $M_s = 1.22$ and $M_s = 1.62$, peaking early. This is followed by variations that show the chaotic character of turbulence and vortex interactions. Conversely, $M_s = 1.22$ shows a substantially slower and more steady increase, with much lower values, suggesting less intense instabilities and less rotational flow. The kinetic energy, or the energy connected to fluid motion, is shown in Figure 8.8(b). Kinetic energy increases quickly for $M_S = 1.62$, peaking at $\tau = 27$, and then gradually decreasing as energy dissipates as a result of turbulence and viscous processes. Kinetic energy for $M_S = 1.22$ increases gradually and stays substantially lower throughout, reflecting the gentler flow dynamics and lower energy input linked to the smaller shock. Mach number affects the flow behavior, as shown in Figure 8.8. Whereas $M_S = 1.62$ produces much more enstrophy and kinetic energy, causing more intense and chaotic flow phenomena, $M_S = 1.22$ results in slower growth and more stable flow patterns. This emphasizes how important shock strength is in controlling the system's energy and vorticity dynamics and intensity.

8.6 Concluding remarks and outlook

This chapter investigates the dynamic evolution of a shock-accelerated V-shaped gas interface, emphasizing the interaction of interface deformation, vorticity generation, and shock-induced instabilities. Using a modal discontinuous Galerkin approach and a compressible two-component Euler equations solver, high-order numerical simulations were used to investigate the impact of two Mach numbers ($M_S = 1.22$ and $M_S = 1.62$) on flow dynamics. The findings show that the flow field is significantly shaped by the Mach number, with higher Mach numbers producing more intense vorticity generation, intense interface deformation, and stronger instabilities. More energy input intensifies the production of complex vortex structures, increases the intensity of mixing, and speeds up the growth of instability for $M_S = 1.62$, as opposed to $M_S = 1.22$. The results show that baroclinic torque and KHI are important processes that drive the creation of vorticity. Greater deformation of the V-shaped interface, larger rolled-up vortex chains, and more variance in the distance between the triple points and Mach stem are all caused by higher Mach numbers. The examination of spatially integrated fields, such as kinetic energy, enstrophy, average vorticity, and baroclinic vorticity generation, highlights the dynamic development of rotational flow and energy transmission. In contrast to $M_S = 1.22$, which showed slower, more stable development with less mixing and turbulence, $M_S = 1.62$ showed faster growth and more chaotic flow character. Future research could concentrate on three-dimensional simulations to capture more intricate mechanisms of turbulence and vorticity formation in shock-accelerated interface, as well as additional flow complexities. The effects of different gas characteristics, initial conditions, and contact geometries may potentially shed more light on the dynamics of these kinds of systems

References

[1] Richtmyer R. D. (1960). Taylor instability in shock acceleration of compressible fluids. Communications on Pure and Applied Mathematics, 13, 297–319.

[2] Meshkov E. E. (1969). Instability of the interface of two gases accelerated by a shock wave. Fluid Dynamics, 4, 101.

[3] Singh S. (2020). Role of Atwood number on flow morphology of a planar shock-accelerated square bubble: A numerical study. Physics of Fluids, 32, 126112.

[4] Zhou Y., Williams R. J., Ramaprabhu P., Groom M., Thornber B., Hillier A., Mostert W., Rollin B., Balachandar S., Powell P. D., Mahalov A., and Attal N. (2021). Rayleigh–Taylor and Richtmyer-Meshkov instabilities: A journey through scales. Physica D: Nonlinear Phenomena, 423, 132838.

[5] Haas J. F., and Sturtevant B. (1987). Interaction of weak shock waves with cylindrical and spherical gas inhomogeneities. Journal of Fluid Mechanics, 181, 41–76.

[6] Ranjan D., Niederhaus J. H. J., Oakley J. G., Anderson M. H., Bonazza R., and Greenough J. A. (2008). Shock-bubble interactions: Features of divergent shock-refraction geometry observed in experiments and simulations. Physics of Fluids, 20, 036101.

[7] Quirk J. J., and Karni S. (1996). On the dynamics of a shock-bubble interaction. Journal of Fluid Mechanics, 318, 129–163.

[8] Singh S., Battiato M., and Myong R. S. (2021). Impact of the bulk viscosity on flow morphology of shock-bubble interaction in diatomic and polyatomic gases. Physics of Fluids, 33, 066103.

[9] Singh S. (2021). Contribution of Mach number to the evolution of the Richtmyer-Meshkov instability induced by a shock-accelerated square light bubble. Physical Review Fluids, 6, 104001.

[10] Singh S., and Battiato M. (2022). Numerical simulations of Richtmyer-Meshkov instability of SF_6 square bubble in diatomic and polyatomic gases. Computers and Fluids, 242, 105502.

[11] Singh S. (2023). Investigation of aspect ratio effects on flow characteristics and vorticity generation in shock-induced rectangular bubble. European Journal of Mechanics - B/Fluids, 101, 131.

[12] Alsaeed S. S., and Singh S. (2024). Numerical study of thermal nonequilibrium effects on Richtmyer-Meshkov flow driven by a heavy forward-triangular bubble. Physical Review E, 110, 065103.

[13] Singh S., and Alsaeed S. S. (2024). High-fidelity simulations of Richtmyer–Meshkov flows triggered by a forward-pentagonal bubble with different Atwood numbers. European Journal of Mechanics - B/Fluids, 108, 151–165.

[14] Fraley G. (1986). Rayleigh–Taylor stability for a normal shock wave–density discontinuity interaction. Physics of Fluids, 29, 376–386.

[15] Mikaelian K. O. (1993). Growth rate of the Richtmyer-Meshkov instability at shocked interfaces. Physical Review Letters, 71, 2903–2906.

[16] Mikaelian K. O. (2005). Richtmyer–Meshkov instability of arbitrary shapes. Physics of Fluids, 17, 034101.

[17] Thornber B., Drikakis D., Youngs D. L., and Williams R. J. R. (2010). The influence of initial conditions on turbulent mixing due to Richtmyer-Meshkov instability. Journal of Fluid Mechanics, 654, 99–139.

[18] Lombardini M., Pullin D. I., and Meiron D. I. (2012). Transition to turbulence in shock-driven mixing: A Mach number study. Journal of Fluid Mechanics, 690, 203–226.

[19] Mohaghar M., Carter J., Pathikonda G., and Ranjan D. (2016). The transition to turbulence in shock-driven mixing: Effects of Mach number and initial conditions. Journal of Fluid Mechanics, 871, 595–635.

[20] Luo X., Dong P., Si T., and Zhai Z. (2016). The Richtmyer–Meshkov instability of a 'V' shaped air/SF6 interface. Journal of Fluid Mechanics, 802, 186–202.

[21] Singh S., Sengupta B., Awasthi M. K., and KumaR V. (2024). Insight on the Flow Physics of Shock-driven Elliptical Gas Inhomogeneity with Different Atwood Numbers. International Journal of Mathematical, Engineering and Management Sciences, 9, 1–22.
[22] Alsaeed S. S., and Singh S. (2024). Numerical Study of Shock Wave Interaction with V-Shaped Heavy/Light Interface. Mathematics, 12, 3131.
[23] Alsaeed S. S., and Singh S. (2023). Numerical investigation of shock Mach number effects on Richtmyer–Meshkov instability in a heavy square bubble. Physica D: Nonlinear Phenomena, 453, 133844.

Sunil* and Vishal Chandel

Chapter 9
Nonlinear and linear analyses of partially ionized plasma

Abstract: This study examines thermal convection in a layer of partially ionized plasma (PIP) heated from below, considering three boundary configurations: free-free, rigid-rigid, and rigid-free. Linear analysis, performed using the normal mode method, and nonlinear analysis, conducted via the energy method, provided comprehensive insights. Exact solutions were derived for free-free boundaries, while the Galerkin-weighted residual method was applied to rigid-rigid and rigid-free boundaries for numerical results. A notable outcome is the agreement between the critical Rayleigh numbers obtained from linear and nonlinear analyses, highlighting global stability and the absence of subcritical regions. The study quantified the role of collisional frequency, demonstrating its significant influence on energy dissipation rates without altering the critical Rayleigh number. Compressibility was found to delay the onset of convection by increasing the Rayleigh number.

In addition to the fundamental investigation of thermal convection in PIP, the effects of rotation and magnetic fields have been separately studied in this chapter. This separation enables a focused exploration of their individual contributions to stabilizing the system. Detailed investigations revealed that at low rotation rates, rigid-rigid boundaries offer the highest stability for PIP, whereas, at high rotation rates, free-free boundaries emerge as the most stable configuration. These findings provide valuable insights into the interplay of compressibility, rotation, magnetic fields, and boundary conditions in governing the thermal convection behavior of PIP.

Keywords: Partially ionized plasma, Thermal convection, Different bounding surfaces, Nonlinear analysis, Linear analysis

9.1 Introduction

Thermal convection is a fundamental phenomenon that governs a wide range of processes in natural and industrial systems. It is the driving mechanism behind large-scale events such as ocean currents, atmospheric circulation, and energy transport in planetary and stellar interiors. On a smaller scale, it underpins essential engineering

***Corresponding author: Sunil**, Department of Mathematics and Scientific Computing, National Institute of Technology Hamirpur, Hamirpur 177005, Himachal Pradesh, India
Vishal Chandel, Department of Mathematics and Scientific Computing, National Institute of Technology Hamirpur, Hamirpur 177005, Himachal Pradesh, India

https://doi.org/10.1515/9783111661674-009

applications like heat transfer, thermal management in electronics, and energy storage systems. The study of thermal convection, often referred to as Bénard convection, can be traced back to the experiments of Henri Bénard in 1900. However, its conceptual foundations were laid much earlier by pioneers like Count Rumford (1797) and James Thomson (1882). A rigorous theoretical framework was introduced by Lord Rayleigh in 1920, who formulated the Rayleigh number – a dimensionless parameter encapsulating the interplay between buoyancy forces and viscous damping in a fluid layer heated from below. This parameter has since become central to the study of convective stability, delineating the conditions under which a fluid transitions from a static state to convective motion.

The analysis of thermal convection typically involves two complementary approaches: linear and nonlinear analysis. Linear analysis identifies the critical thresholds for instability, and nonlinear analysis delves deeper, exploring complex interactions among governing processes and determining sufficient criteria for system stability. This dual approach has driven extensive research into the onset and progression of thermal convection across various fluid systems under diverse external influences, including magnetic fields, rotation, and boundary conditions.

The study of thermal convection has been shaped by key contributions in hydrodynamic and hydromagnetic stability. Chandrasekhar's *Hydrodynamic and Hydromagnetic Stability* [1] provides pivotal insights into fluid instability, while Drazin and Reid's *Hydrodynamic Stability* [2] bridges theory with experimental results on flow transitions. Joseph's Stability of Fluid Motions I [3] explores nonlinear changes in fluid systems, and Straughan's *The Energy Method, Stability, and Nonlinear Convection* [4] advances stability analysis with modern applications like compressible fluids and convection-diffusion, offering essential frameworks for understanding complex convection phenomena.

Partially ionized plasma (PIP), a distinct state of matter composed of both neutral and charged particles, presents unique challenges and opportunities for studying thermal convection. PIP is found in a variety of natural environments, including the solar corona, interstellar medium, cometary tails, and stellar atmospheres, as well as in laboratory settings like fusion devices and plasma processing systems. The dual nature of PIP, with interactions between its neutral and ionized components, introduces complex dynamics that are absent in conventional fluids. These interactions often give rise to a variety of instabilities, including thermal, Rayleigh-Taylor, Kelvin-Helmholtz, and thermosolutal instabilities, which play critical roles in energy transport and structural evolution in both astrophysical and experimental contexts.

The dynamic stability of compressible PIP holds significant importance in astrophysical research, particularly in the context of solar corona heating theories and the magnetic structures of sunspots [5]. To simplify the equations governing compressible fluid flow, Spiegel and Veronis [6] introduced two key assumptions:

(i) The fluid layer's depth is much smaller than the scale height, and
(ii) Motion-induced fluctuations in density, pressure, and temperature are negligible compared to their total static variations.

These approximations effectively bridge the methodologies of compressible and incompressible fluid analyses, facilitating the application of established incompressible techniques to compressible scenarios.

In PIP systems, compressibility effects interact intricately with ion-neutral coupling and collisional dynamics, influencing the onset and progression of thermal convection. Studies by Hans [7], Sharma [8], and more recent works by Zaqarashvili et al. [9] and Soler and Ballester [10] have demonstrated the pivotal role of compressibility in determining PIP stability, with implications for both laboratory and astrophysical plasmas.

Extensive research into thermal convection in PIP has revealed a wide range of instabilities and their potential consequences. Sharma's pioneering work [8] laid the foundation for understanding thermal hydromagnetic convection in PIP. This was later expanded by Sharma and Sunil [11], who studied Rayleigh-Taylor instability in PIP confined in porous media under varying magnetic fields. Further advancements incorporated the Hall Effect, finite Larmor radius corrections, and ion-neutral collisional dynamics. Kaothekar [12] explored thermal instability in viscous PIP, while Kaothekar et al. [13] provided insights into transverse thermal instability in radiative PIP, emphasizing the role of neutral collisions and finite electron inertia. Recent studies by Patidar et al. [14] have explored the effects of magnetized quantum plasma on Jeans instability, extending our understanding of plasma dynamics in various contexts.

Recent studies have also explored the dynamic interactions within PIP systems. Chandel and Sunil [15] and [16] conducted stability analyses of PIP in porous media, considering local thermal nonequilibrium effects as well as the influence of rotation and magnetic fields. Chandel et al. [17, 18] also studied thermal convection in PIP confined within porous media and without porous medium. The work of Galstyan et al. [19] on Hall magnetohydrodynamics has provided valuable insights into large-scale flows in rotating PIP, particularly in space and astrophysical settings. These studies have significantly deepened our understanding of PIP dynamics, particularly the role of magnetic fields, rotation, and plasma properties in shaping convection patterns and stability.

Rotation plays a crucial role in thermal convection in PIP by introducing Coriolis forces that alter the structure and behavior of convective cells. As the system rotates, the conservation of angular momentum generates fluid motion in spiral patterns, which can either enhance or suppress convection depending on the direction and magnitude of rotation. This effect is particularly significant in rotating stars, where convection, rotation, and magnetic fields work together to influence energy transport, stellar evolution, and the synthesis of heavy elements. Similarly, in geophysical fluid dynamics, the Coriolis Effect shapes the dynamics of Earth's mantle and ocean currents. Understanding how rotation influences convection is essential for both astrophysical and engineering applications. Recent work by Chandel et al. [20] has explored the global stability of rotating PIP confined within porous media, providing further insights into the impact of rotation on convection.

Magnetic fields also have a profound effect on thermal convection in PIP. Through magnetohydrodynamics (MHD), magnetic fields interact with fluid motion, stabilizing convection by introducing the Lorentz force, which suppresses convective motions. This stabilization is crucial in astrophysical and geophysical contexts, such as solar flares, sunspots, and Earth's liquid core, where magnetic fields play a central role in energy transport and system stability. The application of MHD extends beyond natural phenomena, offering insights into the dynamics of plasma-based energy systems and thermal management in advanced electronics. Chandel and Sunil [21, 22] analyzed the effects of magnetic fields on the onset of thermal convection in permeable and porous media, showing consistent stability behaviors across both linear and nonlinear regimes.

This chapter explores the dynamics of thermal convection in PIP layers confined by various boundary configurations, including free-free, rigid-free, and rigid-rigid surfaces. Both linear and nonlinear analyses are employed to examine the onset and evolution of convection under these conditions. In addition to the basic study of thermal convection in PIP, the effects of rotation and magnetic fields have been investigated separately. This focused approach highlights the unique contributions of these factors to the stabilization of the system, providing a clearer understanding of their role. The findings presented in this chapter build upon and consolidate our previously published work on thermal convection in PIP, the influence of rotation on convection, and the impact of magnetic fields on thermal convection [21]. These results have broad implications, spanning astrophysical phenomena like stellar dynamics and solar corona behavior, as well as practical applications in plasma-based energy systems and advanced electronics. By combining theoretical insights with practical applications, this chapter seeks to enhance our understanding of thermal convection in PIP systems and its significance in both natural and engineered environments.

9.2 Problem formulation and governing equations

The problem formulation begins with an infinite horizontal layer of compressible PIP of thickness d and density ρ, confined between $z \in (0, d)$. The PIP is permeated by neutral components of density ρ_n. The system is subjected to a gravitational field acting in the negative z-direction, $\boldsymbol{g} = (0, 0, -g)$ and heated from below to maintain a uniform temperature gradient, $\beta\left(=\left|\frac{dT}{dz}\right|\right)$. This setup, illustrated in Figure 9.1, provides the geometric configuration for analyzing thermal convection in PIP.

In this study, the continuum hypothesis is adopted, treating PIP as a continuous fluid medium. Additionally, forces with minimal influence on the primary convection dynamics of PIP are excluded. Thermophoretic forces and neutral fluid drag are disregarded, as the system lacks solid particles. Ion drag and electrostatic forces are also neglected, given the focus on the macroscopic plasma behavior, where such forces

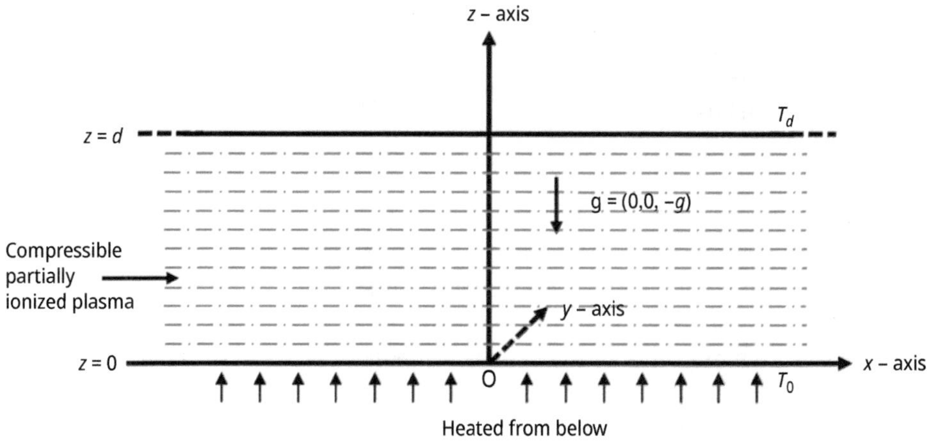

Figure 9.1: Geometrical configuration of the physical system.

have limited impact relative to the bulk dynamics. Also, the two-fluid theory has been employed, where ions and electrons are treated as a single fluid due to their strong coupling, resulting from the high frequency of collisions between them. Consequently, electrons are not explicitly considered in the formulation. Instead, the focus is on the ion-neutral interactions, which are the primary contributors to the dynamics of the system in Rayleigh-Bénard convection. This approach is supported by previous studies, including Sharma [8, 23], Zaqarashvili et al. [9], Sharma and Sharma [24], and Leake et al. [25], and adequately captures the relevant plasma behavior for the problem under consideration.

Based on previous work [5, 8, 13, 23, 26], the governing equations for compressible PIP flow include mass conservation, momentum, and energy equations. The mass conservation equation is given by

$$\frac{\partial \rho}{\partial t} + (\boldsymbol{q} \cdot \nabla)\rho + \rho \nabla \cdot \boldsymbol{q} = 0 \tag{9.1}$$

The momentum equation for the ionized components is

$$\frac{\partial \boldsymbol{q}}{\partial t} + (\boldsymbol{q} \cdot \nabla)\boldsymbol{q} = -\frac{1}{\rho_m}\nabla p + \boldsymbol{g}\frac{\rho}{\rho_m} + \frac{\mu}{\rho_m}\nabla^2 \boldsymbol{q} + \frac{\rho_n \nu_c}{\rho_m}(\boldsymbol{q}_n - \boldsymbol{q}) \tag{9.2}$$

where the term $\nu_c(\boldsymbol{q}_n - \boldsymbol{q})$ represents the frictional effects of the neutral components of PIP on the ionized components [27].

The energy equation, incorporating compressibility effects, is expressed as

$$\frac{\partial T}{\partial t} + (\boldsymbol{q} \cdot \nabla)T + p(\nabla \cdot \boldsymbol{q}) = \kappa \nabla^2 T \tag{9.3}$$

Here, the term $p(\nabla \cdot \boldsymbol{q})$ represents the product of pressure and the velocity divergence, which captures the compressibility effects on the temperature field [18]. This term contributes to the thermodynamic energy balance in compressible fluid dynamics. The density-temperature-pressure relationship is modeled as

$$\rho = \rho_m\{1 - \alpha_m(T - T_m) - k_m(p - p_m)\} \tag{9.4}$$

The Boussinesq approximation [6] is applied to the momentum equation to facilitate analytical simplification. For the neutral component of the PIP model, the momentum equation is further simplified by neglecting the effects of pressure and gravity. This simplification is justified by the predominant influence of collisional interactions, which govern the dynamics of the neutral component in PIP systems. In moderately ionized environments, frequent collisions between charged and neutral species serve as the principal mechanism for momentum exchange, rendering the contributions of pressure gradients and gravitational forces comparatively insignificant. This modeling strategy aligns with established methodologies in PIP research [28], wherein similar assumptions have been adopted to isolate the influence of collisional momentum transfer on the stability and convection behavior of the system. By concentrating on these dominant collisional effects, the model effectively captures the essential physical behavior of PIP under thermal gradients while retaining analytical tractability. This section has been reproduced from the earlier work of Chandel et al. [18] to ensure consistency and to highlight the foundational modeling framework adopted in the present study.

The equation of motion for the neutral components of PIP, which counteracts the motion of the ionized components [8, 18], is expressed as follows:

$$\frac{\partial \boldsymbol{q_n}}{\partial t} + (\boldsymbol{q_n} \cdot \nabla)\boldsymbol{q_n} = -\nu_c(\boldsymbol{q_n} - \boldsymbol{q}) \tag{9.5}$$

Here, $-\nu_c(\boldsymbol{q_n} - \boldsymbol{q})$represents the frictional effects of ionized components on the neutral components of the PIP [27].

Here, $\boldsymbol{q}$ and $\boldsymbol{q_n}$ denote the velocities of the ionized and neutral components of the PIP, respectively, while t is time. The parameters ρ, ρ_m, and ρ_n denote the density at $z = 0$, the constant space-averaged density, and the density of the neutral components, respectively. The variables $p, \mu, \nu_c, \nu, c_p, T,$ and κ represent pressure, coefficient of viscosity, collisional frequency among the composite medium components, kinematic viscosity, specific heat at constant pressure, temperature, and thermal diffusivity, respectively. The temperatures T_0, T_m, and T_d refer to the temperature at $z = 0$, the reference temperature at which $\rho = \rho_m$, and the temperature at $z = d$, respectively.

The thermal expansion coefficient α_m and compressibility parameter k_m are defined as $\alpha_m = -\left(\frac{1}{\rho}\frac{\partial \rho}{\partial t}\right)_m$ (also denoted as α), $k_m = \left(\frac{1}{\rho}\frac{\partial \rho}{\partial p}\right)_m$.

9.3 Basic state and non-dimensionalized perturbation equations

9.3.1 Basic state

The basic state refers to the steady, static configuration of the plasma layer in the absence of motion or any external perturbation. This state represents the thermal and hydrodynamic equilibrium, where the variables – such as velocity, pressure, temperature, and density – depend solely on the vertical coordinate, z, and not on time or the horizontal directions x and y.

The governing equations for the basic state can be derived from the full equations of motion, under the assumption of no flow ($\boldsymbol{q} = \mathbf{0}$) and a linear vertical temperature gradient. The basic state profiles for the velocity, pressure, temperature, and density are expressed as

$$\boldsymbol{q} = \boldsymbol{q_b} = \mathbf{0},\ p = p_b(z),\ \rho = \rho_b(z),\ T = T_b(z),\ \boldsymbol{q_n} = \boldsymbol{q_{n_b}} = \mathbf{0}.$$

Following Sharma [28], the explicit forms of the base state variables are:

$$\begin{aligned} p_b(z) &= p_m - g\int_0^z (\rho_m + \rho_0)dz,\ T_b(z) = -\beta z + T_0, \\ \rho_b(z) &= \rho_m[1 - \alpha_m(T - T_m) - k_m(p - p_m)] \end{aligned} \tag{9.6}$$

where β denotes the imposed vertical temperature gradient. This formulation assumes hydrostatic equilibrium, where the pressure gradient balances the gravitational force, and the temperature decreases linearly with height.

9.3.2 Perturbed equations

Perturbation theory is employed to study the stability of the base state by introducing small disturbances to the system. These perturbations represent deviations from the equilibrium and allow us to examine how the plasma evolves under the influence of external forces or thermal gradients.

Each physical variable in the system is decomposed into a base state component and a perturbation:

$$X(x,y,z,t) = X'(x,y,z,t) + X_b(z)$$

where $X'(x,y,z,t)$ is the small fluctuation about the base state and $X_b(z)$is the steady, spatially varying equilibrium profile.

Consider the small perturbations in the system: $\boldsymbol{q}' = (u, v, w), \theta, \delta\rho, p'$ and $\boldsymbol{q}'_n = (l, r, s)$, which denote the perturbations in $\boldsymbol{q}, T, \rho, p$, and $\boldsymbol{q}_n$, respectively. Using the approximations outlined by Spiegel and Veronis [6], the governing equations for these perturbations are derived. These equations capture the dynamics of small deviations from the basic state and are presented below:

The incompressibility condition for the perturbed velocity field is given by

$$\nabla \cdot \boldsymbol{q}' = 0. \tag{9.7}$$

Despite the plasma's compressibility, assuming a small Mach number ensures that density variations due to sound waves are negligible compared to those arising from thermal effects. This quasi-incompressibility condition is critical for analyzing low-speed, thermally driven convection processes.

The momentum balance for the ionized plasma includes contributions from pressure gradients, buoyancy forces, viscous dissipation, and collisional coupling with the neutral component:

$$\frac{\partial \boldsymbol{q}'}{\partial t} + (\boldsymbol{q}' \cdot \nabla)\boldsymbol{q}' = -\frac{1}{\rho_m}\nabla p' + \alpha\theta g\hat{\boldsymbol{k}} + \nu\nabla^2\boldsymbol{q}' + \frac{\rho_n \nu_c}{\rho_m}(\boldsymbol{q}'_n - \boldsymbol{q}') \tag{9.8}$$

Here, variations in density are assumed to be primarily a function of temperature, while the influence of pressure variations is neglected. This approximation follows the approach of Spiegel and Veronis [6], where the density perturbation is expressed as $\delta\rho = -\rho_m \alpha\theta$. This simplification is valid in thermally driven convection processes, where thermal buoyancy forces dominate over compressibility effects induced by pressure variations.

The temperature perturbation evolves under the influence of advection, thermal diffusion, and source term due to the buoyancy effect:

$$\frac{\partial \theta}{\partial t} + (\boldsymbol{q}' \cdot \nabla)\theta = \kappa\nabla^2\theta + \left(\beta - \frac{g}{c_p}\right)w \tag{9.9}$$

The velocity of the neutral component is primarily influenced by collisional coupling with the ionized component:

$$\frac{\partial \boldsymbol{q}'_n}{\partial t} + (\boldsymbol{q}'_n \cdot \nabla)\boldsymbol{q}'_n = -\nu_c(\boldsymbol{q}'_n - \boldsymbol{q}') \tag{9.10}$$

This equation governs the behavior of the neutral component, which is primarily influenced by collisional interactions with the ionized plasma. The coupling term $-\nu_c(\boldsymbol{q}'_n - \boldsymbol{q}')$ reflects the exchange of momentum and energy between the two plasma constituents.

These perturbation equations provide a comprehensive framework for understanding the interplay of thermal, viscous, and collisional effects in PIP systems. The incompressibility condition ensures consistency in low Mach-number flows. The momentum equations capture the dynamics of the ionized and neutral components, highlighting the effects of buoyancy, viscous damping, and collisional coupling. The thermal energy equation emphasizes the role of vertical motion in heat redistribution. Together, these equations form the foundation for analyzing the stability and evolution of thermal convection in PIP systems, offering insights into how key physical parameters influence convective phenomena.

9.4 Non-dimensionalized perturbed equations

Non-dimensionalization simplifies the governing equations by reducing the number of independent parameters, making the analysis more general and highlighting the key physical mechanisms governing the system. It also ensures that variables have comparable magnitudes, improving numerical stability and enabling systematic exploration of parameter space.

9.4.1 Non-dimensional scales

$$t^* = \frac{\nu}{d^2}t,\ \boldsymbol{q}^* = \frac{d}{\nu}\boldsymbol{q}',\boldsymbol{q}_n^* = \frac{d}{\nu}\boldsymbol{q}_n, z^* = \frac{z}{d}, p^* = \frac{d^2}{\rho_0\nu^2}p', \theta^* = \frac{\sqrt{R}}{\beta d\,\mathrm{Pr}}\theta.$$

9.4.2 Non-dimensional equations

Using these scales, the system is reformulated in terms of dimensionless variables, with the asterisk omitted for clarity:

$$\nabla\cdot\boldsymbol{q} = 0 \tag{9.11}$$

$$\frac{\partial \boldsymbol{q}}{\partial t} + (\boldsymbol{q}\cdot\nabla)\boldsymbol{q} = -\nabla p + R^{1/2}\theta\hat{\boldsymbol{k}} + \nabla^2\boldsymbol{q} + FL(\boldsymbol{q}_d - \boldsymbol{q}) \tag{9.12}$$

$$\mathrm{Pr}\left[\frac{\partial\theta}{\partial t} + (\boldsymbol{q}\cdot\nabla)\theta\right] = \nabla^2\theta + \left(1 - \frac{1}{G}\right)R^{1/2}w \tag{9.13}$$

$$\frac{\partial \boldsymbol{q}_d}{\partial t} + (\boldsymbol{q}_d\cdot\nabla)\boldsymbol{q}_d = -L(\boldsymbol{q}_d - \boldsymbol{q}) \tag{9.14}$$

9.4.3 Non-dimensional parameters and their significance

Each non-dimensional parameter in the system encapsulates key physical aspects of the PIP. The Rayleigh number $\left(R = g\alpha\beta d^4/\nu\kappa\right)$ quantifies the balance between buoyancy forces, which drive convection, and the opposing effects of viscous and thermal diffusion. The collisional frequency parameter $\left(L = \nu_c d^2/\nu\right)$ measures the influence of ion-neutral particle collisions, highlighting the coupling strength between the plasma's ionized and neutral components. The Prandtl number $(\Pr = \nu/\kappa)$ represents the ratio of momentum diffusivity to thermal diffusivity, distinguishing whether the convective flow is dominated by viscous or thermal effects. The density ratio parameter $(F = \rho_n/\rho_i)$ reflects the relative mass densities of the neutral and ionized components, influencing their interaction and dynamics. Finally, the compressibility parameter $\left(G = c_p\beta/g\right)$ characterizes the extent to which compressibility effects arise due to thermal expansion in the presence of gravity. These parameters, derived from established literature [1–4, 24], serve as critical tools for analyzing the interplay of thermal, viscous, and collisional processes in the system.

To examine the influence of the collisional frequency between the ionized and neutral components of PIP on system stability, we simplify eq. (9.14) by neglecting the convective term. This simplification is justified since the convective term does not depend on the collisional frequency. Applying normal mode analysis, we assume a time-dependent solution of the form $f(z).\exp\{i(k_x x + k_y y) + nt\}$, where $n(\equiv \partial/\partial t)$ is the frequency of the harmonic disturbances and k_x and $k_y(a^2 = k_x^2 + k_y^2)$ are horizontal wave numbers [24]. Substituting this into eq. (9.14), we derive the relationship:

$$\boldsymbol{q}_n = \left(\frac{L}{n+L}\right)\boldsymbol{q}$$

Here, the neutral component velocity $(\boldsymbol{q}_n)$ is expressed in terms of the ionized component velocity $(\boldsymbol{q})$ and the collisional frequency parameter L. Incorporating this result into eq. (9.12), we obtain a modified momentum equation for the ionized component:

$$\left[1 + \frac{FL}{n+L}\right]\frac{\partial \boldsymbol{q}}{\partial t} + (\boldsymbol{q}\cdot\nabla)\boldsymbol{q} = -\nabla p + R^{1/2}\,\theta\hat{\boldsymbol{k}} + \nabla^2\boldsymbol{q} \tag{9.15}$$

This equation captures the dynamics of the ionized component, now accounting for its coupling with the neutral component through the collisional parameter L. The presence of the term FL/($n + L$) quantifies the impact of ion-neutral interactions on the system's behavior.

9.4.4 Boundary conditions

The boundary conditions (BCs) associated with eqs. (9.13) and (9.15) are as follows:

$$\boldsymbol{q} = 0, \ \theta = 0, \ \text{on}\, z = 0, \ 1 \tag{9.16}$$

ensuring no-slip conditions for velocity and vanishing temperature perturbations at the horizontal boundaries. Additionally, plane tiling periodicity in the x- and y directions is assumed for both $\mathbf{q}$ and $\boldsymbol{\theta}$, consistent with standard theoretical frameworks [4]. These formulations provide the foundation for analyzing the stability and dynamics of convection in PIP systems.

9.5 Nonlinear analysis

To investigate the nonlinear stability of the system, we employ the energy method, a powerful approach detailed in Straughan's work on thermal convection and stability theory [4]. The nonlinear analysis relies on the evolution equations for kinetic and thermal energy, providing insights into the balance between energy production and dissipation in PIP. In this analysis, $|.|$ denotes the L^2 (V) norm, $\langle . \rangle$ represents integration over V, and V corresponds to the three-dimensional periodic cell.

The derivation begins by multiplying eq. (15) by $\boldsymbol{q}$ and eq. (13) by θ, followed by integration over V. Applying the divergence-free condition, eq. (11), and the boundary conditions, we arrive at the following evolution equations for the kinetic and thermal components:

$$\frac{1}{2}\left[1 + \frac{FL}{n+L}\right]\frac{\mathrm{d}}{\mathrm{d}t}||\boldsymbol{q}||^2 = R^{1/2}\langle \omega\theta \rangle - ||\boldsymbol{q}||^2 \tag{9.17}$$

$$\frac{Pr}{2}\frac{d}{d}||\theta||^2 = \frac{G-1}{G}R^{1/2}\langle \omega\theta \rangle - ||\nabla\theta||^2 \tag{9.18}$$

Equations (9.17) and (9.18) describe the evolution of kinetic and thermal energy in the system, highlighting the interplay between buoyancy-driven energy input and dissipation. Together, these equations quantify the competition between destabilizing buoyancy forces and stabilizing dissipative processes.

To study the system's stability, an L^2 energy function, $E(t)$, is constructed as a combination of the kinetic and thermal energy components. The total energy and its evolution are expressed as

$$\mathrm{E} = \frac{\lambda}{2}\left[1 + \frac{FL}{n+L}\right]||\boldsymbol{q}||^2 + \frac{Pr}{2}||\theta||^2 \tag{9.19}$$

The total energy $E(t)$ is defined as a weighted sum of kinetic and thermal energy. Its evolution is described by the balance of energy production (I_0) and dissipation (D_0) terms:

$$\frac{dE}{dt} = I_0 - D_0 \tag{9.20}$$

where

$$I_0 = R^{1/2}\left(\lambda + \frac{G-1}{G}\right)\langle\omega\theta\rangle \tag{9.21}$$

$$D_0 = \lambda||\nabla \boldsymbol{q}||^2 + ||\nabla\theta||^2 \tag{9.22}$$

Here, λ is the coupling parameter. To maintain stability, the dissipation must outweigh the production, i.e., $m < 1$ where m = $\max_G(I_0/\mathrm{D}_0)$. This ensures:

$$\frac{dE}{dt} \leq a_0 D_0 \tag{9.23}$$

where $a_0 = 1 - m(> 0)$

By applying the Poincare inequality, $(i.e.||\aleph||^2 \leq \frac{1}{\pi^2}||\nabla\aleph||^2)$, the dissipation term is bounded below as follows:

$$D_0 \geq K^* E \tag{9.24}$$

where $K^* = 2 + \pi^2 min\left\{Pr, \left[1 + FL(L+n)^{-1}\right]\right\}$. Substituting this into eq. (9.23),we find:

$$\frac{dE}{dt} \leq a_0 K^* E$$

which integrates to:

$$\mathrm{E(t)} \leq e^{-a_0 K^* E(0)} \tag{9.25}$$

The analysis confirms that the energy $E(t)$ decays exponentially over time, ensuring that as $t \to \infty$, $E(t) \to 0$. This exponential decay guarantees the stability of the system for any initial energy $E(0)$. When the Prandtl number satisfies $\mathrm{Pr} > 1 + \mathrm{FL}(n+L)^{-1}$, the collisional frequency parameter L significantly accelerates energy dissipation, enhancing system stability. This highlights the crucial role of collisional interactions in regulating the stability dynamics of PIP. The results emphasize that energy dissipation mechanisms, such as viscosity and thermal diffusion, counterbalance energy production driven by buoyancy. If dissipation dominates, the system efficiently dissipates energy, ensuring a return to stability following perturbations. Thus, the nonlinear analysis verifies the system's stability and underscores the pivotal influence of collisional dynamics and thermal properties in controlling stability.

9.6 Variational problem

To derive the maximum of $m = \max_{G}(I_0/D_0)$ at the critical argument $m = 1$, we apply the calculus of variations. By transforming the variables as $\hat{q} = \sqrt{\lambda}q$ and $\hat{w} = \sqrt{\lambda}w$ (with caps omitted for simplicity), the Euler-Lagrange equations for the problem are obtained a:

$$\frac{R^{1/2}}{\sqrt{\lambda}}\left(\lambda + 1 - \frac{1}{G}\right)\theta\hat{\boldsymbol{k}} + 2\nabla^2\boldsymbol{q} + 2\nabla\mathcal{P} = 0, \tag{9.26}$$

$$\frac{R^{1/2}}{\sqrt{\lambda}}\left(\lambda + 1 - \frac{1}{G}\right)w + 2\nabla^2\theta = 0, \tag{9.27}$$

where $\mathcal{P}$ is Lagrange multiplier.

By applying the operator $\hat{\boldsymbol{k}} \cdot$ curlcurl to eq. (9.26), we obtain:

$$\frac{R^{1/2}}{\sqrt{\lambda}}\left(\lambda + 1 - \frac{1}{G}\right)\nabla_1^2\theta + 2\nabla^4\omega = 0 \tag{9.28}$$

Assuming a plane tiling solution for velocity and temperature perturbations of the form:

$$(\omega, \theta) = [W(z), \Theta(z)]f(x, y) \tag{9.29}$$

where $f(x,y)$ satisfies:

$$\nabla_1^2 f + a^2 f = 0 \tag{9.30}$$

with a being the wave number, the equations reduce to the following system:

$$2(D^2 - a^2)^2 W - \frac{R^{\frac{1}{2}}}{\sqrt{\lambda}}\left(\lambda + 1 - \frac{1}{G}\right)a^2\ominus = 0 \tag{9.31}$$

$$2(D^2 - a^2)\ominus - \frac{R^{\frac{1}{2}}}{\sqrt{\lambda}}\left(\lambda + 1 - \frac{1}{G}\right)W = 0 \tag{9.32}$$

9.6.1 Boundary conditions

The stability of the system is analyzed under three boundary condition cases:

1. **Free-free boundaries:** Both surfaces are free, imposing W = 0, $\ominus = 0$, and D^2W = 0 at $z = 0, 1$. These conditions allow for unconstrained flow and are mainly of theoretical interest.
2. **Rigid-free boundaries:** The lower boundary is rigid (no-slip condition), while the upper boundary is free. This configuration imposes W = 0, $\ominus = 0$, DW = 0 at the rigid surface, and D^2W = 0 = 0 at the free surface. It represents systems with a par-

tially constrained flow, commonly encountered in practical scenarios such as atmospheric or astrophysical plasmas.

3. **Rigid-rigid boundaries:** Both surfaces are rigid, enforcing no-slip conditions. The constraints $W = 0$, $\ominus = 0$, and $DW = 0$ apply at $z = 0, 1$. This configuration is prevalent in engineering and laboratory plasmas, where physical containment restricts flow at the boundaries, ensuring zero velocity at the interface.

While free-free boundaries offer insights into idealized scenarios, rigid-rigid and rigid-free boundaries are more relevant to real-world applications. The no-slip condition in rigid-rigid boundaries is particularly significant for modeling systems like plasma containers or laboratory experiments, as it enforces a complete suppression of plasma flow at the walls. These boundary conditions, combined with the stability equations, are critical for understanding how PIP systems respond to perturbations and how their stability is influenced by factors such as wave number, buoyancy, and collisional interactions.

9.7 Linear analysis

To assess the threshold of instability, the governing equations are linearized by assuming infinitesimal perturbations. This involves ignoring the nonlinear terms in eqs. (9.15) and (9.13), yielding the linearized perturbation equations:

$$\left[1+\frac{FL}{(n+L)_-}\right]\frac{\partial \mathbf{q}}{\partial \mathrm{t}} = \nabla p + R^{1/2}\theta\hat{\boldsymbol{k}} + \nabla^2 \boldsymbol{q} \tag{9.33}$$

$$Pr\frac{\partial \theta}{\partial t} = \nabla^2\theta + \left(1-\frac{1}{G}\right)R^{1/2}\omega \tag{9.34}$$

By applying the operator $\hat{\boldsymbol{k}}\cdot$ curlcurl to eliminate pressure, eq. (9.33) simplifies to:

$$\left[1+\frac{FL}{(n+L)_-}\right]\frac{\partial}{\partial t}\left(\nabla^2\omega\right) = \sqrt{R}{\nabla_1}^2\theta + \nabla^4\omega \tag{9.35}$$

Assuming perturbations take the normal mode form:

$$\{\omega, \theta\} = \{\mathrm{W}(z), \ominus(z)\}\exp\left[i\left(k_x + k_y y\right) + nt\right] \tag{9.36}$$

where $\sqrt{k_x^2 + k_y^2} = a$ is the wave number.

Using eq. (9.36), eqs. (9.35) and (9.34) reduce to

$$n\left[1+\frac{FL}{(n+L)_-}\right]\left(D^2 - a^2\right)W = \left(D^2 - a^2\right)^2 W - \sqrt{R}a^2\ominus \tag{9.37}$$

$$nPr\ominus = \sqrt{R}\left(1 - \frac{1}{G}\right)W + (D^2 - a^2)\ominus \tag{9.38}$$

This set of coupled differential equations, along with appropriate boundary conditions, forms an eigenvalue problem where the eigenvalue is the Rayleigh number (R). The value of R determines the system's stability.

9.7.1 Principle of exchange of stabilities

The stability problem is analyzed for oscillatory modes using the principle of exchange of stabilities (PES). Multiplying eq. (9.37) by W^* (complex conjugate of W) and eq. (9.38) by $\ominus^*$, integrating over the domain with boundary conditions, yields:

$$-n\left[1 + \frac{FL}{(n+L)_{-}}\right]I_1 = I_2 - \sqrt{R}a^2\int_0^1 W^*\ominus dz \tag{9.39}$$

$$nPrI_3 + I_4 = \sqrt{R}\left(\frac{G-1}{G}\right)\int_0^1 \ominus^* W dz \tag{9.40}$$

where I_1, I_2, $I_{3,}$ and I_4 are positive integrals defined as:

$$I_1 = \int_0^1 (|DW|^2 + a^2|W|^2)dz$$

$$I_2 = \int_0^1 (|D^2W|^2 + 2a^2|W|^2 + a^4|W|^2)dz \tag{9.41}$$

$$I_3 = \int_0^1 (|\ominus|^2 dz, I_4 = \int_0^1 (|D\ominus|^2 + a^2|\ominus|^2)dz$$

We derive the following result by subtracting the product of a^2 with the complex conjugate of eq. (9.40) from the product of $\left(\frac{G-1}{G}\right)$ and eq. (9.39):

$$\left(\frac{G-1}{G}\right)\left\{n\left[1 + \frac{FL}{n+L}\right]I_1 + I_2\right\} - a^2 n^* PrI_3 - a^2 I_4 = 0 \tag{9.42}$$

where n^* is the complex conjugate of n. By substituting $n = \sigma + i\omega$ into eq. (9.42), we obtain the imaginary part of the resulting equation as

$$\omega\left[1 + \frac{FL^2}{(L+\sigma)^2 + \omega^2}\right]\left(\frac{G-1}{G}\right)I_1 + \omega a^2 PrI_4 = 0 \tag{9.43}$$

For $G > 1$, eq. (9.43) indicates that $\omega = 0$, confirming the presence of PES. This guarantees that convection occurs through stationary modes, while oscillatory modes are not present. In the marginal state ($n = 0$), eqs. (9.37) and (9.38) reduce to:

$$(D^2 - a^2)^2 W - \sqrt{R} a^2 \ominus = 0 \tag{9.44}$$

$$(D^2 - a^2)\ominus - \sqrt{R}\left(1 - \frac{1}{G}\right) W = 0 \tag{9.45}$$

These equations, along with the boundary conditions, form the eigenvalue problem for determining the critical Rayleigh number (R_c), which characterizes the onset of convection.

9.8 Numerical method

The Galerkin-weighted residual method [29] is employed for the numerical analysis to solve the eigenvalue problem in both linear and nonlinear cases. This method transforms the governing equations into a system of algebraic equations by representing the approximate solutions as linear combinations of basis functions. These basis functions, also known as trial functions, are chosen to satisfy the boundary conditions of the problem. In the Galerkin method, the residual is minimized in a weighted sense, where the weights are the same as the trial functions. For the velocity (W) and temperature ($\ominus$), the approximate solutions are expressed as

$$\mathrm{W} = \sum_{i=1}^{N} A_i W_i \text{ and } \ominus = \sum_{i=1}^{N} B_i \ominus_i \tag{9.46}$$

where A_i and B_i are unknown coefficients, and W_i (z) and $\ominus_i$ (z) are base functions. Substituting these expressions into the eigenvalue problems and applying the Galerkin procedure results in a system of linear algebraic equations. The basis functions are carefully chosen to satisfy the specific BCs for different bounding surfaces.

For free-free boundaries, the exact solutions are available [1], and the trial functions satisfying the BCs are sinusoidal:

$$W_i = \sin(i\pi z) \text{ and } \ominus_i = \sin(i\pi z) \tag{9.47}$$

For rigid-free boundaries, the basis functions [29] are more complex and defined as

$$W_i = z^2(1 - z)\left[(i + 2) - 2z^{i}\right] \text{ and } \ominus_i = z^i - z^{i+1} \tag{9.48}$$

For rigid-rigid bounding surfaces, the base functions [29] satisfying the BCs are:

$$W_i = z^{i+1} - 2z^{i+2} + z^{i+3} \text{ and } \ominus_i = z^i - z^{i+1} \tag{9.49}$$

When the approximate solutions presented in eq. (9.46) are incorporated into the eigenvalue problem and the Galerkin procedure is employed, the problem is effectively transformed into a linear system of homogeneous equations. These equations are expressed as:

$$C_{ji}A_i + D_{ji}B_i = 0 \tag{9.50}$$

$$E_{ji}A_i + F_{ji}B_i = 0 \tag{9.51}$$

For a nontrivial solution to exist, the determinant of the coefficient matrix formed by these equations must be zero. This condition is expressed as

$$\begin{vmatrix} C_{ji} & D_{ji} \\ E_{ji} & F_{ji} \end{vmatrix} = 0 \tag{9.52}$$

This determinant equation is solved to compute the eigenvalues of the system, with the Rayleigh number (R) being the key eigenvalue. To solve the determinant equation, MATHEMATICA software is used for symbolic manipulation and numerical computation. The software efficiently handles the complexity of the determinant and provides precise values for the Rayleigh number. It has been observed that R obtained through both linear and nonlinear analyses match, confirming the consistency and accuracy of the Galerkin method. This agreement reinforces the robustness of the methodology and its application to analyzing thermal convection problems.

9.8.1 Exact solutions for free-free bounding surfaces

An exact analytical solution for free-free boundaries is obtained using the Galerkin method. R for free-free surfaces, with the optimal λ, is

$$R = \frac{G\left(a^6 + 3a^4\pi^2 + 3a^2\pi^4 + \pi^6\right)}{a^2(G-1)} \tag{9.53}$$

This expression for R aligns with results by Sharma and Misra [30] in the absence of a magnetic field, confirming the accuracy of our results.

To find the critical wave number a_c, set $dR/da = 0$ yielding $a_c = \pi/\sqrt{2} = 2.2214$. The corresponding critical Rayleigh number is $R_c = (27G)/(4(G-1))\pi^4$, which for $G = 3$ gives $R_c = 986.267$. These results were originally derived and presented in the earlier work of Chandel et al. [18], and are reproduced here to maintain analytical continuity and coherence in the discussion.

9.8.2 Numerical solutions for rigid-rigid and rigid-free bounding surfaces

For rigid-rigid and rigid-free boundaries, where exact solutions are infeasible, numerical solutions are computed using higher-order Galerkin approximations. By iteratively refining the basis functions and increasing the number of terms, the values of R_c and a_c converge. The numerical implementation of the Galerkin method is carried out using MATHEMATICA. Results for rigid-rigid and rigid-free boundaries at $G = 3$ are presented in Table 9.1, showing consistent convergence with increasing iterations. This demonstrates the reliability of the method and its capability to handle complex boundary conditions. These results were originally reported in the earlier work of Chandel et al. [18] on thermal convection in partially ionized plasma with different boundary configurations and are reproduced here to ensure completeness and coherence within the present analysis.

Table 9.1: Critical wave number (a_c) and critical Rayleigh number (R_c) for rigid-rigid and rigid-free bounding surfaces at $G = 3$.

Iteration	Rigid-rigid		Rigid-free	
	a_c	R_c	a_c	R_c
1	**3.1165**	**2624.96**	**2.6697**	**1708.05**
2	**3.1165**	**2624.96**	**2.6885**	**1674.69**
3	**3.1164**	**2562.82**	**2.6885**	**1651.78**
4	**3.1163**	**2562.82**	**2.6823**	**1651.14**
5	**3.1163**	**2561.64**	**2.6823**	**1650.97**
6	**3.1163**	**2561.64**	**2.6823**	**1650.97**
7	**3.1163**	**2561.64**	**2.6823**	**1650.97**
8	**3.1163**	**2561.64**	**2.6823**	**1650.97**

Data reproduced from Chandel et al. [18].

9.9 Results and discussion

The analysis yields important insights into the stability characteristics of thermal convection in PIP under varying boundary configurations and compressibility effects. The key outcome is that the critical Rayleigh number R_c remains consistent across both linear and nonlinear analyses, signifying the nonexistence of subcritical instability and thereby confirming the global stability of the system. The results demonstrate that R_c can be determined exactly for free-free boundaries, while numerical methods are required for rigid-rigid and rigid-free boundaries. Despite variations in boundary configurations, collisional effects, which influence energy dissipation, do not impact the computed values of R_c. Figure 9.2 illustrates the variation of the Rayleigh number

(R) with the wave number (a) for different boundary conditions at a fixed compressibility parameter ($G = 3$). The values of R_c and the corresponding a_c for each bounding surface are marked. For $G = 3$, the results indicate that PIP confined between rigid-rigid bounding surfaces exhibits the greatest thermal stability, as evidenced by the highest R_c. In contrast, free-free boundaries are the least stable, with the lowest R_c, allowing convection to occur at a lower threshold. This trend reflects the stabilizing role of rigid boundary conditions, which enforce stronger temperature gradients and delay the onset of convection. The intermediate behavior observed for rigid-free boundaries further supports this stability hierarchy. Additionally, Figure 9.2 reveals that R decreases with increasing a until a_c is reached. This decline corresponds to convection modes dominated by large-scale cellular structures. Beyond a_c, R increases, signifying the emergence of convective oscillations characterized by smaller wave patterns. This interplay between R and a underscores the transition from large-scale cellular convection to finer wave structures as the system evolves.

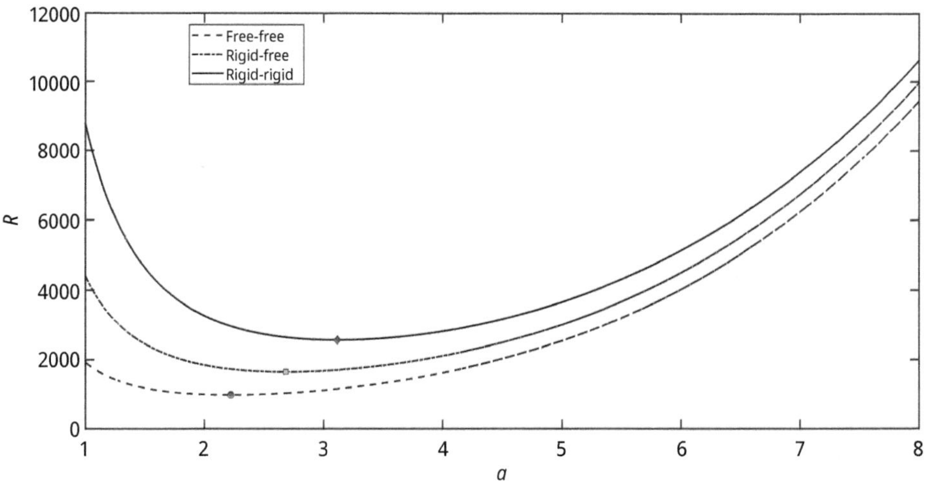

Figure 9.2: Variation of Rayleigh number (R) with wave number (a) for different boundary configurations at a fixed compressibility parameter $G = 3$. Critical values (a_c, R_c) are indicated: free-free (red: 22214986267), rigid-free (green: 26823165097), and rigid-rigid (blue: 31163256164). Results reproduced from Chandel et al. [18].

Figure 9.3 highlights the impact of the compressibility parameter (G) on the critical Rayleigh number (R_c). A clear inverse relationship is observed: as G increases, R_c decreases. This indicates that higher compressibility enhances thermal instability, hastening the onset of convection. Notably, for very high values of G, the PIP begins to exhibit behavior akin to incompressible fluids, as the effects of compressibility diminish. Conversely, for $G \leq 1$, R either diverges or becomes irrelevant, consistent with findings from previous studies [26]. The analysis emphasizes the interplay between

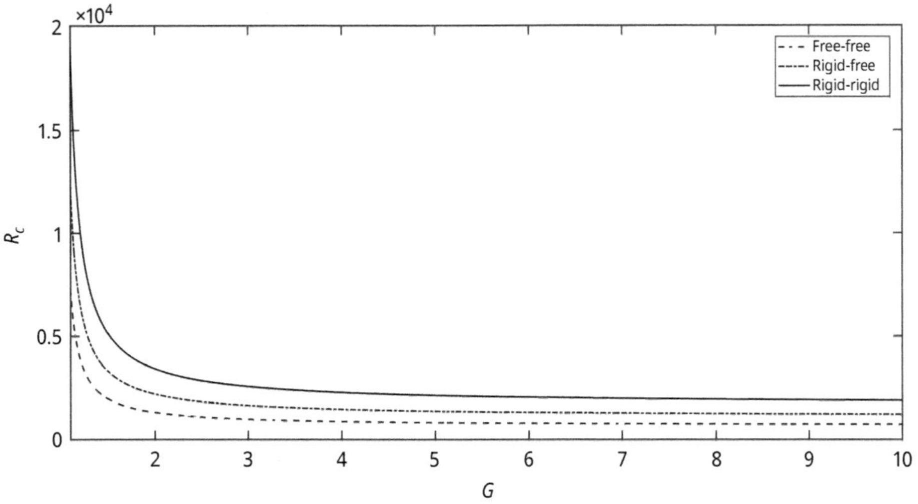

Figure 9.3: Variation of critical Rayleigh number R_c with compressibility parameter G for various boundary configurations. The observed decrease in R_c with increasing G highlights the destabilizing influence of compressibility. Results reproduced from Chandel et al. [18].

compressibility and boundary conditions in governing thermal convection. The rigid-rigid configuration emerges as the most stable, offering maximum resistance to convection onset. In contrast, the free-free configuration represents the least stable case. This comparative stability underlines the importance of boundary conditions in designing systems involving PIP, particularly in applications requiring precise thermal control. The results provide a robust framework for understanding and predicting thermal convection in compressible plasmas, with implications for both theoretical studies and real-world applications.

9.10 Effect of rotation

We now consider the problem of thermal convection in a horizontal layer of PIP that is rotating with an angular velocity $\boldsymbol{\Omega} = (0, 0, \Omega)$. Rotation introduces the Coriolis force into the equations of motion, which fundamentally alters the dynamics of convection. The inclusion of the Coriolis force modifies the governing equation of motion for the ionized components of PIP. While the nondimensional temperature equation remains the same as eq. (9.9), the equation of motion becomes

$$\left[1 + \frac{FL}{n+L}\right]\frac{\partial \boldsymbol{q}}{\partial t} + (\boldsymbol{q} \cdot \nabla)\boldsymbol{q} = -\nabla p + \sqrt{R}\theta\hat{\boldsymbol{k}} + \nabla^2\boldsymbol{q} + Ta^{1/2}\left(\boldsymbol{q} \times \hat{\boldsymbol{k}}\right) \tag{9.54}$$

where the Taylor number ($Ta = 4\Omega^2 d^4/\nu^2$) quantifies the relative importance of rotational forces compared to viscous forces. A higher Ta signifies a stronger stabilizing influence of rotation on the convective system.

By taking the curl and curlcurl of eq. (9.54), and then isolating the vertical components of these expressions, we obtain the governing equations for vertical vorticity ζ and vertical velocity ω:

$$\left[1+\frac{FL}{n+L}\right]n\zeta+\hat{\boldsymbol{k}}\cdot\operatorname{curl}[(\boldsymbol{q}\cdot\nabla)\boldsymbol{q}]=\nabla^2\zeta+Ta^{1/2}w_z \tag{9.55}$$

$$\left[1+\frac{FL}{n+L}\right]n\nabla^2 w-\hat{\boldsymbol{k}}\cdot\operatorname{curlcurl}[(\boldsymbol{q}\cdot\nabla)\boldsymbol{q}]=R^{1/2}\nabla_1^2\theta+\nabla^4 w-Ta^{1/2}\zeta_z \tag{9.56}$$

where $\zeta=\hat{\boldsymbol{k}}\cdot\operatorname{curl}\boldsymbol{q}$ is the vertical vorticity, and $\nabla_1^2\theta=\frac{\partial^2\theta}{\partial x^2}+\frac{\partial^2\theta}{\partial y^2}$. From eq. (9.9), the temperature equation becomes

$$Pr\left[\frac{\partial\theta_z}{\partial t}+(\boldsymbol{q}_z\cdot\nabla)\theta+(\boldsymbol{q}\cdot\nabla)\theta_z\right]=\nabla^2\theta_z+\left(1-\frac{1}{G}\right)R^{1/2}w_z \tag{9.57}$$

By conducting linear and nonlinear analyses as outlined in Sections 4 and 5, the following eigenvalue problems are formulated:

9.10.1 Eigenvalue problem for nonlinear analysis

$$2(D^2-a^2)Z+Ta^{1/2}\left(\frac{\lambda_1+\lambda_2}{\lambda_1{}^{1/2}\lambda_2{}^{1/2}}\right)DW=0 \tag{9.58}$$

$$2(D^2-a^2)^2W+R^{1/2}\left(\frac{G-1}{G}\frac{D^2}{\lambda_1{}^{1/2}}-\lambda_1{}^{1/2}a^2\right)\ominus-Ta^{1/2}\left(\frac{\lambda_1+\lambda_2}{\lambda_1{}^{1/2}\lambda_2{}^{1/2}}\right)DZ=0 \tag{9.59}$$

$$2(D^2-a^2)D^2\ominus+\left(\frac{G-1}{G}\right)\frac{R^{1/2}}{\lambda_1{}^{1/2}}D^2W-\lambda_1^{\frac{1}{2}}R^{\frac{1}{2}}a^2W=0 \tag{9.60}$$

where λ_1 and λ_2 are two positive coupling parameters.

9.10.2 Eigenvalue problem for linear analysis

$$(D^2-a^2)Z+Ta^{1/2}DW=0 \tag{9.61}$$

$$(D^2-a^2)^2W-R^{1/2}a^2\ominus-Ta^{1/2}DZ=0 \tag{9.62}$$

$$(D^2-a^2)D^2\ominus+\left(\frac{G-1}{G}\right)R^{1/2}D^2W=0 \tag{9.63}$$

9.10.3 Boundary conditions

The boundary conditions for the eigenvalue problems depend on the type of bounding surface:

$$W = 0, \Theta = 0, \ D^2\Theta = 0 \ \text{ for } z = 0, 1$$

and

$$DW = 0, \ Z = 0 \ \text{ for rigid surface} \tag{9.64}$$

and

$$D^2W = 0, \ DZ = 0 \ \text{ for free surface}$$

9.10.4 Discussion

Figure 9.4 demonstrates the relationship between Ta and the critical Rayleigh number (R_c) for three combinations of bounding surfaces under the stationary mode of convection. It is observed that as Ta increases, R_c also increases for all cases, indicating the stabilizing effect of rotation. Rotational forces suppress vertical fluid motion by reducing the velocity of the plasma within the boundary layer. This suppression limits the rate of heat transfer from the lower to the upper layer, delaying the onset of convection.

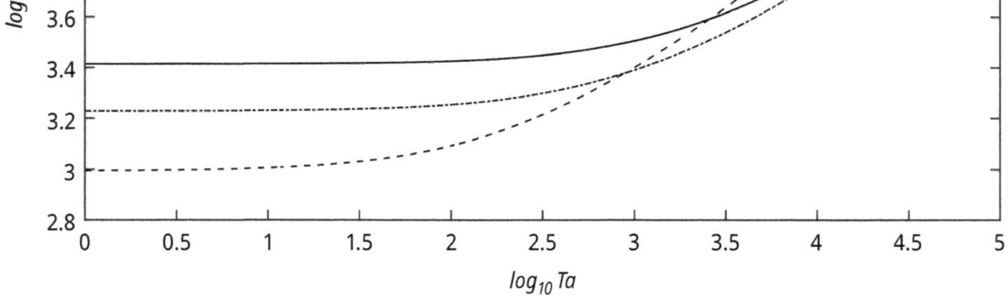

Figure 9.4: Variation of critical Rayleigh number (R_c) with Taylor number (Ta) for a fixed compressibility parameter ($G = 3$) across different boundary configurations, considering the stationary convection mode.

Additionally, rotation encourages the formation and persistence of structured convective cells, leading to stronger stability and necessitating higher R_c values for convection to initiate. The type of bounding surface plays a significant role in the convective behavior. For free-free surfaces at low *Ta*, susceptibility to deformation and temperature fluctuations disrupt the stability of convective cells, resulting in weaker convection and lower R_c. However, as *Ta* increases, these surfaces enhance the formation and interaction of convective cells, improving mixing efficiency and increasing R_c. The flexibility of free-free surfaces allows for dynamic adaptation to rotational effects, resulting in a significant rise in convective stability at higher *Ta* values. Rigid-rigid surfaces, in contrast, provide greater initial stability due to their resistance to deformation. At low *Ta*, this fosters the development of well-defined convective cells, leading to stronger convection and higher R_c. However, at high *Ta*, the rigidity restricts interaction between the larger convective cells that emerge, reducing the system's mixing efficiency. This limitation leads to lower heat transfer rates and a slower increase in R_c, eventually allowing free-free surfaces to surpass rigid-rigid surfaces in stability. Rigid-free surfaces exhibit intermediate behavior, combining the stability of a rigid boundary with the adaptability of a free boundary. They balance the effects of rigidity and flexibility, providing moderate stability at all *Ta* levels.

9.11 Effect of magnetic field

To examine the effect of a magnetic field on the convection problem described in Section 2, we consider a horizontal porous layer of PIP subjected to a uniform vertical magnetic field, **H**. In accordance with the quasi-static MHD approximation introduced by Galdi and Straughan [31], the governing momentum equation is modified by incorporating the Lorentz force term, $\mathcal{L} = \mathbf{j} \times \mathbf{B_0}$, where **j** is the current density and $\mathbf{B_0} = (0,0, B_0)$ denotes the magnetic induction field ($\mathbf{B} = \mu_m \mathbf{H}$), aligned solely in the vertical direction [1]. This modification captures the resistive influence exerted by the magnetic field on plasma motion. Accordingly, the dimensionless form of the perturbed momentum equation, originally given by eq. (9.15), is altered to include this Lorentz force contribution and is written as follows:

$$\left[1 + \frac{FL}{n+L}\right]\frac{\partial \boldsymbol{q}}{\partial t} + (\boldsymbol{q} \cdot \nabla)\boldsymbol{q} = -\nabla p + \sqrt{R}\theta\hat{\boldsymbol{k}} + \nabla^2\boldsymbol{q} + M^2\left[\left(\boldsymbol{q} \times \hat{\boldsymbol{k}}\right) \times \hat{\boldsymbol{k}}\right] \tag{9.65}$$

where $M^2 = (B_0^2\, d^2\sigma_1)/(\rho_m \nu)$ is the square of the Hartmann number, quantifying the relative importance of magnetic forces to viscous forces. Here, σ_1 is the electrical conductivity. Using both nonlinear and linear analyses as outlined in Sections 9.4 and 9.5, we derived the corresponding eigenvalue problems:

9.11.1 Eigenvalue problem for nonlinear analysis

$$2(D^2 - a^2)^2 W - \sqrt{\frac{R}{\lambda}}\left(\lambda + \frac{G-1}{G}\right) a^2 \ominus - 2M^2 D^2 W = 0 \tag{9.66}$$

$$2(D^2 - a^2)\ominus + \sqrt{\frac{R}{\lambda}}\left(\lambda + \frac{G-1}{G}\right) W = 0 \tag{9.67}$$

9.11.2 Eigenvalue problem for linear analysis

$$2(D^2 - a^2)^2 W - a^2\theta - \sqrt{R} M^2 D^2 W = 0 \tag{9.68}$$

$$\sqrt{R}\left(1 - \frac{1}{G}\right) W + (D^2 - a^2)\ominus = 0 \tag{9.69}$$

9.11.3 Boundary conditions

The boundary conditions associated with these equations depend on the type of bounding surface:

$$W = 0, \quad \ominus = 0 \quad \text{for } z = 0, 1$$

and

$$DW = 0 \quad \text{for rigid surface} \tag{9.70}$$

and

$$D^2 W = 0 \quad \text{for free surface}$$

9.11.4 Discussion

Table 9.2 presents the variations in the critical wavenumber (a_c) and the critical Rayleigh number (R_c) with increasing values of the square of the Hartmann number (M^2) for three different combinations of bounding surfaces. These results were originally reported in the prior work of Chandel and Sunil [21], which investigated the influence of magnetic fields on thermal convection in partially ionized plasma, and are reproduced here to provide continuity and contextual relevance to the current analysis. The corresponding trends are visualized in Figure 9.5, showing that as M^2 increases, R_c also increases for all boundary configurations.

This trend emphasizes the stabilizing effect of the magnetic field, which delays the onset of thermal convection. The Lorentz force generated by the magnetic field

Table 9.2: Variations of critical wavenumber (a_c) and critical Rayleigh number (R_c) with the square of Hartman number (M^2) for three different boundary configurations.

M^2	Free-free		Rigid-free		Rigid-rigid	
	a_c	R_c	a_c	R_c	a_c	R_c
0	2.22144	986.267	2.66384	1686.93	3.10957	2576.97
1,000	5.68421	22810.5	5.81465	27357.2	5.88248	28710.5
2000	6.45316	41549.9	6.59793	49690.3	6.66358	51747.7
3,000	6.94451	59601.3	7.09979	71233.5	7.16667	73991.3
4,000	7.31299	77276.8	7.47657	92344.4	7.54515	95798.0
5,000	7.61070	94703.2	7.78115	113169	7.85144	117314

Data reproduced from Chandel and Sunil [21].

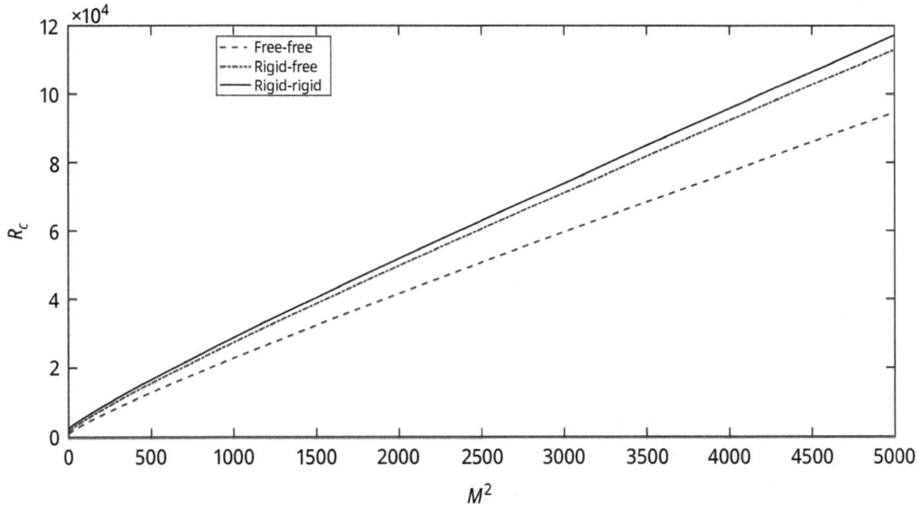

Figure 9.5: Variations of critical Rayleigh number (R_c) with the square of Hartmann number (M^2) for three distinct combinations of bounding surfaces. Results reproduced from Chandel and Sunil [21].

opposes plasma motion, thereby suppressing convective instabilities. As the plasma tries to initiate convective motion, the magnetic field serves as a resistive force, dissipating energy and impeding fluid movement. As a result, the threshold for the onset of convection, denoted by R_c, increases with the strength of the magnetic field. Furthermore, both Table 9.2 and Figure 9.5 demonstrate that the thermal stability of PIP is influenced by the boundary conditions. The plasma layer confined between rigid boundaries exhibits the highest R_c, indicating greater stability compared to configurations with free-free or rigid-free boundaries. This increased stability can be attributed to the enhanced restriction of motion imposed by the rigid boundaries, which further suppresses the development of convective cells. These findings emphasize the significant role of magnetic fields in modulating convection thresholds and highlight their potential for controlling convective instabilities in PIP systems across various boundary conditions.

9.12 Conclusions

This study delves into thermal convection in a PIP layer heated from below and bounded by three distinct surface configurations: rigid-rigid, rigid-free, and free-free. Employing both linear and nonlinear analyses, the investigation yields the following key findings:

- The study emphasizes the critical influence of the collisional frequency parameter in dissipating energy, which governs the thermal dynamics of the system.
- The thresholds for the onset of convection determined through linear and nonlinear analyses are identical across all boundary conditions, highlighting the global stability of the system.
- Compressibility plays a significant role in delaying the onset of convection by elevating the Rayleigh number. Conversely, an increase in the compressibility parameter leads to a reduction in the critical Rayleigh number, thus promoting the initiation of convection. It is important to note that the collisional frequency parameter does not exert any influence on stability limits.
- Rotation suppresses the velocity of PIP within the boundary layer, reducing heat transfer from the lower to the upper layer. This diminished heat transfer delays the onset of convection, thereby enhancing system stability.
- At low rotation rates, rigid-rigid bounding surfaces offer the highest stability, while free-free surfaces are the least stable. At higher rotation rates, this trend reverses, with free-free surfaces becoming the most stable and rigid-rigid surfaces the least.
- The presence of a magnetic field significantly improves the system's stability by effectively suppressing convective motions due to the opposing effects of the Lorentz force.

The findings of this study open avenues for several future research directions. One particularly exciting area is the investigation of surface tension's impact on thermal convection in PIP, especially across diverse bounding surface configurations. This can be effectively explored using both linear and nonlinear analyses. Additionally, the influence of advanced physical mechanisms, such as varying ionization levels, finite Larmor radius effects, and Hall currents, on stability and convection thresholds warrants detailed exploration. Another exciting direction is the application of machine learning techniques to study and control Rayleigh Bénard convection in PIP. Such approaches could provide advanced strategies for optimization and control in thermal management systems. These future studies will not only enhance the understanding of thermal convection in PIP but also broaden their real-world applications in various scientific and industrial domains.

9.13 Applications

The insights gained from this study not only advance the theoretical understanding of thermal convection in PIP but also have practical implications for astrophysical, geophysical, and industrial systems. These findings lay a strong foundation for future investigations, enabling broader applications in energy, environment, and technology domains where PIP and thermal convection play a critical role. Below are some key applications:

- **Stellar atmospheres**: The findings enhance our understanding of thermal convection in stellar atmospheres, such as those of the sun, where PIP dynamics, rotation, and magnetic fields govern energy transport and solar activity.
- **Planetary interiors**: Insights into the effects of rotation and compressibility apply to the study of convection in the molten cores of planets, which influence magnetic field generation through the geodynamic process.
- **Nuclear fusion reactors**: The stabilizing effects of magnetic fields and rotation are directly relevant to plasma confinement in tokamaks and stellarators, aiding in the design of more stable and efficient fusion reactors.
- **Space weather prediction**: Modeling thermal convection in PIP layers is vital for predicting space weather phenomena, such as solar flares and coronal mass ejections, which can impact satellite operations and communication systems.
- **Atmospheric modeling**: The role of rotation and boundary conditions in convection helps improve climate models, particularly for regions where PIP, such as the ionosphere, interact with atmospheric dynamics.
- **Plasma processing**: Understanding convection in PIP can optimize industrial processes such as plasma etching, coating, and semiconductor fabrication, which rely on controlled plasma behavior.
- **Energy storage**: The knowledge of convection thresholds and stability mechanisms can inform the design of plasma-based thermal storage systems, which are critical for renewable energy technologies.

References

[1] Chandrasekhar S. (1981). Hydrodynamic and Hydromagnetic Stability, New York: Dover.
[2] Drazin P. G., and Reid W. H. (2004). Hydrodynamic Stability, Cambridge: Cambridge University Press.
[3] Joseph D. D. (1976). Stability of Fluid Motions I, Heidelberg: Springer Berlin.

[4] Straughan B. (2004). The Energy Method, Stability, and Nonlinear Convection, New York: Springer-Verlag, doi: 10.1007/978-0-387-21740-6.
[5] Ogbonna N., and Bhatia P. K. (1984). Effect of magnetic resistivity on the stability of a partially ionized compressible Hall plasma. Astrophysics and Space Science, 103, 9–19. doi: 10.1007/BF00650038.
[6] Spiegel E. A., and Veronis G. (1960). On the Boussinesq approximation for a compressible fluid. The Astrophysical Journal, 131, 442–447. doi: 10.1086/146849.
[7] Hans H. K. (1968). Larmor radius and collisional effects on the combined Taylor and Kelvin instabilities in a composite medium. Nuclear Fusion, 8(2), 89–92. doi: 10.1088/0029-5515/8/2/002.
[8] Sharma R. (1976). Thermal hydromagnetic instability of a partially ionized medium. Physica B+C, 81, 199–204. doi: 10.1016/0378-4363(76)90259-X.
[9] Zaqarashvili T. V., Khodachenko M. L., and Rucker H. O. (2011). Magnetohydrodynamic waves in solar partially ionized plasmas: Two-fluid approach. Astronomy & Astrophysics, 529, A82. doi: 10.1051/0004-6361/201016326.
[10] Soler R., and Ballester J. L. (2022). Theory of fluid instabilities in partially ionized plasmas: An overview. Frontiers in Astronomy and Space Sciences, 9, 789083. doi: 10.3389/fs pas.2022.789083.
[11] Sharma R. C., and Sunil (1992). Rayleigh-Taylor instability of a partially ionized plasma in a porous medium in presence of a variable magnetic field. Z Naturforsch A, 47(12), 1227–1231. doi: 10.1515/zna-1992-1208.
[12] Kaothekar S. (2018). Thermal instability of partially ionized viscous plasma with Hall effect FLR corrections flowing through porous medium. Journal of Porous Media, 21, 679–699. doi: 10.1615/JPorMedia.2018017559.
[13] Kaothekar S., Mishra S., and Phadke S. (2023). Transverse thermal instability of partially ionized radiative plasma with impact of neutral collisions and finite electron inertia effect in ISM. Materials Today: Proceedings, 89, 24–33. doi: 10.1016/j.matpr.2023.05.596.
[14] Patidar A. K., Joshi H., Patidar S., et al. (2023). The effect of magnetized quantum plasma on Jeans instability. Z Naturforsch A, 78(9), 833–840. doi: 10.1515/zna-2023-0084.
[15] Chandel V., and Sunil (2025). Stability analysis of partially ionized plasma in a porous medium with local thermal non-equilibrium effects. Journal of Plasma Physics, 91(1), E1. doi: 10.1017/S0022377824001272.
[16] Chandel V., and Sunil (2024). Effects of local thermal non-equilibrium and magnetic fields on the stability of rotating partially ionized plasma heated from below. IEEE Trans Plasma Sci, 52(11), 5315–5326. doi: 10.1109/TPS.2024.3505538.
[17] Chandel V., Sunil, and Devi R. (2024). Stability analysis of thermal convection in partially-ionized plasma saturating a porous medium. Journal of Applied Nonlinear Dynamics, 14(4), 835–846. doi: 10.5890/JAND.2025.12.006.
[18] Chandel V., Sunil, and Devi R. (2025). Stability and convection in compressible partially ionized plasma layers: Nonlinear and linear analysis. Z Naturforsch A, 80(4), 323–333. doi: 10.1515/zna-2024-0207.
[19] Galstyan T. V., Koshkina D. A., Klimachkov D. A., et al. (2024). A theory of large-scale f lows of rotating partially ionized space and astrophysical plasma in the approximation of Hall magnetohydrodynamics. Plasma Physics Reports, 50(9), 1165–1179. doi: 10.1134/S1063780X24601159.
[20] Chandel V., Sunil, and Sharma P. (2024). Study of global stability of rotating partially-ionized plasma saturating a porous medium. Special Topics & Reviews in Porous Media, 15(6), 27–46. doi: 10.1615/SpecialTopicsRevPorousMedia.2024050815.
[21] Chandel V., and Sunil (2024). Influence of magnetic fields and bounding surface configurations on thermal convection in partially ionised plasmas: Nonlinear and linear stability analyses. Pramana Journal of Physics, 98(3), 107. doi: 10.1007/s12043-024-02805-3.

[22] Chandel V., and Sunil (2024). Nonlinear and linear analysis for thermal convection in partially-ionized plasma saturating a porous medium in the presence of magnetic field. The European Physical Journal Plus, 139(9), 856. doi: 10.1140/epjp/s13360 024-05663-5.

[23] Sharma R. C. (1972). Finite Larmor radius and Hall effects on thermal instability of a rotating plasma. Physics of Fluids, 15, 1822–1826. doi: 10.1063/1.1693781.

[24] Sharma R. C., and Sharma K. C. (1978). Thermal instability of a partially ionized plasma. Australian Journal of Physics, 31(2), 181–188. doi: 10.1071/PH780181.

[25] Leake J. E., Lukin V. S., Linton M. G., et al. (2012). Multi-fluid simulations of chromo spheric magnetic reconnection in a weakly ionized reacting plasma. The Astrophysical Journal, 760, 109. doi: 10.1088/0004-637X/760/2/109.

[26] Sharma R. C., and Sunil (1996). Thermal instability of a compressible finite-Larmor radius Hall plasma in a porous medium. Journal of Plasma Physics, 55, 35–45. doi: 10.1017/S002237780001864X.

[27] El-Sayed M. F., and Mohamed R. A. (2011). Gravitational instability of rotating viscoelastic partially ionized plasma in the presence of an oblique magnetic field and Hall current. International Scholarly Research Notices, 2011, 597172. doi: 10.5402/2011/597172.

[28] Sharma R. C., and Sunil. (1995). Thermal instability of a compressible finite Larmor radius, Hall plasma in porous medium. Physics of Plasmas, 2, 1886–1892. doi: 10.1063/1.871275.

[29] Yadav D., Bhargava R., and Agrawal G. S. (2013). Thermal instability in a nanofluid layer with a vertical magnetic field. Journal of Engineering Mathematics, 80, 147–164. doi: 10.1007/s10665-012-9598-1.

[30] Sharma R. C., and Misra J. N. (1986). Thermal instability of a compressible and partially ionized plasma. Z Naturforsch A, 41, 729–732. doi: 10.1515/zna-1986-0506.

[31] Galdi G. P., and Straughan B. (1985). Exchange of stabilities, symmetry, and nonlinear stability. Archive for Rational Mechanics and Analysis, 89(3), 211–228. doi: 10.1007/BF00276872.

Ashok K. Barik*, Prafulla K. Swain, and Mohamed M. Awad

Chapter 10
Thermo-fluid behavior of electroosmotic flow in a hydrophobic microchannel under Joule heating and external fields

Abstract: In this chapter, electromagnetohydrodynamic characteristics of a thermally fully developed flow of an electrolyte solution through micro-parallel plates under the influence of the external magnetic, electric, and radiation fields have been numerically investigated. The finite difference method has been implemented in an in-house computer program to solve electric potential, momentum, and energy equations with velocity slip and temperature jump conditions at the wall. The effects of dimensionless pressure gradient, Joule heating, viscous heating, and other parameters such as radiation, slip, and temperature-jump on the normalized velocity field, temperature field, flow rate, and Nusselt number have been critically analyzed. It has been observed that the velocity of the flow increases with the slip on the wall. Also, the flow rate is observed to increase with the slip parameter and decreases with the Hartmann number. For electromagnetohydrodynamic flow, the Nusselt number is found to increase rapidly at the low value of the electric double layer (EDL) parameter (κ), which slows down at the high value of κ. We also observe that Nusselt number increases with the Hartmann number (Ha), EDL parameter (κ), and radiation parameter (Nr). We believe that the present study may be helpful to design microfluidic devices such as micro-heat pipes, micro-heat exchangers, and microelectronics. Moreover, this study may provide valuable insights into magnetic and radioactive therapeutics such as muscle pain relief and tumor treatment.

Keywords: interfacial slip, microchannel, Nusselt number, magnetohydrodynamics, thermal radiation

***Corresponding author: Ashok K. Barik,** Department of Mechanical Engineering, National Institute of Technology, Jamshedpur 831014, India, e-mails: ashokbarik.mech@gmail.com
ashokbarik.me@nitjsr.ac.in
Prafulla K. Swain, Department of Mechanical Engineering, GITAM Institute of Technology, Visakhapatnam 530045, Andhra Pradesh, India
Mohamed M. Awad, Mechanical Power Engineering Department, Faculty of Engineering, Mansoura University, Mansoura 35516, Egypt

https://doi.org/10.1515/9783111661674-010

10.1 Introduction

In recent years, there have been tremendous technological improvements in micro–fabrication processes, which lead to the development of various microfluidic devices such as micromixtures and micropumps [1–3], heat exchangers [4], heat sinks [5], medical diagnostic devices at lab-on-chip level [6], and intravenous drug delivery systems [7–10], to name a few. Microscale transport processes have been applied in various engineering fields such as in micropower generation and conversion [11–15], microelectromechanical systems (MEMS), chemical processes, biochemical processes, polymeric and colloidal science, etc. All these applications require a clear understanding of the transport process at the microscale level, revealing the dynamics of the microfluidic flow phenomena like electroosmosis. Generally, a surface acquires electric charge, on its contact with the electrolyte solution/aqueous medium. The polarity and the strength of the surface charge primarily depend on the pH of the electrolyte solution and the condition of the surface charge group (acidic or basic). The charged surface usually influences the distribution of the free ions in the electrolyte solution in a way that the oppositely charged ions, called the counter ions, are attracted to the charged surface, while the same charged ions, called the coions, are repelled by the surface, thereby developing an electric double layer (EDL) [16]. The charges in the EDL will move by the application of the external electric field and drag the viscous fluid along with it to set the motion of the bulk fluid. Nevertheless, electroosmosis is a typical surface phenomenon under the influence of the external electric field. So, the liquid slip [17] and other parameters, due to the interplay between electric and magnetic fields, would be crucial to control the transport process in the microfluidic devices. For the last two decades, a large number of studies have been carried out on the electro kinetic flows in micro-parallel plates. Research works on the electro kinetic flow of Newtonian electrolyte in a narrow channel are abundant [18–21]. The effect of Joule heating on the fluid flow and energy transport characteristics of electroosmotic flow (EOF) through a parallel channel has been studied numerically by Yang et al. [22]. They reported that both temperature and Nusselt number of the flow depend on the velocity profile as well as the Joule heating for a microchannel, subjected to a constant heat flux condition. Mukhopadhyay et al. [23] carried out an analytical solution for a fully developed pressure-driven EOF in a microchannel, subjected to both asymmetric zeta potential and wall heat flux, to show that the thermo-fluid characteristics were influenced by asymmetries in the wall boundary condition, which could be beneficial for controlling the flow.

Recently, the filling dynamics of a soft microchannel with grafted polyelectrolyte layer on its inner wall and carrying a Newtonian electrolyte through it was studied by Gorthi et al. [24]. They observed an oscillatory filling length that depends on the intriguing interplay between the surface tension force and electroviscous force that controls the filling rate. Heat transfer and entropy generation of the electrokinetic flow of non-Newtonian fluid through a microchannel have been investigated by Shamshiri et al. [25]. They studied the effects of applied pressure gradient, Joule heating, and viscous heating on the fluid flow and heat transfer characteristics. The mixing efficiency of an electroosmotic micro-mixer has been numerically investigated by Dallakehnejad et al. [26] to demonstrate that an increased mixing efficiency due to a high transverse magnetic field applied normal to the main flow that could generate the vortices in the flow field. The effects of heterogeneous surface potential and slip boundary condition on the heat transfer characteristics of electroosmotic nanofluid flow in a microchannel have been studied numerically by Sheikhizad and Kalteh [27]. Zade et al. [28] analytically investigated a thermally fully developed EOF in microchannels, demonstrating the increase in the Nusselt number with the wall heat flux and the decrease with the channel height for a constant pressure gradient. Also, it has been reported by them that the Nusselt number may become negative when the magnitude of the bulk temperature exceeds the wall temperature due to high Joule heating.

The electrokinetic flow of a dilute suspension of spherical particles over a hydrophobic surface has been comprehensively reviewed by Ohshima [29]. They presented the general expressions for the electrohydrodynamic phenomena, such as electrophoretic mobility, sedimentation velocity, potential, and electrical conductivity. They also reported that the sedimentation potential and electrophoretic mobility increase with the slip length. Yavari et al. [30] studied the EOF in a circular pipe under the influence of Joule heating and viscous dissipation. They considered the temperature-dependent electrical resistivity and dynamic viscosity to demonstrate that the Nusselt number of the temperature-dependent fluid properties is lesser than that of the constant fluid properties. They explained that in the case of the temperature-dependent fluid property, the electrical resistivity reduces with temperature, leading to a higher local as well as bulk fluid temperatures, which in turn, decreases the Nusselt number. The hydrodynamics of slipping off the EOF of the Powell-Eyering fluid through a circular micro-tube has been analytically investigated using the homotopy perturbation method (HAM) by Dutta et al. [31] to demonstrate a higher flow rate for the Powell-Eyering fluid as compared to the Newtonian fluid.

Chen et al. [32] carried out a numerical investigation on the thermo-fluid characteristics of a thermally fully developed electrokinetic flow in a microtube, with interfacial velocity slip, using finite difference method (FDM). They reported that velocity and temperature fields are strongly influenced by the slip, the wall potential, and other parameters. Although their study considers the external electric field and magnetic fields, the effects of the temperature jump and radiation on the electromagnetohydrodynamics of the flow were not included. The electroosmotic flow in a microchannel slit, subjected to constant heat flux and under the influence of varying surface concentration and pH, has been studied by Mo and Hu [33] to show that the flow rate increases with ion concentration and slip, whereas it decreases with pH. The magnetohydrodynamics and electrokinetics transport characteristics of a slip-mediated thermally developing flow in a narrow channel have been studied analytically by Sarkar et al. [34], taking the effects of the external electric and magnetic fields on velocity profile, temperature profile, and the Nusselt number. The effect of slip on the velocity field of a microchannel has been studied by Jamaati et al. [35], demonstrating that the slip could increase the velocity and the induced voltage, ascertaining the benefits of hydrophobic surfaces in microfluidic applications, such as power generations/batteries.

External thermal radiation finds application in several medical and therapeutic procedures, such as treatment of myalgia (muscle pain), muscle spasm, chronic pain (fibromyalgia), shortening of muscles (contracture), tumor, and cancer [36] by the laser source, to name a few. In this context, several experimental [37–39] and numerical [40–42] investigations have been perused by various investigators to study the velocity and temperature fields of blood flow using thermal radiation. Later, the pressure-driven electrokinetic flow of blood has been analytically studied by Shina and Shit [43] under constant heat flux and external thermal radiation imposed on a micro-capillary tube. The effects of different nondimensional parameters, such as the Joule heating parameter, radiation parameter, Hartmann number, velocity slip parameter, and electroosmotic parameter, on the velocity and temperature fields have been investigated by them. Recently, the EOF of non-Newtonian fluid has been numerically studied by Nayak et al. [44], considering the effects of the external magnetic field, electric field, viscous dissipation, and external radiation.

From the above literature survey, it could be said that there exists abundant research works on the electromagnetohydrodynamic characteristics of a plane microchannel either under constant wall temperature/constant wall heat flux, with/without wall slip. A little work has been devoted toward the study of a microchannel flow, considering the effects of the external magnetic field, electric field, and radiation field along with the velocity slip and temperature jump conditions at the wall. Furthermore, external radiation is very important from the medical/biological application of view. Similarly, the slip at the wall is also equally important for the flows in micro/nanoscales. Hence, both these effects may not be eliminated completely in microscale flows, in particular, electrokinetic flows. Therefore, in this work, we aim at studying the combined effects of the external magnetic, electric, and thermal radiation fields, along with applied pressure gradient in a microchannel with wall-slip and temperature-jump conditions. The interplay among the external pressure gradient, Joule heating, and viscous heating on the overall velocity and temperature fields of a Newtonian flow in the parallel microchannel has been numerically studied to unravel their dependencies on different parameters of practical significance. We also report the parametric dependency of the volume flow rate and the Nusselt number of both the EOF and electromagnetohydrodynamic flow (EMHF). Furthermore, we believe that the present results may help in better control and designing of microfluidic devices for various engineering and bio-engineering applications. The results of the present study may directly be used for magnetic therapeutic and radiation therapeutic processes for the treatment of different diseases.

10.2 Physical problem and mathematical formulation

The physical configuration of a hydrophobic microchannel consisting of two parallel plates separated by a distance of $2H$ and a length of L is shown in Figure 10.1.

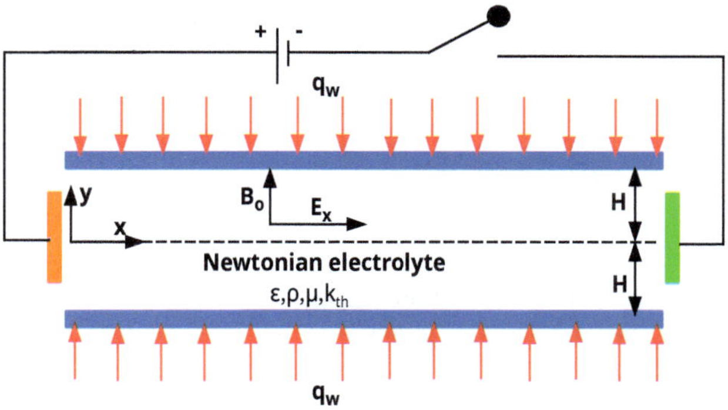

Figure 10.1: Diagram showing the application of external electric and magnetic fields to a microparallel plate under constant heat flux. The plates contain Newtonian electrolyte with length (L) and separation of $2H$.

The EMHF occurs in the microchannel due to the combined action of the external pressure gradient and an external electric field along the direction of motion of the fluid. A magnetic field of constant strength is applied normal to the flow direction. The walls are subjected to a constant wall heat flux, with a sudden jump in the temperature condition. External thermal radiation is also taken into consideration in the present flow situation. The flow is assumed thermal and hydrodynamically fully developed. The other assumptions relevant to the present situation will be discussed later on.

10.2.1 Equation for electric double layer (EDL) potential

When an electrolyte is put inside a microchannel, with its surfaces charged by an external electric field, its free ions are either attracted to or repelled from the charged surface based on the polarity, so that the charges will separate and redistribute to give rise to a phenomenon called the EDL in the close proximity of the charged surface [45, 46]. As we apply an external electric field to the charged surface, the free charges/ions move in the ED layer so as to drag the liquid in the buffer layer due to a cohesive hydrogen bond, to set up a flow called the EOF. In this work, the thermo-fluid characteristics of EOF in a narrow microchannel have been investigated under the influence of the radiation, magnetic as well as electric fields. The FDM has been adopted to solve the governing equations for EDL potential, velocity, and temperature fields. A steady and incompressible fluid of a hydrodynamically and thermally fully developed flow of symmetric electrolyte solution having constant thermophysical properties has been considered. The Debye-Hückel approximation is assumed to be valid in the EDL layer because of the small zeta potential. The charged ions are considered as point charges with constant permittivity of the electrolyte. We also assume that the electrokinetic potential (ψ) is related to the net charged density through the Poisson equation, as given below:

$$\frac{d^2\psi}{dy^2} = \frac{-\rho_c}{\varepsilon} \tag{10.1}$$

In eq. (10.1), ρ_c is the charge density of free ions and ε is the permittivity of the electrolyte solution. For symmetric electrolyte of the valence $z_v(z_v = z_{v+} = z_{v-})$, the charge density is further simplified using the Boltzmann distribution for ionic number concentration in the EDL [47], as given below:

$$\rho_c = 2n_0 e z_v \sinh\left(\frac{e z_v \psi}{k_B T_{ab}}\right) \tag{10.2}$$

where the ion density of the solution, valence of the ion, proton charge, Boltzmann constant, the absolute temperature and the electrokinetic potential induced in the EDL are denoted as n_0, z_v,e, k_B,T_{ab} and ψ, respectively. For a fully developed flow, the electric potential could be the function of only the y-direction, and the effect of EDL overlap may be neglected [48]. For small values of zeta potentials at the surface (i.e., $\zeta \leq 25$ mV), the electric potential is usually small as compared to the thermal potential (i.e., $|e z_v \psi| \leq k_B T_{ab}$). Thus, the right-hand side of eq. (10.2) can be approximated to $e z_v \psi / k_B T_{ab}$ using the well-established Debye-Hückel linearization. The use of the small zeta potential is practically relevant for the fabrication of polydimethysiloxane and polymethyl methacrylate [49], which are quite commonly used in lab-on-chip applications. Now, eq. (10.1) with the help of eq. (10.2) and Debye-Hückel linearization can be written as follows:

$$\frac{d^2\psi}{dy^2} = K^2\psi \tag{10.3}$$

where K is called the Debye length parameter $(=1/\lambda_D)$ and λ_D is called the Debye length ($\lambda_D = \frac{2e^2 {z_v}^2 n_0}{\varepsilon k_B T_{ab}}$). The boundary conditions for eq. (10.3) are given as:

$$\text{At } y = 0, \frac{d\psi}{dy} = 0, \text{ and at } y = H, \ \psi = \xi \tag{10.4}$$

Equations (10.3) and (10.4) are converted to nondimensional form using $y^* = y/H$ and $\psi^* = \psi/\xi$, respectively. The nondimensional EDL potential equation and its boundary conditions are invoked as follows:

$$\frac{d^2\psi^*}{dy^{*2}} = \kappa^2\psi^*, \text{ where } \kappa = KH \tag{10.5}$$

$$\text{At } y^* = 0, \frac{d\psi^*}{dy^*} = 0, \text{ and at } y^* = 1,\ \psi^* = 1 \tag{10.6}$$

10.2.2 Equation for momentum transport

The transport equations for charged ions in microfluidic confinement subjected to thermal radiation, external electric, and magnetic fields constitute the continuity and Navier-Stokes equations of the following forms:

$$\vec{\nabla} \cdot \vec{V} = 0 \tag{10.7}$$

$$\rho\left(\vec{V} \cdot \vec{\nabla}\right)\vec{V} = -\vec{\nabla}P + \mu\nabla^2\vec{V} + \vec{F} \tag{10.8}$$

where ρ is the density of the fluid, $\vec{V} = u\bar{i} + v\bar{j} + w\bar{k}$ is the velocity vector, $\vec{\nabla}P$ is the applied pressure gradient, μ is the dynamic viscosity of the fluid, and $\vec{F}$ is the body force due to the externally imposed electric and magnetic fields, which are defined [45, 50] as:

$$\vec{F} = \rho_c\vec{E} + \vec{J} \times \vec{B} \tag{10.9}$$

where $\vec{E} = E_x\bar{i}$ is the externally applied electric field, and the current density, $\vec{J} = \sigma_e(\vec{E} + \vec{V} \times \vec{B})$, where σ_e is the electrical resistivity of the electrolyte. Since the external magnetic field is applied along the y-direction of the microchannel, so $\vec{B} = B_y\bar{j} = B_0$. Since the microchannel that constitutes two parallel plates are separated by a distance of $2H$ and the channel length (L) is sufficiently long along the x-direction, it can be safely assumed that $2H \ll L$. Moreover, since the conduit is infinitely long along the z-direction, the gradients of the physical quantities along the z-direction can safely be neglected. Also, we assume that $w \approx 0$, without losing the generality of the present problem. Under all these aforementioned assumptions and arguments, the scalar form of the momentum transport equation of the present EMHF is given as follows:

$$0 \approx -\frac{dP}{dx} + \mu\frac{d^2u}{dy^2} + \rho_c E_x - \sigma_e B_0^2 u \tag{10.10}$$

The last two terms of eq. (10.10) imply the body forces (per unit volume) due to the external electric as well as the magnetic fields, whereas the first two terms are the pressure force and viscous force per unit volume, respectively. The velocity slip conditions have been imposed at both the top and bottom walls of the channel, while the symmetry condition has been used at the centerline of the conduit.

The boundary conditions are given as: $u|_{y=H} = -b\frac{du}{dy}|_{y=H}$ and $u|_{y=-H} = b\frac{du}{dy}|_{y=-H}$. However, one can use any of these velocity boundary conditions and the symmetry condition (i.e., $\frac{du}{dy}|_{y=0} = 0$) to solve for the velocity profile. The momentum equation and its boundary conditions are made dimensionless using a nondimensional velocity, $u^* = u/u_{HS}$, where $u_{HS} = \frac{-\varepsilon\zeta E_X}{\mu} = \frac{-\varepsilon E_X k_B T_{ab}}{\mu z_v e}$ is known as Helmholtz-Smoluchowski velocity [47]. The nondimensional momentum equation and its boundary conditions are given as follows:

$$\frac{d^2u^*}{dy^{*2}} - Ha^2u^* + \Gamma + \kappa^2\psi^* = 0 \tag{10.11}$$

$$At\ y^* = 1, u^* = -\beta\frac{du^*}{dy^*} \text{ and at } y^* = -1, u^* = \beta\frac{du^*}{dy^*} \tag{10.11}$$

The velocity boundary condition at the centerline is $u^* = \frac{du^*}{dy^*}\Big|_{y^*=0} = 0$. One can use any one of the conditions of eq. (10.12), along with the symmetry condition, to solve for the velocity of a half channel, or eq. (10.12) can be used to solve for velocity in the entire channel. In the present study, we have used the entire channel to find the velocity. Thus, eqs. (10.11) and (10.12) are solved numerically using the FDM. In eq. (10.11), the Hartmann number is given as $Ha = B_0H\sqrt{\frac{\sigma_e}{\mu}}$ and the ratio of the pressure-driven velocity (u_{PD}) to the electroosmotic velocity (u_{HS}) is given as $\Gamma = u_{PD}/u_{HS} = \frac{-H^2}{\mu u_{HS}}\frac{dP}{dx}$. The slip parameter is given as $\beta = b/H$. The flow is said to be purely EOF when $\Gamma = 0$, and the flow is said to be EMHF when $\Gamma \neq 0$, provided there would be externally imposed electric and magnetic fields. Nevertheless, in our present study, the effects of the external electric and magnetic fields along with the external radiation have been considered.

10.2.3 Equation for thermal energy transport

The conservation of the energy equation in the presence of viscous dissipation, axial conduction, Joule heating, external thermal radiation, and the electromagnetic effects is written in dimensional form as follows:

$$\rho c_p u_m\frac{\partial T}{\partial x} = k_{th}\frac{\partial^2 T}{\partial y^2} - \frac{\partial q_r}{\partial y} + \mu\left(\frac{du}{dy}\right)^2 + \sigma_e E_x^2 + \sigma_e B_0^2 u^2 \tag{10.13}$$

where c_p and k_{th} are the specific heat and thermal conductivity of the medium. The mean velocity in dimensional and nondimensional forms are given as $u_m = \frac{1}{H}\int_0^H u dy$

and $u_m^* = \frac{u_m}{u_{HS}} = \int_0^1 u^* dy^*$, respectively. Rosseland approximation [43–44, 51–52] has been used to approximate the radiative heat flux (q_r) of the form

$$q_r = -\frac{4\sigma_{SB}}{3k_a}\frac{\partial T^4}{\partial y} \tag{10.14}$$

where σ_{SB} ($=5.67\times10^{-8}$W/m^2 K^4) is the Stefan Boltzmann constant and k_a ($=10^{-1}$m^{-1}) is the absorption coefficient [44], respectively. We assume that the temperature of the fluid in the microchannel is small so that the nonlinear term in eq. (10.14) (i.e., T^4) can be expressed as a linear combination of the temperature difference between the local (T) and wall temperatures via Taylor's series expansion around the wall temperature (T_w), as follows:

$$T^4 = T_w^4 + 4T_w^3(T - T_w) + 6T_w^2(T - T_w) + \cdots \tag{10.15}$$

The higher powers of $(T - T_w)$ can be neglected as the temperature in the channel is small. By doing this, T^4 can be further simplified to

$$T^4 = T_w^4 + 4T_w^3(T - T_w) \tag{10.16}$$

Now, taking the derivative of eq. (10.16), and its substitution in eq. (10.14) takes the form as

$$q_r = -\frac{16\sigma_{SB}T_w^3}{3k_a}\frac{\partial T}{\partial y} = -k_{th}N_r\frac{\partial T}{\partial y} \tag{10.17}$$

where the thermal radiation parameter $N_r = \frac{16\sigma_{SB}T_w^3}{3k_a k_{th}}$. Equation (10.17) is substituted into eq. (10.13) to get a new dimensional form of the energy transport equation, as follows:

$$\rho c_p u_m \frac{\partial T}{\partial x} = k_{th}(1 + N_r)\frac{\partial^2 T}{\partial y^2} - \frac{\partial q_r}{\partial y} + \mu\left(\frac{du}{dy}\right)^2 + \sigma_e E_x^2 + \sigma_e B_o^2 u^2 \tag{10.18}$$

The thermal boundary conditions are:

$$\text{At } y = \pm H,\ T = T_w + S_t\frac{\partial T}{\partial y} \text{ and } y = -H, T = T_w - S_t\frac{\partial T}{\partial y} \tag{10.19}$$

where S_t is the thermal jump factor at the wall.

$$\text{At the centerline of the channel: } \left.\frac{\partial T}{\partial y}\right|_{y=0} = 0 \tag{10.20}$$

For our analysis, we assume the flow in the microchannel is fully developed both thermally and hydrodynamically, so that $\frac{\partial}{\partial x}\left(\frac{T - T_w}{T_m - T_w}\right) = 0$ and $\frac{\partial T}{\partial x} = \frac{dT_m}{dx} = \frac{dT_w}{dx} = \text{const.}$ Therefore, to get the value of $\frac{\partial T}{\partial x} = \frac{dT_m}{dx}$, it is needed to carry out an energy balance over an elemental control volume of infinitesimal length dx [53].

$$\rho c_p u_m H dT_m = 2q_w dx + \sigma_e E_x^2 H dx + \sigma_e \int_0^H B_o^2 u^2 dx dy + q_r dx \tag{10.21}$$

One can get the nondimensional energy equation after the substitution of $\frac{dT_m}{dx}$ from eq. (10.21) and then nondimensionalizing the resulting energy equation, using $\theta = \frac{T - T_W}{q_w H / k_{th}}$. For the sake of the brevity, the nondimensional energy equation and the boundary conditions are only presented here:

$$\begin{aligned} &\frac{u*}{u_m^*}\left(2 + N_r + S + Ha^2 Br \beta_{21}\right) = (1 + N_r)\frac{\partial^2 \theta}{\partial y*^2} + S \\ &+ Br\left(\frac{du*}{dy*}\right)^2 + HaBru*^2 \quad N_r \frac{u*}{u_m^*} N_r \frac{\partial \theta}{\partial y*}\bigg|_{y*=1} \end{aligned} \tag{10.22}$$

Where $S = \frac{\sigma_e E_x^2 H}{q_w}$ is the ratio of the volumetric Joule heating to the applied heat flux [54, 55], the Brinkmann number $Br = \frac{\mu u_{HS}^2}{q_w H}$, and $\beta_{21} = \int_0^1 u*^2 dy*$, respectively. The corresponding nondimensional boundary conditions are given as follows:

$$\text{At } y* = 1,\ \theta = S_t \frac{\partial \theta}{\partial y*} \text{ and at } y* = -1,\ \theta = -S_t \frac{\partial \theta}{\partial y*} \tag{10.23}$$

where S_t is the dimensionless thermal jump parameter.

The thermal boundary condition in the centerline of the duct is

$$\frac{\partial \theta}{\partial y*}\bigg|_{y*=0} = 0. \tag{10.24}$$

10.3 Numerical solution procedure

The finite difference formulation has been adopted to solve for the electric potential (i.e., eqs. (10.3)–(10.6)), the velocity field (i.e., eqs. (10.11) and (10.12)), and the temperature field (i.e., eqs. (10.22)–(10.24)) using our in-house developed computer program in Engineering Equation Solver (EES) platform. The computational domain has been discretized by taking n_i ($n_i = 1, 2 ... 20$) number of grid points in the y-direction, such that each discrete interval size is $\Delta y = 1/n_i$. The governing transport equations, in particular, the electric potential (i.e., eq. (10.5)), momentum (i.e., eq. (10.11)), and energy equations (i.e., eq. (10.22)) are first discretized over the entire computational domain, except at the boundaries. For inner grids in the domain, we use the central difference scheme to discretize both the first- and second-order derivatives that appeared in the transport equations. Different numerical difference schemes [56, 68] have been employed to discretize the first-order derivatives that appeared in the boundary conditions. The discretized forms of eqs. (10.5), (10.11), and (10.22) are given as follows:

$$\frac{\psi_{i+1}^* - 2\psi_i^* + \psi_{i-1}^*}{\Delta y^{*2}} - \kappa^2\psi_i^* = 0 \tag{10.25}$$

$$\frac{u_{i+1}^* - 2u_i^* + u_{i-1}^*}{\Delta y^{*2}} - Ha^2u_i^* + \Gamma + \kappa^2\psi_i^* = 0 \tag{10.26}$$

$$\frac{u_i^*}{u_m^*}\left(2 + N_r + S + Ha^2Br\beta_{21}\right) = (1+N_r)\frac{\theta_{i+1}^* - 2\theta_i^* + \theta_{i-1}^*}{\Delta y^{*2}} + S + Br\left(\frac{u_{i+1}^* - u_{i-1}^*}{2\Delta y^*}\right)^2 + HaBru_i^{*2} \tag{10.27}$$

The mean velocity $u_m^* = \int_0^1 u^* dy^*$ is implemented in our computer program using the trapezoidal rule of integration [44] as

$$u_m^* = \sum_{i=1}^{n_i} \frac{u_i^* + u_{i+1}^*}{2}\Delta y^* \tag{10.28}$$

To calculate the value of $\beta_{21}(= \int_0^1 u^{*2} dy^*)$, the same trapezoidal integration rule has been adopted. We first solved the upper-half of the channel, and then the lower half of the channel to get the values of ψ_i^*, u_i^*, and θ_i^*, respectively. While doing so, we used the symmetry condition for both the halves of the channel. Next, we solved the discretized equations with the relevant boundary conditions in order to close the governing equations. The grids at the centerline and on the wall of the channel are called the boundary nodes. We used the first three nodes (the counting of the node being done from the centerline and is implanted through a counter-loop implemented in our computer code) to discretize the boundary condition at the centerline (i.e., $n_i = 1$), as given below:

$$\frac{4\psi_2^* - \psi_3^* - 3\psi_1^*}{2\Delta y^*} = 0 \tag{10.29}$$

Similarly, the velocity and the temperature conditions at the centerline are discretized as follows:

$$u_{n_i+1}^* = -\beta\left(\frac{u_{n_i-1}^* - 4u_{n_i}^* + 3u_{n_i+1}^*}{2\Delta y^*}\right) \text{ and } \theta_{n_i+1}^* = S_t\left(\frac{\theta_{n_i-1}^* - 4\theta_{n_i}^* + 3\theta_{n_i+1}^*}{2\Delta y^*}\right) \tag{10.30}$$

The discretized equations and their boundary conditions yield a set of coupled algebraic equations, which are solved using Newton's methods with appropriate guess value in EES. EES essentially converts the systems of algebraic equations into the Jacobian matrix, and then the blocking algorithm of Tarjan [57] is implemented to convert the entire problem into smaller blocks for ease with the solution procedure. The Jaco-

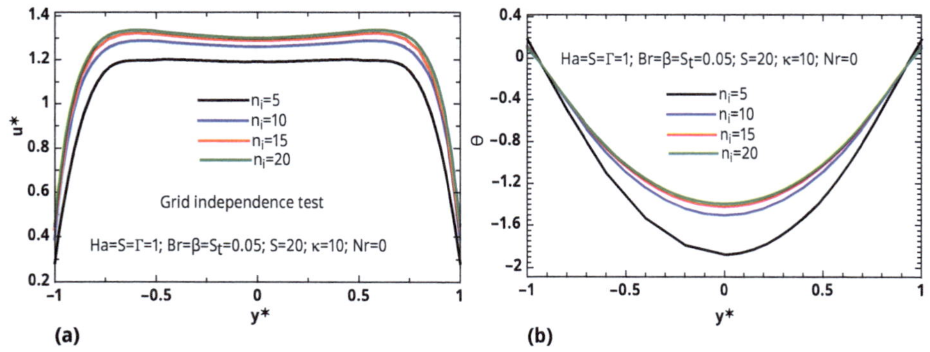

Figure 10.2: Left: Diagram shows the evaluation of the nondimensional velocity profile with the nondimensional height of the channel with increasing number of nodes in the domain of interest. The velocity profile shows insignificant change when twenty number nodes are taken. Right: Non-dimensional temperature variation with the number of nodes.

bian matrix is evaluated at the end of each iteration until the user-specified relative tolerance is satisfied. In this present study, we have requested relative tolerances of 10^{-6} for each of the dependent variables and the maximum number of iterations as 100. A grid independence study has been executed to test the influence of the number of nodes in the computation on the dependent variables, especially, on the velocity and temperature of the electrolyte. The grid independence study is depicted in Figure 10.2 (left) and 10.2 (right). It is quite apparent from these plots that both the temperature and velocity show negligible changes at 20 grids. In our further analysis, we have taken 20 grids (n_i) in the domain.

10.4 Validation of numerical code

The validation for the nondimensional zeta potential (ψ^*) has been carried out with the Ref. [45]. For microchannel flows, it is believed that the electric potential (ψ^*) is a function, normal to the flow direction (here, the y-direction). We substantiate the present methodology with analytical expression for ψ^*, as depicted in Figure 10.3 (top left). We construct 20 node points along the y-direction to get the numerical values of the electric potential. To validate the velocity profiles, we select Ref. [49], which deals with an EMHF in a narrow channel under the influence of the Joule heating and viscous dissipation.

Since the present problem investigates the effects of Joule heating, viscous dissipation, and external radiation with wall slips, we decided to validate the numerical methods with Ref. [49], in which the external magnetic and electric fields were considered. Figure 10.3 (top left) illustrates the validation for the velocity profile at low val-

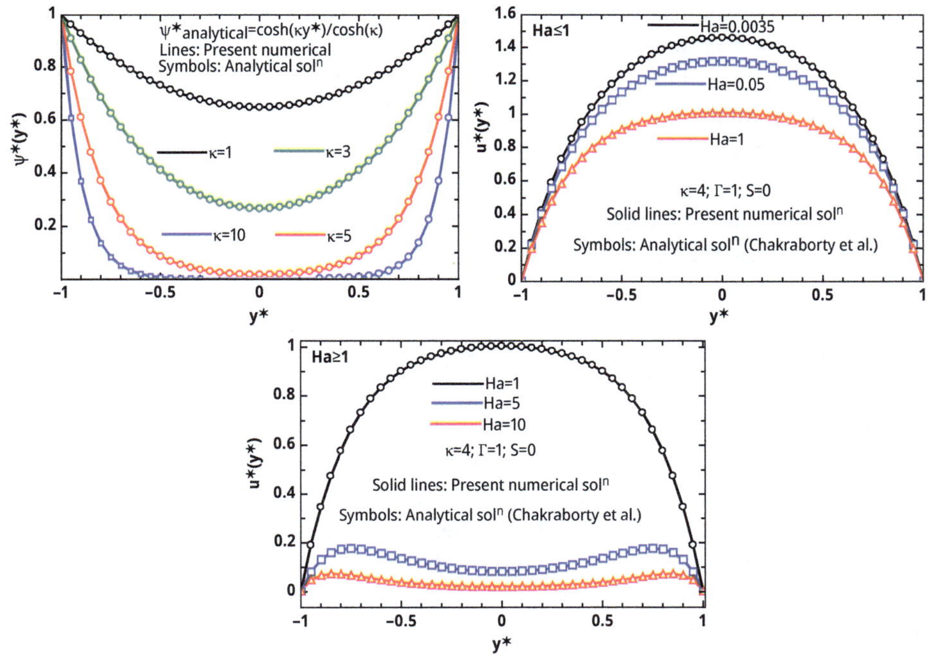

Figure 10.3: Top left: Evaluation of the dimensionless zeta potential along the channel height at various Debye-Hückel parameters [45]. Higher Debye-Hückel parameter shows an almost flat zeta potential having strong gradients adjacent to the wall. The top right diagram shows the variation of velocity profile (from the present study and Ref. [49]) with Hartmann number for $Ha \leq 1$ and the bottom-middle figure illustrates its variation when $Ha \geq 1$.

ues of Hartmann number ($Ha \leq 1$), whereas Figure 10.3 (bottom) shows its variation at relatively high values of Hartmann number ($Ha \geq 1$). It has been observed from Figure 10.3 that the velocity computed using the present FDM matches well with the analytical results of Ref. [49].

10.5 Results and discussion

10.5.1 Effect of different parameters on the velocity profile

First, the effects of different nondimensional parameters on the velocity profile and volume flow rate of the pure EOF and EMHF have been delineated. Then, critical discussion on the influences of these parameters along with additional parameters that arise from the nondimensional energy equations on the temperature distribution and Nusselt number have been discussed. The effect of the slip parameter (i.e., β and S_t)

on the velocity profile of purely EOF (i.e., $\Gamma \to 0$) has been depicted in Figure 10.4 (top left). Several other parameters (i.e., radiation parameter (*Nr*), Joule heating parameter [53, 55] ($S = \sigma_e E_x^2 H / q_w$= Joule heating/applied heat flux), Brinkmann number (*Br*), and temperature jump parameter (S_t)) do not explicitly appear in the nondimensional momentum equation, and hence, the velocity profiles are not much affected by these parameters. However, these parameters directly appear in the nondimensional form of the energy equation and significantly influence the temperature distribution in EMHF flows, which is the central focus of our study. It can be seen from Figure 10.4 (top left) that the fluid velocity increases with the higher values of slip parameter at the wall [58]. The increased flow velocity (due to a higher slip parameter) is attributed to a longer slip length, resulting in a higher slip velocity at the wall, as explained by Gaikwad et al. [59]. Similar consequences have been observed in fluid velocity as β increases from 0 (i.e., no-slip condition) to 0.1 (i.e., partial wall-slip condition). For the EMHF, the nondimensional pressure gradient ($\Gamma = u_{PD}/u_{HS}$) is usually taken as nonzero. Figure 10.4 (top right) represents the EMHF flow in a microchannel at different Hartmann numbers, with and without slips at the walls. All other parameters required for the simulation of the present EMHF may be read from Figure 10.4 (top right). At $Ha = 1$, the velocity gradually increases, from zero at the wall to attain the maximum value, little away from the walls, with the no-slip condition. The velocity becomes almost flat away from the wall, and this velocity profile covers most of the region of the channel. At the same Hartmann number and when the slip is considered, the velocity of the fluid, both on the wall and in the inner region of the channel, significantly increases. The velocity of the flow gradually reduces with the Hartmann number. In a typical EMHF, there exist flow-assisting and flow-opposing forces due to the external transverse electric and magnetic fields, respectively. In our case, the flow is not applied to a transverse electric field (i.e., $E_z = 0$), and hence, the flow-assisting force due to the transverse electric field does not exist. But the flow-opposing force (i.e., $\sim Ha^2 u^*$) due to the external magnetic field still exists, which decelerates the flow as the value of the Hartmann number is increased. Therefore, the velocity diminishes at higher value of *Ha*, as shown in Figure 10.4 (top right). For both slip and nonslip walls, the flat velocity profile is distorted and reduced more in the central region of the channel, than on the wall. The higher velocity near the wall is because of the inertia of the flow due to the slip imposed at the walls. Furthermore, the flow-opposing force is proportional to the velocity and the square power of the Hartmann number.

For a fully developed flow, the velocity in the central region is more as compared to at the wall. Thus, the retarding force is expected to be higher in the central region than that at the wall region. Thus, at a high Hartmann number, the fluid velocity is strongly retarded in the channel center than on the wall, resulting in a lower velocity at the center of the channel. The present results are consistent with the results reported by Chakraborty et al. [49], Chu and Jian [58], and Liu et al. [60]. Also, it has been observed that the difference of the velocities for the slip-wall case and the no-slip-wall case is reduced at higher values of the Hartmann number (i.e., Figure 10.4

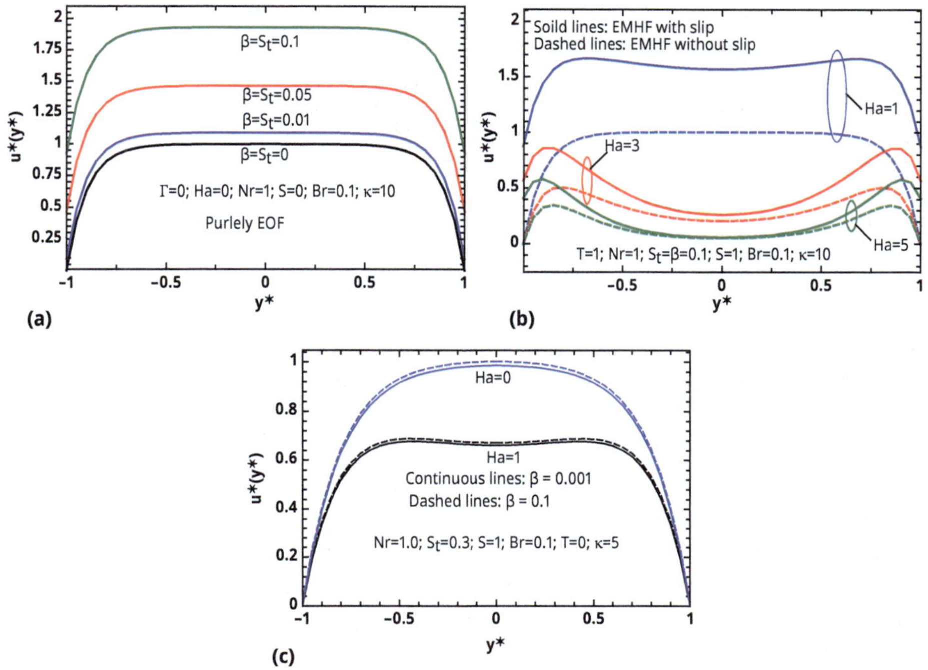

Figure 10.4: Diagram in the left top shows the variation in velocity profile with the parameter β, illustrating a higher velocity with this parameter. The right top plot shows the velocity variation with the Hartmann number when $Ha \geq 1$, whereas the bottom one depicts the velocity change with Ha for $Ha \leq 1$.

(top right)). For $Ha = 5$, we observed the centerline velocity of the non-slip- and slip-walled channels becomes almost the same at the channel centerline. For high electric potential (ψ^*), the EMHF velocity increases significantly with the wall-slip, as observed from Figure 10.4 (top right) and (bottom). The convective velocity increases with higher velocity slip [61, 62]. At a constant β (i.e., 0.1), the variation of the EMHF velocity with respect to κ is illustrated in Figure 10.5 (left). The fluid velocity increases with κ. At the low κ $(=1)$, we observe a parabolic velocity profile. However, at a high κ (i.e., high value of inverse Debye length), the velocity of EMHF looks like an M-shape because of the retarding force. The inverse of κ $(= H/\lambda_D)$ is the characteristic Debye length (λ_D), which physically signifies the thickness of the EDL, and depends on the concentration of ions. At low value of κ, the EDL covers almost the entire width of the channel. So the charged ions are distributed across the channel, resulting in a decreased charged density and the Helmholtz-Smoluchowski velocity (u_{HS}) as well. Also, the induced EDL potential (ψ^*) is decreased due to the low value of κ, which in turn decreases the flow velocity (u^*).

At low $\kappa(=1)$ value, the velocity profile takes the parabolic profile because the EDL is very thick and probably reaches the centerline of the channel. But at high value of κ, the EDL becomes very thin and is restricted to near the wall. Due to the

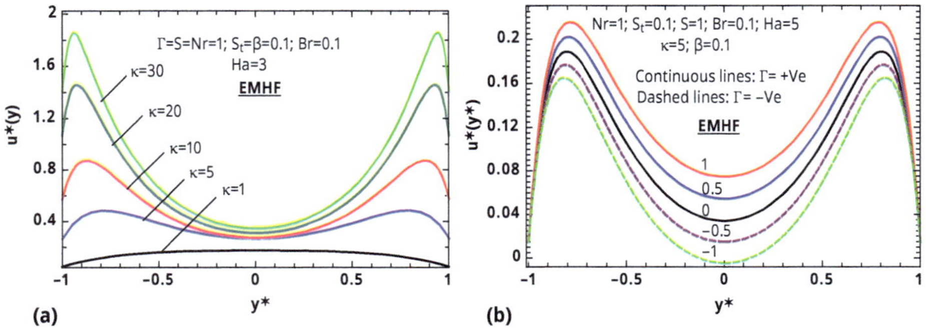

Figure 10.5: Variation in velocity profile with κ (left) and Γ (right). Magnitude of velocity increases with κ and Γ depicting a distinct peak near the wall.

high κ values, the charge density is enhanced in the thin EDL, leading to an increased flow velocity [59, 63, 64]. For practical microchannel applications, thin EDL layers are formed. Thus, for a practical application, one needs to take a high value of κ. The value of κ can be controlled by controlling the ion concentration of the electrolyte solution and the pH value of the solution. Another advantage of taking a high value of κ is the validity of the Debye-Hückel approximation, which greatly simplifies the EDL potential distribution equation. The dimensionless velocity distribution across the channel width is depicted in Figure 10.5 (right) for different values of nondimensional pressure gradient; $\frac{dP}{dx} = 1, 0$ and -1. Other parameters can be read from the same plot. For a positive Γ, the fluid velocity increases, whereas it decreases with a negative Γ. For a positive Γ, the axial pressure gradient becomes negative, i.e., $\frac{dP}{dx} < 0$, which is commonly known as the favorable pressure gradient in classical fluid dynamics. So the fluid velocity increases with positive Γ. However, for negative Γ, an adverse pressure gradient (i.e., $\frac{dP}{dx} > 0$) acts against the fluid motion, reducing its velocity. A comparative analysis of velocity profiles with and without slip for different cases such as: (i) $\Gamma = 1$ and $Ha = 0$, (ii) $\Gamma = Ha = 1$, and (iii) $\Gamma = Ha = 0$ is shown in Figure 10.6. In all three cases, the slip at the wall increases the fluid velocity. For the combined EOF (i.e., $\Gamma \neq 0$), a lower velocity is obtained when an external magnetic field is applied, as compared to a case when the magnetic field is withdrawn.

This is due to the fact $\beta = 0$ and 0.02 that fluid flow is resisted β by the Lorentz force (i.e., $\vec{J} \times \vec{B}$), which appears as a body force in the momentum equation and acts against the fluid motion. When the wall slip is withdrawn, the velocity of a purely EOF (i.e., $\Gamma = 0$ and shown in solid red line), without an external magnetic field, is found to be the same with the velocity of EMHF (shown in solid blue line), with the application of the magnetic field ($\Gamma = Ha = 1$). Furthermore, in the absence of an external magnetic field (i.e., $Ha = 0$), the EOF with flow-assisting (i.e., $\Gamma > 0$) nondimensional pressure gradient shows a higher velocity as compared to a pure (i.e., $\Gamma = 0$) EOF.

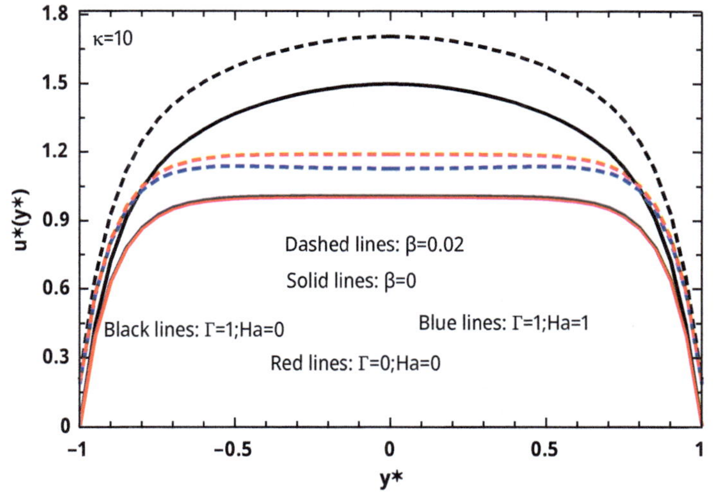

Figure 10.6: The plot shows the velocity variation across the channel height for $\beta = 0$ and 0.02. Higher value of β implies a higher velocity in the channel.

10.5.2 Effect of different parameters on the volume flow rate

Before presenting a discussion on the effects of different parameters on the volume flow rate of EMHF, the mathematical equation to compute the volume flow rate has been invoked as follows:

$$\dot{Q} = 2\int_0^H u_m dA$$

where

$$u_m = \frac{1}{H}\int_0^H u dy \tag{10.31}$$

Using previously defined dimensionless quantities, the dimensionless volume flow rate is given as $\dot{Q}^* = 2 \times u_m^*$, where $u_m^* = \int_0^1 u^* dy^*$. The computational procedure to calculate nondimensional mean velocity has been given in eq. (10.28). The effect of dimensionless slip parameter (β) on the volume flow rate of a purely EOF at different values of the inverse Debye layer thickness (κ) is depicted in Figure 10.7 (left).

At a particular κ, the volume flow rate increases linearly with the wall slip parameter. It is evident from the velocity profile (see Figure 10.4 (top left)) that a higher slip increases the fluid velocity, which in turn, increases the flow rate. Moreover, the

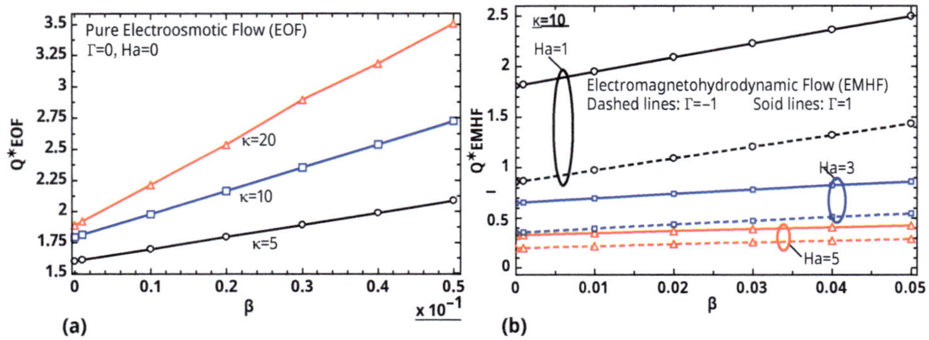

Figure 10.7: The variation in dimensionless volume flow rate for electroosmotic flow (Q^*_{EOF}) with β (Left) and volume flow rate of electromagnetohydrodynamic flow (Q^*_{EMHF}) (Right). For both flow cases, the volume flow rate increases with β. But the flow rate in the latter case is lower than in the former case.

volume flow rate increases with higher value of κ, which is attributed to the higher charge density in a very thin EDL layer near the charged wall, promoting the electroosmotic body force that acts on the fluid mass. In this case (i.e., EOF), no external transverse magnetic field has been imposed to the flow. Both the velocity and the flow rate have been increased due to increased κ value, which is altered either by varying the ion concentration or the channel size, as explained by Yang and Kwok [65]. Figure 10.7 (right) illustrates the volume flow rate of EMHF with respect to velocity slip parameter at different values of both the Hartmann number (Ha) and the dimensionless pressure gradient (Γ) at a constant EDL parameter ($\kappa = 10$). It has been observed that the volume flow rate of EMHF increases with β. However, the increase in the volume flow rate rather reduces with the externally imposed magnetic field, which is because of the opposing Lorentz force. For example, when $Ha = 5$, the volume flow rate increases marginally, as depicted in Figure 10.7 (right). Moreover, a higher volume flow rate is obtained for flow-assisting pressure gradient (i.e., $\Gamma = 1$) than that of the flow-opposing pressure gradient (i.e., $\Gamma = -1$).

10.5.3 Effect of different parameters on the temperature profile

Figure 10.8 shows the variation of the nondimensional temperature (θ) across the channel for different values of Debye parameters, in the range of $1 \le \kappa \le 50$. The slip and the temperature jump parameters are taken as 0.1 and 0.5, respectively. All other parameters are given in Figure 10.8. Since we considered the nonzero values for all other parameters, such as Γ, S, Ha, Br, and Nr, the temperature field is for the case of an EMHF with radiation effect. At a low value of κ (=1), we observed a higher temperature difference between the wall and the centerline of the electrolyte solution (i.e., $\theta_w - \theta_c$). This temperature difference is gradually decreased as the Debye parameter increases. For example, at the constant values of all other parameters, as given in Figure 10.8, the values of the temperature difference, i.e., $\theta_w - \theta_c$, are obtained as 0.9286 K, 0.8318 K, and 0.7698 K at $\kappa = 1$, 5, and 10, respectively. However, the mean temperature (θ_m) of fluid is improved with the higher values of κ. For example, the mean temperatures at $\kappa = 1$, 5, and 10 are found to be 0.06432 K, 0.1616 K, and 0.2221 K, respectively. The reduction in the temperature difference and the improvement in the mean temperature suggest higher convection of EMHF [62] at the high value of κ. The enhancement in the electroosmotic convection is attributed to the improvement in the flow velocity and charge density in a very thin EDL layer at the high value of κ. The increased fluid velocity might be beneficial to microfluidic devices, such as micro mixtures. It has also been observed from Figure 10.8 that the wall temperature gradually decreases with higher values of κ, owing to the fluid convection. Haddout and Lahjomri [66] reported similar observations in their study of gas flow through micro-pipe and micro-parallel plates with wall slip under the influences of axial conduction, pressure work, and viscous dissipation. The effects of Hartmann number on the temperature profile for two different cases (i.e., $Ha \leq 1$ and $Ha \geq 1$) are shown in Figure 10.9 (left) and Figure 10.9 (right). For $Ha \leq 1$, the temperature differences decrease with increasing value of Hartmann number. The same trend in the temperature profile has been observed for $Ha \geq 1$. For $Ha \leq 1$, the temperature profile seems to be the parabolic shape. The temperature on the wall is the maximum on the wall and the minimum at the center of the channel, indicating transfer of heat from the wall to the bulk fluid.

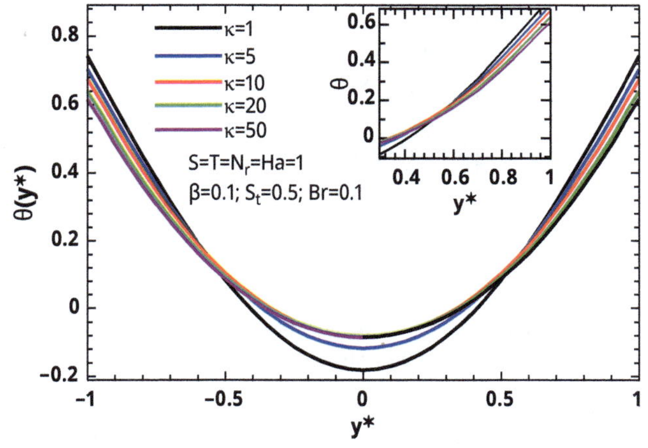

Figure 10.8: Nondimensional temperature (θ) variation across channel heights at different κ values. A high κ value shows a low temperature at the center of the channel.

The effect of the nondimensional pressure ratio (i.e., the ratio of the pressure and electroosmotic velocity scales) on the temperature profile has been shown in Figure 10.10 (left).

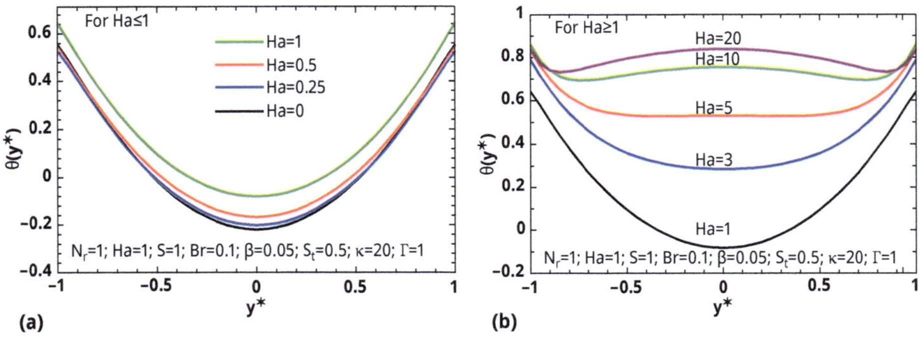

Figure 10.9: Variation in dimensionless temperature profile for low ($Ha \leq 1$) (left) and high ($Ha \geq 1$) (right) Hartmann numbers. High Hartmann number increases the centerline temperature of the channel.

For $\Gamma > 0$, a favorable pressure gradient improves the fluid convection, exhibiting an increase in the magnitude of the dimensionless temperature. However, for $\Gamma < 0$, the fluid advection is impended to lower the dimensionless temperature. A higher temperature difference between the wall and the fluid mass at the channel center has been noticed for $\Gamma < 0$.

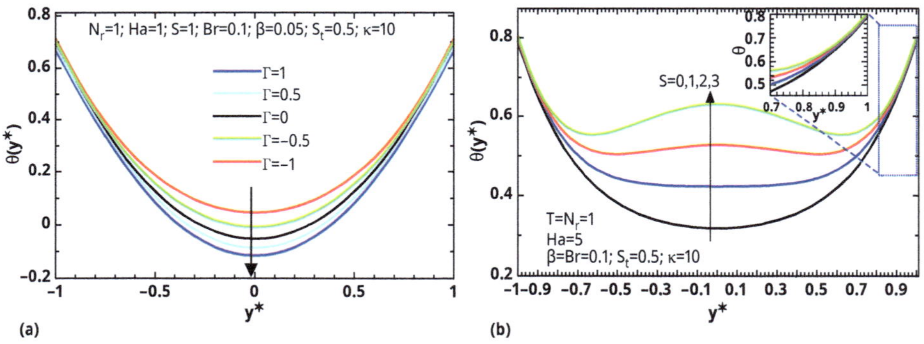

Figure 10.10: Nondimensional temperature variation across the channel with Γ (left) and *S* (right). The direction of the arrow shows the increase in a parameter in that direction.

Physically, the advection of the thermal energy from the wall rather slows down for $\Gamma < 0$, and speeds up for $\Gamma > 0$. The effect of the dimensionless Joule heating parameter on the temperature profile is shown in Figure 10.10 (right). The dimensionless temperature significantly increases with the Joule heating parameter (S). In the absence of Joule heating, a large temperature difference between the wall and the center of the channel is observed. This difference gradually reduces as one goes on increasing the value of S. For $S = 2$ and 3, a significant increase in the temperature has been observed at the channel center, which is mainly attributed to a high electromagnetic force [49].

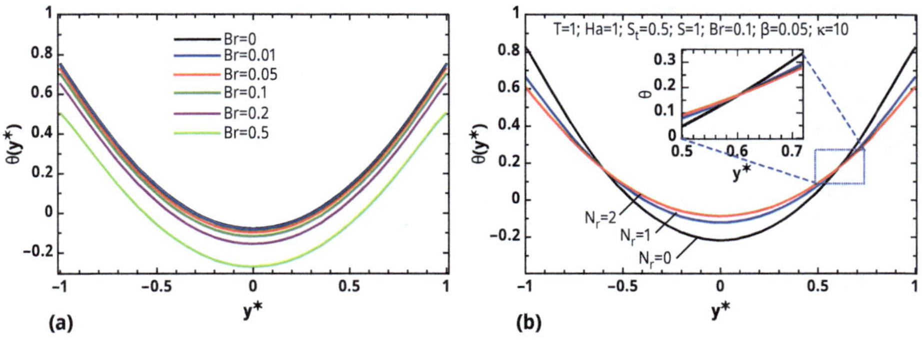

Figure 10.11: Variation of dimensionless temperature (θ) with Brinkmann number (Br) (left) and radiation parameter (Nr) (right).

The variation in θ with Br is illustrated in Figure 10.11 (left). The Brinkmann number (Br) represents the ratio of the viscous dissipations to the applied heat flux. With the increase in Br, the dimensionless temperature decreases, leading to a large temperature difference [67]. The increase in viscous dissipation, which acts as an energy source, exhibits a higher temperature near the wall [30], which could increase the temperature difference. The effect of the radiation parameter is shown in Figure 10.11 (right). As can be observed, θ increases with Nr at the channel center, and it reduces with Nr toward the wall. As the temperature gradient near the wall falls with Nr, so does θ reduce with Nr. Moreover, the radiation heats up the bulk fluid to enhance its temperature at the channel center. So, the temperature profiles at different Nr cross at a region between the wall and the center of the channel. Similar variations in temperature profiles have been observed by Nayak et al. [44] in their study of an EOF of a non-Newtonian fluid in a hydrophobic channel with radiation and Joule heating. The effect of the thermal jump parameter on the temperature profile has been depicted in Figure 10.12. A higher temperature jump parameter enhances the dimensionless temperature. The enhancement rate is found more at the center of the channel than on the wall.

10.5.4 Effect of different parameters on the Nusselt number

Before the discussions on the parametric effects on the Nusselt number, we present its formulation, as follows:

$$Nu = \frac{hD_h}{k_{th}} = \frac{q_w D_h}{k_{th}(T_w - T_m)} \tag{10.32}$$

where h is the convective heat transfer coefficient, $D_h(=4H)$ is the hydraulic diameter of the channel, and T_m is the mean temperature. Equation (10.32) can be expressed in the dimensionless form [32] as:

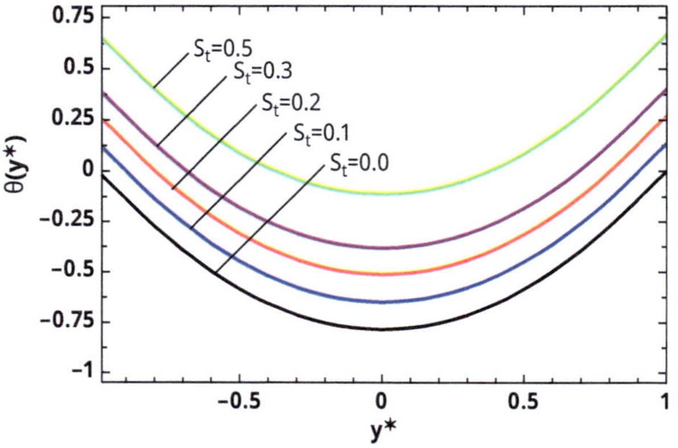

Figure 10.12: Variation of dimensionless temperature across the channel heights with thermal jump parameter (S_t). A higher value of this parameter makes the electrolyte to acquire a uniform temperature.

$$Nu = -\frac{4}{\theta_m}$$

where the mean temperature,

$$\theta_m = \frac{\int_0^1 u^* \theta \, dy^*}{\int_0^1 u^* \, dy^*} \tag{10.33}$$

First, the effects of different parameters such as κ, S_t, Ha, and Nr on the Nusselt number of the EOF are presented. Then, the influences of these parameters on EMHF are discussed thoroughly. Finally, a comparative analysis of the Nusselt number has been given for EOF and EMHF, with and without velocity slip and temperature jump condition. The variation in the Nusselt number with the EDL potential difference (κ), and with the temperature jump parameter has been shown in Figure 10.13 (left) and Figure 10.13 (right), respectively. In this case, purely EOF ($\Gamma = 0$) has been analyzed in the absence of transverse magnetic field ($Ha = 0$), Joule heating ($S = 0$), viscous heating ($Br = 0$), and the thermal radiation ($Nr = 0$). For a particular value of the velocity-slip parameter (β), we observe a rapid increase in Nu with κ, when $\kappa \leq 20$.

For $\kappa \geq 20$, the increase in Nu is found to be insignificant or approximately constant. As κ increases, the nondimensional temperature difference (i.e., $\theta_w - \theta_m$) decreases, owing to the aiding effects of the electromagnetic effects. For example, at $\beta = 0$, the temperature difference, $(\theta_w - \theta_m)$, decreases by 30.26% as κ increases from 1 to 20, registering an increase in the Nusselt number by 30.7%. However, at higher κ, the EDL thickness is very small in the proximity of the wall, where a strong temperature gradient prevails. The outer flow is not much affected by the electromagnetic force, so the temperature difference is reduced to cause a slow rate of increment in the Nusselt number. For instance, an increase in the Nusselt number of only 4.72%

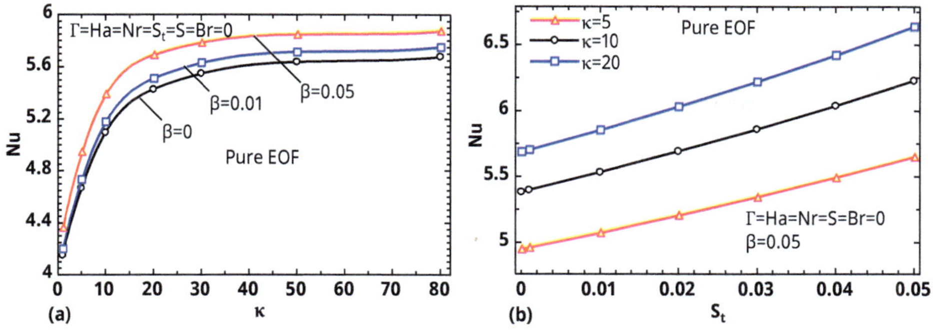

Figure 10.13: These plots demonstrate the variation in the Nusselt number (*Nu*) with κ at different β (left) and with S_t at different κ(right). The Nusselt number rapidly increases with κ before becoming constant at high κ values (left). But *Nu* shows an increasing trend with S_t(right).

has been registered as the EDL parameter (κ) increases from 20 to 80. This is attributed to a lower increase in $\theta_w - \theta_m$ (i.e., 6.55%), owing to a reduction in θ_m and an enhancement in the centerline dimensionless temperature (θ_c), as shown in Figure 10.5 (left). The high slip at the wall further increases fluid advection [22] and enhances the Nusselt number. For different values of κ, the effect of temperature jump parameter (S_t) on *Nu* is depicted in Figure 10.13 (right). At a particular κ, the Nusselt number monotonously increases with S_t for an EOF. As the temperature jump parameter increases, the temperature difference ($\theta_w - \theta_m$) is reduced, exhibiting a higher Nusselt number. For an EOF with slip condition, the effect of Hartmann number (*Ha*) on the Nusselt number is shown in Figure 10.14 (left) at different values of Γ(=0, 1, and −1). One can observe that *Nu* increases with *Ha* for all the considered values of dimensionless pressure gradients (Γ). However, at a particular Hartmann number, especially at a low value, a marginal increase in the Nusselt number with respect to different Γ has been noticed.

But at a thigh value of the Hartmann number, a significant increase in the Nusselt number has been obtained. At a high Hartmann number, a strong opposing Lorentz force is imposed on the flow field, thereby reducing the velocity of the flow and the temperature difference ($\theta_c - \theta_m$), which can increase the Nusselt number. But the magnitude of the channel centerline temperature (θ_c) gradually increases with the Hartmann number (see Figure 10.9 (right)) because of the continuous transfer of thermal energy from the wall to the slow-moving fluid mass. At a particular *Ha*, a high Nusselt number is obtained for $\Gamma < 0$ than for $\Gamma > 0$. The temperature difference ($\theta_w - \theta_m$) gradually increases as the dimensionless pressure gradient changes its sign from negative to positive (i.e., the adverse pressure gradient to favorable pressure gradient), and as a result of this, we get a high and a low Nusselt number for $\Gamma < 0$ and $\Gamma > 0$, respectively. The effect of the radiation parameter on the Nusselt number has been shown in Figure 10.14 (right).

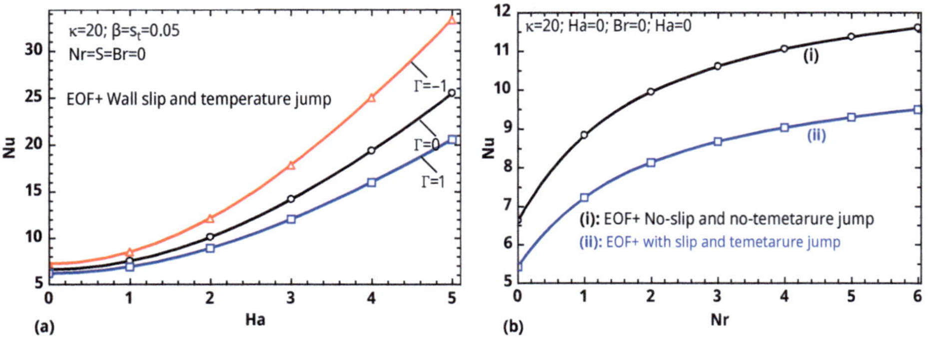

Figure 10.14: Variation in *Nu* with the Hartmann (*Ha*) number at different Γ for an EOF under both wall slip and temperature jump condition (left) and *Nu* variation with *Nr* for the cases, as shown in right-side plot.

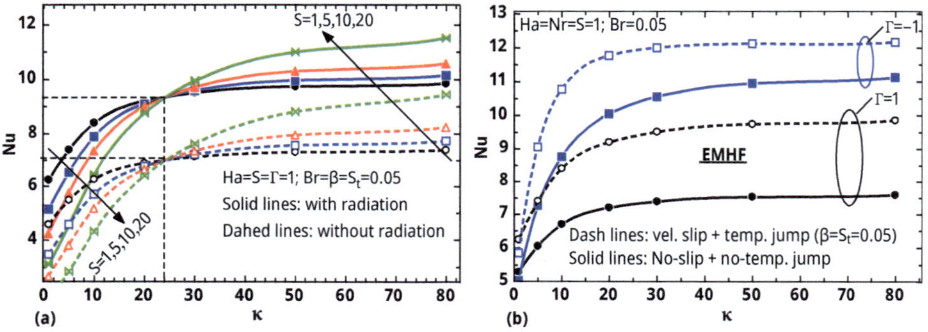

Figure 10.15: Variation of *Nu* with *κ* at different *S* for EMHF (left); and with *κ* at different Γ for EMHF (right). In both cases, the Nusselt number initially increases at a faster rate and then, the rate of increment slows down.

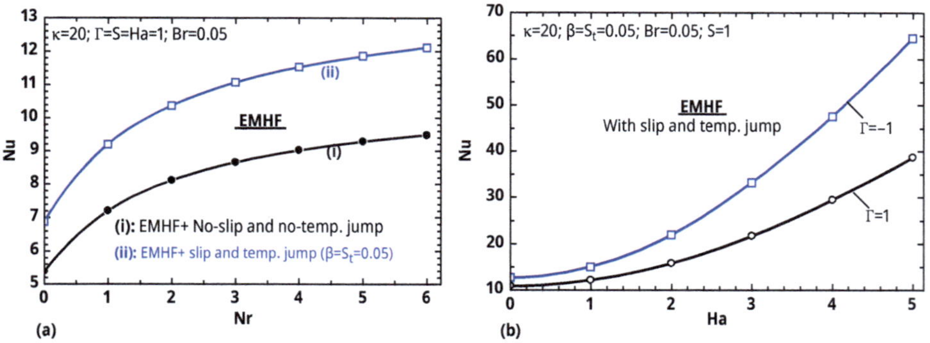

Figure 10.16: Variation of *Nu* with *Nr* of an EMHF for the slip and no-slip conditions (left) and with *Ha* of an EMHF for the slip-wall condition (right). EMHF flow with slip and temperature jump shows a higher Nusselt number (left).

It can be seen that the Nusselt number increases with *Nr* because of the addition of thermal energy due to the radiation effect. For FOF, we observed a strong influence of the radiation parameter on the Nusselt number, in particular, at a low range of the radiation parameter ($1 \le Nr \le 4$). Also, a higher Nusselt number is found for EOF with wall slip and temperature jump, as compared to the EOF without the wall slip. This is attributed to the high value of $\theta_w - \theta_m$ for an EOF with hydrophobic/wall slip, as compared to an EOF without the wall slip, and as a consequence, the Nusselt number of an EOF without slip is higher than an EOF with slip. Since the objective of the present study is to carry out a detailed parametric investigation on an EMHF and its comparison with EOF under externally imposed radiation heat transfer, the details of different parameters on Nusselt of an EMHF are enlightened in the next part of our discussion. Therefore, the electromagneto and thermofluid characteristics and their dependencies on different significant parameters are only discussed. The variation of the Nusselt number with the EDL potential difference the case with external radiation is applied and the case without external radiation has been depicted in Figure 10.15 (left). At a particular value of *S*, the Nusselt number, as expected, increases with the EDL potential difference (κ). For $\kappa \ge 25$, the Nusselt number for both the cases (i.e., with and without wall slips, Figure 10.15 (left)) either remains constant or increases slightly.

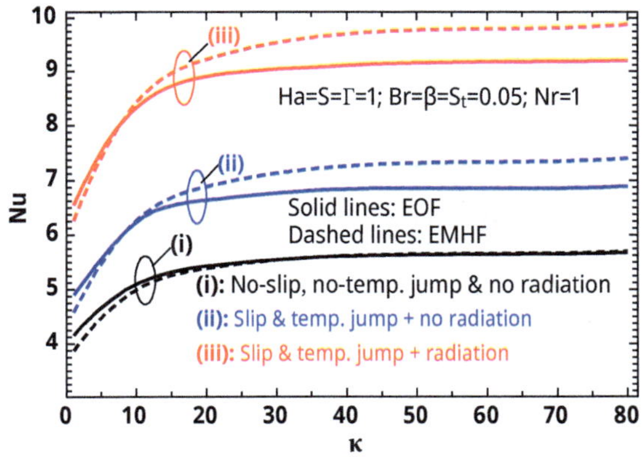

Figure 10.17: Variation of *Nu* with *κ* for an EMHF for different flow situations. The comparison of various cases shows a high Nusselt number for the EMHF with wall slip and temperature jump condition, coupled with external radiation.

In the same plot, it has been observed that the Nusselt number of the non-radiative EMHF is more than the radiative EMHF. When the radiation is imposed, the fluid temperature increases in order to improve the mean temperature and bring down the wall and mean temperature difference ($\theta_w - \theta_m$). According to eq. (10.32), the Nusselt number will decrease with mean temperature. Figure 10.15 (right) illustrates the variation of the Nusselt number with the EDL parameter as the function of the pressure gradient parameter for the cases with and without the wall slip. The rate of enhancement in the Nusselt number is faster in the low range of the EDL parameter ($\kappa \leq 20$) than at the high range of the EDL parameter ($\kappa \geq 20$). The Nusselt number for the case with $\Gamma = -1$ is more than the case with κ. The slip on the wall further improves the Nusselt number. For the case with $\Gamma = -1$, the adverse pressure gradient suppresses the fluid convection, because of which, the difference between the wall and the mean temperature is reduced to strengthen the Nusselt number. However, for $\Gamma = 1$, the favorable pressure gradient increases convection and the mean temperature of the fluid so as to get a low Nusselt number. Moreover, at a high κ, the electromagnetic effect becomes prominent in the thin EDL layer, and in combination with the Joule heating parameter is primarily responsible for a high Nusselt number. Figure 10.16 (left) shows the variation of Nu with radiation parameter (Nr). The effect of the radiation parameter on the Nusselt number of two different flow situations: (i) EMHF without slip wall and temperature jump (hereafter, this case is treated as "EMHF-no-slip-wall") and (ii) EMHF with slip wall and temperature jump are compared (hereafter, it is treated as EMHF-slip-wall). It is quite apparent that the Nusselt number for both the flow situations increases with the radiation parameter. The fluid temperature increases further on enforcing the radiation, giving rise to a lower value of $\theta_w - \theta_m$, which in turn, augments the Nusselt number. As explained earlier, the mean temperature of the fluid is increased by virtue of the enhanced convection due to the higher value of the slip parameters. Thus, the Nusselt number of the EMHF-slip-wall is found more than that of the EMHF-no-slip-wall. The rate of the enhancement in the Nusselt number of both the EMHF-slip-wall and EMHF-no-slip-wall is gradually increased as a result of the increase in the radiation parameter. For example, the Nusselt number of the EMHF-slip-wall is increased by 27.43% and 27.89% over the EMHF-no-slip-wall at $Nr = 1$ and 6, respectively. The efficacy of the Hartmann number (Ha) on the Nusselt number of EMHF-slip-wall at different values of Γ has been depicted in Figure 10.16 (right). It is observed that the Nusselt number increases with the Hartmann number. A significant increment in the Nusselt number has been noticed at higher values of the Hartmann number. This could be stemmed from the fact that the flow is more decelerated by the strong Lorentz force so that the wall and mean temperature difference ($\theta_w - \theta_m$) is reduced to elevate the Nusselt number. After due deliberations of the implications of the different parameters that regulate the flow characteristics of EMHF in the circumstances of the present case, we finally present the variation of the Nusselt number with the EDL parameter for three different cases designated as Case-I, Case-II, and Case-III, as depicted in Figure 10.17. It has been observed from Figure 10.17 that the Nusselt number for all the three cases increases with the EDL parameter (κ). At a particular value of the EDL parameter (κ), the Nusselt number for an EMHF with wall slip, the temperature jump and the thermal radiation are higher as compared to an EMHF either without radiation or without the wall slip. Thus, it may be argued that the wall slip, temperature jump along with the external radiation have strong influences on the thermo-fluid characteristics of an EMHF that quite frequently finds application in microfluidics and in medical fields such as the treatment of muscle pain, tumor, etc. We believe that the present work might furnish further understanding on the dynamics of electrothermofluid interactions.

10.6 Conclusions

In this study, a numerical investigation has been carried out to study the implications of the Joule heating, viscous heating, and external radiation on the convection heat transfer characteristics, relevant to electromagnetohydrodynamically and pressure-driven flows in hydrophobic microchannels. An in-house finite-difference program has been developed to solve the governing equations pertaining to the transport of the EDL potential, momentum, and energy, revealing the intricacies of the microchannel flows, especially, under the velocity slip and temperature jump conditions. The present numerical solution has been carefully corroborated with well-established results of previous studies. Some of the major conclusions obtained from this study are summarized as follows

(i) For a pure EOF, the velocity increases significantly with higher slip at the wall. This increment in the velocity is found even more in the case of EMHF. The velocity of EMHF increases with the inverse-Debye parameter. At low values of the Hartmann number, the flow velocity remains flat across most part of the channel, whereas at high values of the Hartmann number, the flow velocity decreases in the central region of the channel due to an increased magnetic field. The velocity increases for $\Gamma > 0$ and decreases for $\Gamma < 0$ due to the flow-assisting and flow-resisting forces, respectively, acted on the fluid flow.

(ii) Both the dimensionless flow rates of EOF (Q^*_{EOF}) and EMHF (Q^*_{EMHF}) increase with the velocity slip parameter (β). However, at a particular κ and β, $Q^*_{\mathrm{EOF}} > Q^*_{\mathrm{EMHF}}$, which is attributed to the external magnetic field that opposes the flow. The flow rate is further reduced at higher magnetic field strength.

(iii) The parameters, such as the EDL (κ), the Hartmann number (Ha), the nondimensional pressure gradient (Γ), the Joule heating parameter (S), and the Brinkmann number (Br) have influence on the temperature distribution. The radiation (Nr) as well as the wall-slip parameters (i.e., β and S_t) have strong influence on both the velocity and temperature fields.

(iv) The Nusselt number of both the EOF and EMHF increases with the DEL parameter (κ). The increment is more prominent at low values of κ, and the rate of increment in the Nusselt number is found be less for Nu. It has been observed that the Nusselt number of EMHF with the wall slip is more than that of the no-slip-wall. The Nusselt number is observed to increase with the Joule heating parameter and the Hartmann number. At a high Hartmann number, the enhancement in the Nusselt number is attributed to the enhanced opposition force, called the Lorentz force.

(v) The Nusselt number of the EMHF with an adverse pressure gradient is found to be more as compared to the favorable pressure gradient. Also, the radiation parameter significantly improves the Nusselt number of EMHF at a particular value of κ. We observed that at $Nr = 6$, the Nusselt number of EMHF with the slip-wall is 27.89% more than the Nusselt number of the EMHF without wall slip.

(vi) The Nusselt number of the EMHF with slip-wall condition is found to be significantly increased when an external radiation is imposed on the flow. This suggests that external radiation has a significant effect on the EMHFs, in particular, when the slip on the walls of the microchannel is taken into consideration.

Conflict of interest: There is no potential conflict of interest among the authors.

Nomenclature

A	Cross-sectional area (m^2)
b	Slip length (m)
B_o	Magnetic field
B_y	*y*-direction magnetic field
Br	Brinkmann number
c_p	Specific heat (kJ/kg K)
D_h	Hydraulic diameter (m)
e	Proton charge (C)
E_x	Axial electric field (V/m)
F	Body force vector
H	Half height of the channel (m)
h	Heat transfer coefficient (W/m^2 K)
J	Ion current density
k_a	Absorption coefficient (m^{-1})
k_B	Boltzmann constant (J/K)
k_{th}	Thermal conductivity (W/m K)
k_j	Temperature jump factor
K	Zerta potential (V)
n	Number of nodes
n_0	Ion density (m^{-3})
Nr	Radiation parameter $\left(= \frac{16\sigma_{SB}T_w^3}{3k_a k_{th}}\right)$
Nu	Nusselt number
P	Pressure (Pa)
$\dot{Q}$	Volume flow rate (m^3/s)
Q^*	Nondimensional flow rate
q_w	Wall-heat flux (W/m^2)
q_r	Radiative heat flux (W/m^2)
S	Joule heating parameter $\left(= \frac{\sigma_e E_x^2 H}{q_w}\right)$
S_t	Temperature jump parameter $\left(= k_j/H\right)$
T	Temperature (K)
T_{ab}	Absolute temperature (K)

T_m Mean temperature (K)
T_w Wall temperature (K)
u Axial velocity (m/s)
u_{HS} Helmholtz-Smoluchowski velocity $\left(=\frac{-\varepsilon\zeta E_x}{\mu}\right)$
u_{PD} Pressure-driven velocity $\left(=\frac{-H^2}{\mu}\frac{dP}{dx}\right)$
u_m Mean velocity (m/s)
x Axial direction (m)
y y-Direction (m)
z_v Ion valence

Greek symbols

β Velocity slip parameter $(=b/H)$
ε Electrolyte permittivity (C/V m)
σ_e Electrical resistivity (Ωm)
σ_{SB} Stefan Boltzmann constant (W/m^2 K^4)
Γ Dimensionless pressure gradient $(=u_{PD}/u_{HS})$
ζ Zeta potential (V)
θ Nondimensional temperature
κ Nondimensional Debye-Hückel parameter ($=KH$)
λ_D Debye length (m) ($=1/\kappa$)
μ Dynamic viscosity (kg/m s)
ρ Density (kg/m^3)
ρ_c Net charge density (C/m^3)
ψ EDL potential (V)
ψ^* Nondimensional EDL potential

Subscripts

a Absorption
ab Absolute
B Boltzmann
c Centerline
D Debye
e Electric
HS Helmholtz-Smoluchowski
j Jump
PD Pressure-driven
r Radiation
SB Stefan Boltzmann
th Thermal
v Valence
w Wall
m Mean

References

[1] Song Y. J., and Zhao T. S. J. (2001). Modelling and test of a thermally-driven phase-change nonmechanical micropump. Micromechanics and Microengineering, 11, 713–719.

[2] Abhari F., Jaafar H., and Yunus N. A. M. (2012). A comprehensive study of micropumps technologies. International Journal of Electrochemical Science, 7, 9765–80.

[3] Gaikwad H. S., Basu D. N., and Mondal P. K. (2017). Slip driven micro-pumping of binary system with a layer of non-conducting fluid under electrical double layer phenomenon. Colloids and Surfaces A, 518, 166–172.

[4] Hunt C. E., Desmond C. A., Ciarlo D. R., and Benett W. J. (1991). Direct bonding of micromachined silicon wafers for laser diode heat exchanger applications. Journal of Micromechanics and Microengineering, 1(3), 152–156.

[5] Pfrahler J. N. (1990). Advances in Thermal Modeling of Electronic Components and Systems, vol. II, New York, USA: SME Press.

[6] Harrison D. J., Fluri K., Seiler K., Fan Z., Effenhauser C. S., and Manz A. (1993). Micromachining a miniaturized capillary electrophoresis-based chemical analysis system on a chip. Science, 261, 895–897.

[7] Blackshear P. J. (1979). Implantable Drug-delivery Systems. Scientific American, 241, 52–59.

[8] Penn R. D., Paice J. A., Gottschalk W., and Ivankovich A. D. (1984). Cancer pain relief using chronic morphine infusion: Early experience with a programmable implanted drug pump. Journal of Neurosurgery, 61, 302–306.

[9] Lavan D. A., McGuire T., and Langer R. (2003). Small-scale systems for in vivo drug delivery. Nature Biotechnology, 21, 1184–1191.

[10] Bruno G., Geninatti T., Hood R. L., Fine D., Scorrano G., Schmulen J., Hosali S., Ferrari M., and Grattoni A. (2015). Leveraging electrokinetics for the active control of dendritic fullerene-1 release across a nanochannel membrane. Nanoscale, 7, 5240–5248.

[11] Olthuis W., Schippers B., Eijkel J., and van den Berg A. (2005). Energy from streaming current and potential. Sensors and Actuators B, 111, 385–389.

[12] Siria A., Poncharal P., Biance A. L., Fulcrand R., Blase X., Purcell S. T., and Bocquet L. (2013). Giant osmotic energy conversion measured in a single trans membrane boron nitride nanotube. Nature, 494, 455–458.

[13] Tseng S., Li Y. M., Lin C. Y., and Hsu J. P. (2016). Salinity gradient power: Influences of temperature and nanopore size. Nanoscale, 8, 2350–2357.

[14] Chein R., Liao C., and Chen H. (2009). Electrokinetic energy conversion efficiency analysis using nanoscale finite-length surface-charged capillaries. Journal of Power Sources, 187, 461–470.

[15] Goswami P., and Chakraborty S. (2010). Energy transfer through streaming effects in time periodic pressure-driven nanochannel flows with interfacial slip. Langmuir, 26, 581–590.

[16] Bera S., and Bhattacharyya S. (2013). On mixed electroosmotic-pressure driven flow and mass transport in microchannels. International Journal of Engineering Science, 62, 165–176.

[17] Tretheway D. C., and Meinhart C. D. (2002). Apparent fluid slip at hydrophobic micro channel walls. Physics of Fluids, 14, 9–12.

[18] Levine S., Marriott J. R., and Robinson K. (1975). Theory of electrokinetic flow in a narrow parallel-plate channel. Journal of the Chemical Society, Faraday Transactions II, 71, 1–11.

[19] Mala G. M., Li D. Q., and Dale J. D. (1997). Heat transfer and fluid flow in microchannels. International Journal of Heat and Mass Transfer, 40, 3079–3088.

[20] Yang C., Li D. Q., and Masliyah J. H. (1998). Modeling forced liquid convection in rectangular microchannels with electrokinetic effects. Int. J. Heat Mass Transfe, 41, 4229–4249.

[21] Levine S., Marriott J. R., Neale G., and Epstein N. (1975). Theory of electrokinetic flow in fine cylindrical capillaries at high zeta-potentials. Journal of Colloid and Interface Science, 52, 136–149.

[22] Yang Z., Peng X. F., and Wang B. X. (2006). Fully developed electroosmotically and hydrodynamically induced convection between two parallel plates. Numerical Heat Transfer Part A, 50(10), 905–926.
[23] Mukhopadhyay A., Banerjee S., and Gupta C. (2009). Fully developed hydrodynamic and thermal transport in combined pressure and electrokinetically driven flow in a microchannel with asymmetric boundary conditions. International Journal of Heat and Mass Transfer, 52, 2145–2154.
[24] Gorthi S. R., Gaikwad H. S., Mondal P. K., and Biswas G. (2020). Surface Tension Driven Filling in a Soft Microchannel: Role of Streaming Potential. Industrial & Engineering Chemistry Research, 59(9), 3839–3853.
[25] Shamshiri M., Khazaeli R., Ashrafizaadeh M., and Mortazavi S. (2012). Heat transfer and entropy generation analyses associated with mixed electrokinetically induced and pressure-driven power-law microflows. Energy, 42, 157–169.
[26] Dallakehnejada M., Mirbozorgi S. A., and Niazmand H. (2019). A numerical investigation of magnetic mixing in electroosmotic flows. Journal of Electrostatics, 100, 103354.
[27] Sheikhizad M., and Kalteh M. (2020). Heat transfer investigation of combined electroosmotic/ pressure driven nanofluid flow in a microchannel: Effect of heterogeneous surface potential and slip boundary condition. European Journal of Mechanics / B Fluids, 80, 13–25.
[28] Zade A. Q., Manzari M. T., and Hannani S. K. (2007). An analytical solution for thermally fully developed combined pressure– Electroosmotically driven flow in microchannels. International Journal of Heat and Mass Transfer, 50, 1087–1096.
[29] Ohshima H. (2019). Electrokinetic phenomena in a dilute suspension of spherical solid colloidal particles with a hydrodynamically slipping surface in an aqueous electrolyte solution. Advances in Colloid and Interface Science, 272, 101996.
[30] Yavari H., Sadeghi A., Saidi M. H., and Chakraborty S. (2012). Combined influences of viscous dissipation, non-uniform Joule heating and variable thermophysical properties on convective heat transfer in microtubes. International Journal of Heat and Mass Transfer, 55, 762–772.
[31] Dutta S., Mondal P. K., and Goswami P. (2018). Slipping hydrodynamics of Powell-Eyring fluid in a cylindrical microchannel under electrical double layer phenomenon. Physica Scripta, 94, 025002.
[32] Chen X., Jian Y., Xie Z., and Ding Z. (2018). Thermal transport of electromagnetohydrodynamic in a microtube with electrokinetic effect and interfacial slip. Colloids and Surfaces A, 540, 194–206.
[33] Mo X., and Hu X. (2019). Electroviscous effect on pressure driven flow and related heat transfer in microchannels with surface chemical reaction. International Journal of Heat and Mass Transfer, 130, 813–820.
[34] Sarkar S., Ganguly S., and Dutta P. (2016). Thermally developing combined magnetohydrodynamic and electrokinetic transport in narrow confinements with interfacial slip. International Journal of Heat and Mass Transfer, 100, 451–463.
[35] Jamaati J., Niazmand H., and Renksizbulut M. (2010). Pressure-driven electrokinetic slip-flow in planar microchannels. International Journal of Thermal Sciences, 49, 1165–1174.
[36] Szasz A. (2007). Hyperthermia, a modality in the wings. Journal of Cancer Research and Therapeutics, 3(1), 56–66.
[37] Kobu Y. (1999). Effects of infrared radiation on intraosseous blood flow and oxygen tension in rat tibia. Kobe Journal of Medical Sciences , 45, 27–39.
[38] Inoue S., and Kobaya M. (1989). Biological activities caused by far infrared radiation. International Journal of Biometeorology, 33, 145–150.
[39] Nishimoto C., Ishiura Y., Kuniasu K., and Koga T. (2006). Effects of ultrasonic radiation on cutaneous blood flow in the paw of decerebrated rats. Kawasaki Journal of Medical Welfare , 12, 13–18.
[40] Ogulu A., and Bestman A. R. (1994). Blood flow in a curved pipe with radiative heat transfer. Acta Physiologica Hungarica , 74, 189–201.

[41] He Y., Shirazaki M., Liu H., Himeno R., and Sun Z. (2006). A numerical coupling model to analyze the blood flow, temperature, and oxygen transport in human breast tumor under laser irradiation. Computers in Biology and Medicine, 36, 1336–1350.
[42] Pal D., and Mondal H. (2013). Influence of thermophoresis and Soret–Dufour on magnetohydrodynamic heat and mass transfer over a non-isothermal wedge with thermal radiation and Ohmic dissipation. Journal of Magnetism and Magnetic Materials , 331, 250–255.
[43] Sinha A., and Shit G. C. (2015). Electromagnetohydrodynamic flow of blood and heat transfer in a capillary with thermal radiation. Journal of Magnetism and Magnetic Materials , 378, 143–151.
[44] Nayak A. K., Haque A., and Weigand B. (2018). Analysis of electroosmotic flow and Joule heating effect in a hydrophobic channel. Chemical Engineering Science, 176, 165–179.
[45] Masliyah J. H., and Bhattacharjee S. (2006). Electrokinetic and Colloid Transport Phenomena, New Jersey, Hoboken, USA: John Wiley & Sons, Inc. Publication.
[46] Karniadakis G., Beskok A., and Aluru N. (2005). Microflows and Nanoflows: Fundamentals and Simulation, USA, Springer Science Business Media, Inc.
[47] Probstein R. F. (1994). Physicochemical Hydrodynamics, 2nd ed. New York, USA: Wiley.
[48] Shamshiri M., Khazaeli R., Ashrafizaadeh M., and Mortazavi S. (2012). Heat transfer and entropy generation analyses associated with mixed electrokinetically induced and pressure-driven power law microflows. Energy, 42, 157–69.
[49] Chakraborty R., Dey R., and Chakraborty S. (2013). Thermal characteristics of electromagnetohydrodynamic flows in narrow channels with viscous dissipation and Joule heating under constant wall heat flux. International Journal of Heat and Mass Transfer, 67, 1151–1162. 50.
[50] Griffiths D. J. (1999). Introduction to Electrodynamics, 3rd ed. Upper saddle River, New Jersey USA: Prentice Hall Press.
[51] Barik A. K., Mishra S. K., Mishra S. R., and Pattnaik P. K. (2020). Multiple slip effects on MHD nanofluid flow over an inclined, radiative, chemically reactive stretching sheet by means of FDM. Heat Transfer Asian Research, 49(1), 447–501.
[52] Shit G. C., Mondal A., Sinha A., and Kundu P. K. (2016). Electro-osmotic flow of power-law fluid and heat transfer in a micro-channel with effects of Joule heating and thermal radiation. Physica A, 462, 1040–1057.
[53] Xie Z., and Jian Y. (2017). Entropy generation of two-layer magnetohydrodynamic electroosmotic flow through microparallel channels. Energy, 139, 1080–1093.
[54] Yavari H., Sadeghi A., Saidi M. H., and Chakraborty S. (2012). Combined influences of viscous dissipation, non-uniform Joule heating and variable thermophysical properties on convective heat transfer in microtubes. International Journal of Heat and Mass Transfer, 55, 762–772.
[55] Xie Z., and Jian Y. (2018). Entropy generation of magnetohydrodynamic electroosmotic flow in two-layer systems with a layer of non-conducting viscoelastic fluid. International Journal of Heat and Mass Transfer, 127, 600–615.
[56] Dash S. K. (2014). Engineering Equation Solver: Application to Engineering and Thermal Engineering Problems, Daryaganja, New-Delhi, India: Norosa Publishing House Pvt. Ltd.
[57] Tarjan R. (1972). Depth-First Search and Linear Graph Algorithms. SIAM Journal on Computing, 1, 146–160.
[58] Chu X., and Jian Y. (2019). Magnetohydrodynamic electro-osmotic flow of Maxwell fluids with patterned charged surface in narrow confinements. Journal of Physics D: Applied Physics , 52, 405003.
[59] Gaikwad H., Basu D. N., and Mondal P. K. (2016). Electroosmotic transport of immiscible binary system with a layer of non-conducting fluid under interfacial slip: The role applied pressure gradient. Electrophoresis, 37(14), 1998–2009.

[60] Liu Y. P., Jian Y. J., Liu Q. S., and Li F. Q. (2015). Alternating current magnetohydrodynamic electroosmotic flow of Maxwell fluids between two micro-parallel plates. Journal of Molecular Liquids , 211, 784–791.
[61] Ngoma G. D., and Erchiqui F. (2007). Heat flux and slip effects on liquid flow in a microchannel. International Journal of Thermal Sciences , 46, 1076–1083.
[62] Silkina E. F., Asmolov E. S., and Vinogradova O. I. (2019). Electro-osmotic flow in hydrophobic Nanochannels. Physical Chemistry Chemical Physics, 21, 23036.
[63] Dhinakaran S., Afonso A. M., Alves M. A., and Pinho F. T. (2010). Steady viscoelastic fluid flow between parallel plates under electro-osmotic forces: Phan-Thien–Tanner model. Journal of Colloid and Interface Science, 344, 513–520.
[64] Sarma R., Deka N., Sarma K., and Mondal P. K. (2018). Electroosmotic flow of Phan-Thien–Tanner fluids at high zeta potentials: An exact analytical solution. Physics of Fluids, 30, 062001.
[65] Yang J., and Kwok D. Y. (2003). Effect of liquid slip in electrokinetic parallel-plate microchannel flow. Journal of Colloid and Interface Science, 260, 225–233.
[66] Haddout Y., and Lahjomri J. (2015). The extended Graetz problem for a gaseous slip flow in micropipe and parallel-plate microchannel with heating section of finite length: Effects of axial conduction, viscous dissipation and pressure work. International Journal of Heat and Mass Transfer, 80, 673–687.
[67] Jian Y. (2015). Transient MHD heat transfer and entropy generation in a microparallel channel combined with pressure and electroosmotic effects. International Journal of Heat and Mass Transfer, 89, 193–205.
[68] Barik A. K., and Swain P. K. (2024). Effects of axial electric and transverse magnetic fields on a rotating electro-osmotic flow in micro-parallel plates. European Journal of Mechanics-B/Fluids, 107, 148–164.

Mohammad Hassan*, Dhananjay Yadav, and Lham Tashi

Chapter 11
The study of oscillating water column energy device in a two-layer fluid system of finite impermeable depth

Abstract: An upright floating hollow cylindrical arrangement is considered in a two-layer flow with impermeable depth. Under the linear water waves theory, the interaction of a vertical floating hollow cylindrical structure with water waves is examined in the surface and inner wave modes. Since the oscillating water column is primarily characterized by a hollow cylindrical arrangement, the proposed device may be regarded as an oscillating water column (OWC), which makes up one among the key wave energy devices and it can transform wave energy into electrical energy. The separation of variables technique is utilized to tackle the boundary value problems for the diffracted potentials in the identified regions and the matched eigenfunction expansion and MATLAB tools are used to determine the unknown coefficients appearing in the obtained diffracted potentials.

The chapter explains how hydrodynamic forces induced by the device are derived with the help of obtained diffracted velocity potentials in both the surface and the inner wave modes. A number of results are plotted for the hydrodynamic forces with different parameters of the device. When the cylinder is located either at the interface, the upper layer, or the lower layer, the exciting forces in the surface and inner wave modes show the significant behavior with different parameters of the device. There are remarkable effects observed with a two-layer flow system compared to a one-layer flow. The oscillations of exciting forces can be seen in the inner wave mode, and peak values can be seen in the surface wave mode.

Note: MSC Classification 2000: 76B07, 76B15

Acknowledgment: The authors are grateful to MMMUT Gorakhpur and NERIST by the authors for giving him access to the facilities and support needed to carry out his research.

---

*Corresponding author: **Mohammad Hassan**, Mathematics and Scientific Computing Department, Madan Mohan Malaviya University of Technology, Gorakhpur, 273010, Uttar Pradesh, India, e-mail: mdhassan000@gmail.com
Contributing author: Dhananjay Yadav, Department of Mathematical and Physical Sciences, University of Nizwa, Nizwa, Oman, e-mail: dhananjayadav@gmail.com; 519014@nerist.ac.in
Lham Tashi, Department of Mathematics North Eastern Regional Institute of Science and Technology, Nirjuli 791109, Arunachal Pradesh, India

https://doi.org/10.1515/9783111661674-011

Keywords: surge radiation, eigenfunction expansion, two-layer, added mass, damping coefficient, wavenumber, floating cylinder, bottom-mounted cylinder

11.1 Introduction

The density of water in some offshore regions is not uniform throughout the depth due to the mixing of fresh and salty water. Therefore, it is widely investigated how hydrodynamic loads of water waves are predicted in two-layer flows. Following are some examples of two-layer flows as subsurface hydrological flows: stratified fluid flow in glacial material, related problems from the biological and industrial worlds. It was Stokes who first looked into the interplay of water waves in a two-layer flow [1]. Subsequently, many researchers investigated the problem of internal waves in which an upper fluid is either encircled by a rigid lid or unbounded. Recently, a number of researchers have studied the internal solitary waves in two-layer fluid, such as Benjamin [2], Keulegan [3] and Long [4]. The solution to the diffraction problem related to hollow cylindrical structure in water of finite depth was given by Zhu and Mitchell [5]. Sarah et al. [6] studied the effect of two-layer fluid on towing a horizontally placed circular cylinder in a horizontal direction perpendicular to its axes. The vibrations caused by waves on the cylinder were calculated. The hydrodynamic force of a single hollow cylinder in finite depth was given by Hassan et al. [7]. Various works have been done in two-layer flow with uneven bottom, for example, Bains [8] has studied two-layer flow over an uneven bottom.

Also, Chanda and Bora [9] investigated wave scattering in two-layer flow with elastic sea bed and upper fluid covered by an ice-sheet. Linton and McIver [10] explained a theory for the propagation of two-dimensional water waves. They also calculated the transmission and reflection coefficients. The integral equation method is used to solve the three-dimensional radiation and scattering problems by the floating bodies in a two-layer flow of limited depth [11]. The investigation of radiation and scattering problems of a sphere in fluid of two-layer flow was performed [12]. The investigation of scattering due to the oblique waves was discussed in a two-layer flow [13]. The interaction problems of water waves with a floating thin barrier and a rectangular box in a two-layer flow were studied by [14] and [15]. The work on the wave dispersion by a cylinder in a homogeneous fluid of limitless depth with an ice cover, later on extended the case for two-layer fluid [16]. Kundu and Cohen [17] explained the significance of the two-layer fluid in the different density's ratios in the two-layer fluid. They also represented the importance of the upper and lower layers of two-layer flow. Various works on break waters across seashores have also been done recently. The cost of setting up wave energy converters based on oscillating water columns (OWCs) in the open water is high, so a combination of these devices with offshore structures like breakwater is cost-saving. Some recent works are as follows: Barman and Bora [18] investigated the scattering of slanted waves with a single chamber of rectangular box type water break. David et al. [19] studied internal wave force by a rectangular bar in a two-layer fluid. Perfectly absorbed wave energy converter of water waves by using perfectly matched layer method was developed by Porter [20]. Multi-scattering technique is used to model wave energy converter due to the hydrodynamic interactions with the geometrical structure [21]. A study of trapped flexural waves was performed by taking two cylinders in the two-layer fluid system [22]. Analysis of hydrodynamic forces by a cylindrical structure in the two-layer fluid system is shown in [23]. The study of instability of nanofluid layer with rotation was performed in [24]. Champak and Hassan [25] studied the surge hydrodynamic force by a floating cylinder in the presence of bottom-mounted obstacle in two-layer fluid.

The role of renewable energies is increasing during this era of energy transition. Therefore, hydro power, solar, and wind energy are the well-known examples of source of renewable energy. Moreover ocean waves are also a good source of renewable energy. In the sea surface, mechanical structures are subject to waves moving up and down. Using an electrical generator installed internally, by this setup as an energy device (oscillating water column) energy from waves is converted into electrical voltage. Most of the vital characteristics of oscillating water columns can be found in the cylindrical hollow structure. The suggested device can therefore be categorized as an oscillating water column (OWC), a type of wave energy technology. When the water wave interacts with a structure, some energy may be transferred from one mode to another mode for arbitrary but stable density ratio. Various methods have been applied to study the interaction of water waves with a horizontal cylinder in the infinite or finite fluid domain, which may be divided into analytical and numerical methods. Less work has been done on the scattering and radiation of water waves by vertical hollow cylinder in two-layer flow. The main objective in this work is to find out the hydrodynamic wave loads results of a vertical hollow cylinder in a two-layer flow. Such work is more complicated and different than a single layer fluid as here there are two possible wave modes of motion.

11.2 Formulation

In order to solve such diffraction problem as shown in Figure 11.1, we consider that the fluid is incompressible, homogeneous and inviscid and the motion is small in amplitude and irrotational in nature. Let us take the whole depth of the two-layer flow is h and h_1 is the depth of the upper layer having density ρ_1, and h_2 is the depth of the lower layer having density ρ_2. The floating vertical hollow cylinder is semi-submerged in the upper layer and its draft is at $z = d$. The bottom is considered flat and impermeable. The direction of wave propagation is regarded along the x-axis while the y-axis is perpendicular in the right-handed Cartesian coordinate system $Oxyz$, with the origin O in the uninterrupted free surface and the z-axis coincident with the cylinder's axis measured positively upwards. We divide the whole region into two regions; the first region is defined by $\Omega_1 =$

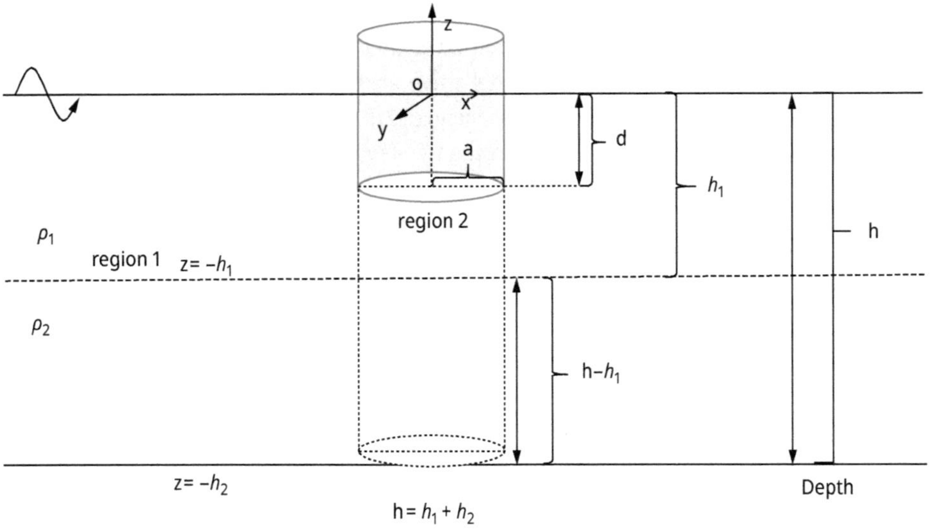

Figure 11.1: Diagram of the device (side view).

$[r > a,\ 0 < \theta \le 2\pi,\ -h \le z \le 0]$ and the second region is defined by $\Omega_2 = [r < a,\ 0 < \theta \le 2\pi,\ -h \le z \le 0]$ as shown in Figure 11.1.

In these physical regions, the corresponding boundary value problems are obtained and subsequently, solve the boundary value problem in each region for the diffracted velocity potentials. Here, we write the total potential as $\Phi^s = Re\left[\phi^s \phi_d^s(r,\ \theta,\ z)e^{-i\omega t}\right]$, where the upper and the lower fluid layers are indicated by the superscript $s = 1,\ 2,$ respectively, and $d = 1,\ 2$ denote the diffracted potentials in region Ω_1 and Ω_2, respectively. Hence, the spatial component of the complex diffracted potential for regions Ω_1 and Ω_2, respectively, are denoted by ϕ_1^s and ϕ_2^s. In the positive x-direction, the expression of the incident velocity potential having angular frequency ω is given by

$$\phi_{inc} = -\frac{i\bar{a}(\mu_n)g\,Z^{(s)}(\mu_n,\ z)}{\omega}\sum_{m=0}^{\infty}\epsilon_m i^m J_m(\mu_n r)\cos m\theta \tag{11.1}$$

where $\epsilon_0 = 1$, $\epsilon_m = 2$ for $m \ge 1$, $i = \sqrt{-1}$, g is the gravitational acceleration, J (.) stands for Bessel function of the first kind having order m and μ_n is given below as the dispersion relation:

$$\omega^2 = \frac{g\,\mu_n\left\{(g_1+g_2) \pm \sqrt{(g_1+g_2)^2 - 4\epsilon\, g_1 g_2(1+\gamma g_1 g_2))}\right\}}{2(1+\gamma g_1 g_2)} \tag{11.2}$$

$\epsilon = 1-\gamma$, $g_1 = \tanh(h_1)$, $g_2 = \tanh(h_2)$ and

$$Z^{(1)}(\mu_n, z) = \frac{\left(\frac{\omega^2}{g\nu_n}\right)\sinh\mu_n z + \cosh\mu_n z}{a(\mu_n)} \tag{11.3}$$

$$Z^{(2)}(\mu_n, z) = \frac{\omega^2}{g\mu_n \sinh \mu_n h_2} \cosh \mu_n (z + h) \tag{11.4}$$

Here $a(\mu_n)$ represents the amplitude defined as:

$$a(\mu_1) = a_0^{(1)}(\mu_1) \cosh \mu_1 h_1 \left(1 - \frac{g\mu_1}{\omega^2} \tanh \mu_1 h_1\right) \tag{11.5}$$

$$a(\mu_2) = a_0^{(2)}(\mu_2), \tag{11.6}$$

where $a_0^{(1)}(\mu_1)$ and $a_0^{(2)}(\mu_2)$ represent the amplitudes of the incident surface and inner wave modes, respectively.

11.3 Boundary value problem and their solution

11.3.1 The governing equation and boundary conditions

The governing equation and the respective boundary conditions for the diffracted potentials in regions Ω_1 $(i = 1)$ and Ω_2 $(i = 2)$ are given by:

$$\nabla^2 \phi_i^{(s)} = 0; \ (-\boldsymbol{h} < \boldsymbol{z} < 0, \ 0 < \boldsymbol{r} < \boldsymbol{a}; \boldsymbol{r} > \boldsymbol{a}; -\infty < \boldsymbol{x}, \boldsymbol{y} < \infty) \tag{11.7}$$

$$\frac{\partial \phi_i^{(1)}}{\partial \boldsymbol{z}} = \frac{\boldsymbol{\omega}^2}{\boldsymbol{g}} \phi_i^{(1)}; \ (\boldsymbol{z} = 0) \tag{11.8}$$

$$\frac{\partial \phi_i^{(2)}}{\partial \boldsymbol{z}} = 0; \ (\boldsymbol{z} = -\boldsymbol{h}) \tag{11.9}$$

$$\frac{\partial \phi_i^{(1)}}{\partial z} = \frac{\partial \phi_i^{(2)}}{\partial z}; \ (\boldsymbol{z} = -\boldsymbol{h}_1) \tag{11.10}$$

$$\gamma \left(\frac{\partial \phi_i^{(1)}}{\partial \boldsymbol{z}} - \frac{\boldsymbol{\omega}^2}{\boldsymbol{g}} \phi_i^{(1)} \right) = \frac{\partial \phi_i^{(2)}}{\partial \boldsymbol{z}} - \boldsymbol{K} \phi_i^{(2)}; \ (\boldsymbol{z} = -\boldsymbol{h}_1) \tag{11.11}$$

and the radiation condition, which is valid in the unbounded region:

$$\lim_{r \to \infty} \sqrt{r} \left(\frac{\partial \phi_1^{(1)}}{\partial r} - ik\phi_1^{(1)} \right) = 0 \tag{11.12}$$

11.3.2 Matching conditions

The undetermined coefficients in the analytical expression of the diffraction potentials are evaluated by maintaining the continuity of pressure and velocity along the virtual and physical boundaries. Whereas the spatial component of the complex velocity potential $\phi_d^{(s)}$ is divided into two potentials denoted by:

$$\phi_d^{(s)} = \begin{cases} \phi_1^{(s)}(r \geq a) \\ \phi_2^{(s)}(r < a) \end{cases} \tag{11.13}$$

Now with the necessary matching conditions on the edge of the boundary $r = a$, we establish:

$$\phi_1^{(s)}(\boldsymbol{r}, \boldsymbol{\theta}, \boldsymbol{z}) = \phi_2^{(s)}(\boldsymbol{r}, \boldsymbol{\theta}, \boldsymbol{z});\ (-\boldsymbol{h} \leq \boldsymbol{z} \leq -\boldsymbol{d}) \tag{11.14}$$

$$\frac{\partial \phi_1^{(s)}}{\partial r} = \begin{cases} \frac{\partial \phi_{inc}^{(s)}}{\partial r};\ (-d \leq z \leq 0) \\ \frac{\partial \phi_2^{(s)}}{\partial r};\quad (-h \leq z \leq -d) \end{cases} \tag{11.15}$$

In order to evaluate the undetermined coefficients that emerge in the expression of diffracted potentials, the aforesaid matching criteria are to be applied.

11.3.3 Methodology to obtain solutions

Due to the fact that the problem is formulated in terms of diffracted potentials, the expansion of matched eigenfunction and variables separation tools are used to determine the diffracted potentials in the identified regions of Ω_1 and Ω_2. On the application of the above methods to the boundary-value problem (eqs. (11.7)–(11.12)) for the region Ω_1, the diffracted potential $\phi_1^{(s)}$ in region Ω_1 can be written as:

$$\begin{aligned} &\phi_1^{(s)}(r, \theta,\ z) = \\ &\frac{-i\, a_n\, (\mu_n) g}{\omega} \sum_{m=0}^{\infty} \epsilon_m i^m \cos m\theta\ \{J_m(\mu_n r)\}\, Z^{(s)}(\mu_n,\ z) + \\ &A_{m0}^{(n,\ 1)} H^*{}_m(\mu_1, r) Z^{(s)}(\mu_1, z) + A_{m0}^{(n,\ 2)} H^*{}_m(\mu_2, r) Z^{(s)}(\mu_2, z) + \\ &\sum_{j=1}^{\infty} A_{mj}^{(n,\ 1)} K^*{}_m\left(\mu_j^{(1)}, r\right) Z^{(s)}\left(\mu_j^{*(1)}, z\right) + \sum_{j=1}^{\infty} A_{mj}^{(n,\ 2)} K^*{}_m\left(\mu_j^{(2)}, r\right) Z^{(s)}\left(\mu_j^{*(2)}, z\right), \end{aligned} \tag{11.16}$$

where the vertical eigenfunctions $Z^{(1)}(\mu_n, z)$ and $Z^{(2)}(\mu_n, z)$ are given by eqs. (11.3) and (11.4), respectively, and $\mu_j^{(1)}$ and $\mu_j^{(2)}$ are roots of the dispersion relations, which are given as:

$$\omega^2 = -\frac{g\, \mu^{(1)} \left\{ (G_1 + G_2) + \sqrt{(G_1 + G_2)^2 - 4\epsilon\, G_1 G_2 (1 - \gamma G_1 G_2)} \right\}}{2(1 - \gamma G_1 G_2)}, \tag{11.17}$$

$$\omega'^2 = -\frac{g\,\mu^{(2)}\left\{(G_1+G_2)-\sqrt{(G_1+G_2)^2-4\epsilon\, G_1G_2(1-\gamma G_1G)}\right\}}{2(1-\gamma G_1G_2)} \tag{11.18}$$

where

$$G_1 = \tanh\mu^{(1)}h_1, G_2 = \tanh\left(\mu^{(2)}h_2\right) \tag{11.19}$$

$$\mu_j^{*(1)} = -i\mu_j^{(1)},\ \mu_j^{*(2)} = -i\mu_j^{(2)} \tag{11.20}$$

$$H_m^*(\mu_n, r) = \frac{H_m^{(1)}(\mu_n r)}{H_m^{(1)}(\mu_n a)} \tag{11.21}$$

$$K_m^*\left(\mu_j^{(n)}, r\right) = \frac{K_m\left(\mu_j^{(n)} r\right)}{K_m\left(\mu_j^{(n)} a\right)} \tag{11.22}$$

where $n = 1$ or 2, $H_m^{(1)}$ $(m = 0,1,2, \ldots)$ are the Hankel functions of order m to satisfy the radiation condition and K_m $(m = 0,1,2, \ldots)$ are modified Bessel functions of second kind of order m.

Similarly, the same methods are applied as indicated above for determining diffracted potential of the boundary-value problem (eqs. (11.7)–(11.12)). Thus, the diffracted potential $\phi_2^{(s)}$ in region Ω_2 can be written as:

$$\begin{aligned}\phi_2^{(s)}(r,\theta,z) = \frac{-ia_n(\mu_n)g}{\omega}\sum_{m=0}^{\infty}\epsilon_m i^m \cos m\theta\Big\{ B_{m0}^{(n,1)}V_m\left(\lambda_0^{(1)}, r\right)Y\left(\lambda_0^{(1)}, r\right) + \\ B_{m0}^{(n,2)}V_m\left(\lambda_0^{(2)}, r\right)Y\left(\lambda_0^{(2)}, r\right) + \sum_{k=1}^{\infty}B_{mk}^{(n,1)}V_m\left(\lambda_k^{(1)}, r\right)Y\left(\lambda_k^{(1)}, r\right) + \\ \sum_{k=1}^{\infty}B_{mk}^{(n,2)}V_m\left(\lambda_k^{(2)}, r\right)Y\left(\lambda_k^{(2)}, r\right)\Big\}\end{aligned} \tag{11.23}$$

where $V_m(\lambda_0^{(1)}, r)$, $V_m(\lambda_0^{(2)}, r)$, $V_m(\lambda_k^{(1)}, r)$, and $V_m(\lambda_k^{(2)}, r)$ are the radial functions, which are given by

$$V_m\left(\lambda_0^{(1)}, r\right) = \frac{J_m\left(\lambda_0^{(1)} r\right)}{J_m\left(\lambda_0^{(1)} a\right)} \tag{11.24}$$

$$V_m\left(\lambda_0^{(2)}, r\right) = \frac{J_m\left(\lambda_0^{(2)} r\right)}{J_m\left(\lambda_0^{(2)} a\right)} \tag{11.25}$$

$$V_m\left(\lambda_k^{(1)}, r\right) = \frac{I_m\left(\lambda_k^{(1)} r\right)}{I_m\left(\lambda_k^{(1)} a\right)} \tag{11.26}$$

$$V_m\left(\lambda_k^{(2)},r\right)=\frac{I_m\left(\lambda_k^{(2)}r\right)}{I_m\left(\lambda_k^{(2)}a\right)} \tag{11.27}$$

and the vertical eigenfunctions $Y\left(\lambda_0^{(1)},r\right)$, $Y\left(\lambda_0^{(2)},r\right)$, $Y\left(\lambda_k^{(1)},r\right)$, and $Y\left(\lambda_k^{(2)},r\right)$, $k\geq 1$ can be expressed as:

$$Y\left(\lambda_0^{(1)},z\right)=\begin{cases}\dfrac{\lambda_0^{(1)}\cosh\lambda_0^{(1)}z+K\sinh\lambda_0^{(1)}z}{K\cosh\lambda_0^{(1)}h_1-\lambda_0^{(1)}\sinh\lambda_0^{(1)}h_1}; & (-h_1<z<0)\\[2ex] \dfrac{\cosh\lambda_0^{(1)}(z+h)}{\sinh\lambda_0^{(1)}h_2}; & (-h<z<-h_1)\end{cases} \tag{11.28}$$

$$Y\left(\lambda_k^{(1)},z\right)=\begin{cases}\dfrac{\lambda_k^{(1)}\cos\lambda_k^{(1)}z+K\sin\lambda_k^{(1)}z}{K\cos\eta_k^{(1)}h_1-\eta_k^{(1)}\sin\eta_k^{(1)}h_1}; & (-h_1\leq z\leq 0)\\[2ex] \dfrac{\cos\lambda_k^{(1)}(z+h)}{\sin\lambda_k^{(1)}h_2}; & (-h<z<-h_1)\end{cases} \tag{11.29}$$

$$Y\left(\lambda_0^{(2)},z\right)=\begin{cases}\dfrac{\lambda_0^{(2)}\cosh\lambda_0^{(2)}z+K\sinh\lambda_0^{(2)}z}{K\cosh\lambda_0^{(2)}h_1-\lambda_0^{(2)}\sinh\lambda_0^{(2)}h_1}; & (-h_1\leq z\leq 0)\\[2ex] \dfrac{\cosh\lambda_0^{(2)}(z+h)}{\sinh\lambda_0^{(2)}h_2}; & (-h\leq z\leq -h_1)\end{cases} \tag{11.30}$$

$$Y\left(\lambda_k^{(2)},z\right)=\begin{cases}\dfrac{\lambda_k^{(2)}\cos\lambda_k^{(2)}z+K\sin\lambda_k^{(2)}z}{K\cos\lambda_k^{(2)}h_1-\lambda_k^{(2)}\sin\lambda_k^{(2)}h_1}; & (-h_1\leq z\leq 0)\\[2ex] \dfrac{\cos\lambda_k^{(2)}(z+h)}{\sin\lambda_k^{(2)}h_2}; & (-h\leq z\leq -h_1).\end{cases} \tag{11.31}$$

In the equations of diffracted potentials represented by eqs. (11.16) and (11.23), it should be noted that $A_{mj}^{(n,1)}$, $A_{mj}^{(n,2)}$, $B_{mk}^{(n,1)}$, and $B_{mk}^{(n,2)}$, are the unknown coefficients that must be calculated by using the proper matching conditions.

11.3.4 Undetermined coefficients

In order to find undetermined coefficients of the diffracted potentials in regions Ω_1 and Ω_2, we utilize the matching situations, which show the consistency of flow velocity and its pressure. Thus, the diffracted potentials represented by eqs. (11.16) and (11.23) are applied to the matching conditions, which are given by eqs. (11.13)–(11.15), we have

$$J_m(\mu_n a)Z^s(\mu_n, z) + \sum_{j=0}^{\infty} A_{mj}^{(n,1)} Z^{(s)}(\mu_j^{*(1)}, z) + \sum_{(j=0)}^{\infty} A_{mj}^{(n,2)} Z^{(s)}(\mu_j^{*(2)}, z) = \sum_{k=0}^{\infty} B_{mk}^{(1)} Y(\lambda_k^{(1)}, z) + \sum_{k=0}^{\infty} B_{mk}^{(n,2)} Y(\lambda_k^{(2)}, z) \tag{11.32}$$

where

$$\mu_0^{*(1)} = \mu_0^{(1)} = \mu_1, \mu_0^{*(2)} = \mu_0^{(2)} = \mu_2, \ \mu_j^{*(1)} = i\,\mu_j^{(1)}, \mu_j^{*(2)} = -i\,\mu_j^{(2)}$$

$$\mu_n J_m'(\mu_n a)Z^{(s)}(\mu_n, z) + \mu_0^{(1)} A_{m0}^{(n,\,1)} H_1^{*'}\left(\mu_0^{(1)}, a\right) Z^{(s)}\left(\mu_0^{(1)}, z\right) + \mu_0^{(2)} A_{m0}^{(n,\,2)} H_m^{*'}\left(\mu_0^{(2)}, a\right) Z^{(s)}\left(\mu_0^{(2)}, z\right)$$
$$+ \sum_{j=1}^{\infty} \mu_j^{(1)} A_{mj}^{(n,\,1)} K_1^{*'}\left(\mu_j^{*(1)}, a\right) Z^{(s)}\left(\mu_j^{(1)}, z\right) + \sum_{j=1}^{\infty} \mu_j^{(2)} A_{mj}^{(n,\,2)} K_1^{*'}\left(\mu_j^{(2)}, a\right) Z^{(s)}\left(\mu_j^{(2)}, z\right) =$$
$$\begin{cases} 0; & (-d \le z \le 0), \\ \sum_{k=0}^{\infty} \lambda_k^{(1)} B_{mk}^{(n,1)} V_m'\left(\lambda_k^{(1)}, a\right) Y\left(\lambda_k^{(1)}, z\right) + & (-h \le z \le -d) \\ \sum_{k=0}^{\infty} \lambda_k^{(2)} B_{mk}^{,(n,\,2)} V_m'\left(\lambda_k^{(2)}, a\right) Y\left(\lambda_k^{(2)}, z\right). \end{cases} \tag{11.33}$$

Equations (11.32) and (11.33) are now multiplied by the orthogonal functions as per their respective regions and then integrated along the corresponding interface boundaries. The system of linear equations $MX = B$ can be obtained. Here, the undetermined coefficients $A_{mj}^{(n,1)}$, $A_{mj}^{(n,2)}$, $B_{mk}^{(n,1)}$, and $A_{mk}^{(n,2)}$ can be obtained by solving obtained system of linear equations after truncating appropriately, say, $j = 0, 1, 2, 3, \ldots, J$ and $k = 0, 1, 2, 3, \ldots, K$. As a results, we construct an algebraic linear system having orders $[2(J + 1) + 2(K + 1)] \times [2(J + 1) + 2(K + 1)]$ and then use MATLAB programming to solve this system of linear equations for the unknown coefficients.

11.4 Wave exciting force

From the linearized Bernoulli's formula, the exciting forces due to the diffracted potential and the incident potential waves are calculated as follows:

$$F_x = i\,a\,\omega \int_{-d}^{0}\int_{0}^{2\pi} \rho(z)\phi_1^{(s)}(a,\ \theta,\ z)(-\cos\theta)d\theta\,dz \tag{11.34}$$

or

$$F_x = i\, a\, \omega \int_0^{2\pi} \left[\int_{-d}^{h_1} \rho_2(z)\phi_1^{(2)}(a,\ \theta,\ z) + \int_{-h_1}^{0} \rho_1(z)\phi_1^{(1)}(a,\ \theta,\ z) \right] (-\cos\theta) d\theta\, dz \tag{11.35}$$

where $\rho(z)$ is the density in the region defined by

$$\rho(z) = \begin{cases} \rho_1, & -h_1 \le z \le 0 \\ \rho_2, & -h \le z \le -h_1 \end{cases} \tag{11.36}$$

11.5 Numerical results

11.5.1 Convergence table

We now studied the convergence tables of the exciting forces in the x-direction due to the incident wave of surface mode, i.e., $F_x^{(1)}$ and for the incident wave of internal mode, i.e., $F_x^{(2)}$ against the various iterations of eigenvalue μ_j for the frequency $\omega\sqrt{h/g}$. From Table 11.1, it can be seen that there is a good convergence in the values of exciting force, which validates our results.

Table 11.1: Convergence study of non-dimensionalized exciting force $F_x^{(1)}/(a\pi\rho_1 h_2)$ due to diffraction for various iterations and different values of nondimensional angular frequencies $\omega\sqrt{h/g}$ with h_1: h_2 = 7: 3, d = 0.4 h, a = 0.2 h.

$\omega\sqrt{\frac{h}{g}}$	$(J+1, K+1)$	$\frac{F_x^{(1)}}{a\,\pi\rho_1 h^2}$	$\omega\sqrt{\frac{h}{g}}$	$(J+1, K+1)$	$\frac{F_x^{(1)}}{a\,\pi\rho_1 h^2}$
0.06	(4, 4)	0.0262413256	1.58	(4, 4)	0.8213474615
	(8, 8)	0.0262414137		(8, 8)	0.8213475368
	(12, 12)	0.0262445210		(12, 12)	0.8213457296
	(16, 16)	0.0262445542		(16, 16)	0.8213465023
	(20, 20)	0.0262445731		(20, 20)	0.8213462071
0.16	(4, 4)	0.1269900412	1.98	(4, 4)	0.6199261479
	(8, 8)	0.1269900532		(8, 8)	0.6199264782
	(12, 12)	0.1269900009		(12, 12)	0.6199262716
	(16, 16)	0.1269895691		(16, 16)	0.6199277374
	(20, 20)	0.1269896128		(20, 20)	0.6199272134
0.26	(4, 4)	0.0985730024	2.11	(4, 4)	0.6063914561
	(8, 8)	0.0985730044		(8, 8)	0.6063916663
	(12, 12)	0.0985730037		(12, 12)	0.6063943256
	(16, 16)	0.0985739963		(16, 16)	0.6063985714
	(20, 20)	0.0985739965		(20, 20)	0.6063988178

Table 11.1 (continued)

$\omega\sqrt{\frac{h}{g}}$	$(J+1, K+1)$	$\frac{F_x^{(1)}}{a\,\pi\rho_1 h^2}$	$\omega\sqrt{\frac{h}{g}}$	$(J+1, K+1)$	$\frac{F_x^{(1)}}{a\,\pi\rho_1 h^2}$
0.52	(4, 4)	0.1906525123	2.64	(4, 4)	0.1566964525
	(8, 8)	0.1906526128		(8, 8)	0.1566978173
	(12, 12)	0.1906524191		(12, 12)	0.1566975608
	(16, 16)	0.1906524188		(16, 16)	0.1566975554
	(20, 20)	0.1906524187		(20, 20)	0.1566979271
1.05	(4, 4)	0.3087812589	3.17	(4, 4)	0.3257021473
	(8, 8)	0.3087814035		(8, 8)	0.3257025261
	(12, 12)	0.3087808341		(12, 12)	0.3257044295
	(16, 16)	0.308783315		(16, 16)	0.3257048495
	(20, 20)	0.3087833186		(20, 20)	0.3257041058

11.5.2 Discussion

After finding the diffracted potentials along with its unknown coefficients in each defined region, our next step is to find the related exciting forces by the structure and it is important to obtain the significant influence of different variables of the device on the exciting forces. In order to observe it, we need to show all the obtained results through suitable graphs with distinct values of the variables such as length of submerged part (d), and depth ratio for two-layer fluid ($\frac{h_1}{h_2}$) of the device. Here, we prescribe the values of following variables all over our evaluations, i.e., the value for the acceleration due to gravity (g), density ratio (γ), ϵ, depth (h), and the radius (a) are taken as $g = 9.8$ m/s^2, $\gamma = 0.97$, $\epsilon = 1 - \gamma = 0.30$, $h = h_1 + h_2 = 10$ m, and $a = 0.5$ h and the exciting forces are non-dimensionalized by $\rho_1 gaha_0^{(n)}$. For the study of the effects of depth ratio (h_1/h_2) on exciting forces of the floating cylinder, we have taken three cases as: *(I)*$\frac{h_1}{h_2} = 3/7$, i.e., the bottom of floating cylinder presence in the lower layer, *(ii)* $\frac{h_1}{h_2} = 7/3$, i.e., the bottom of the cylinder presence in the upper layer, and *(iii)* $\frac{h_1}{h_2} = 5/5$, i.e., when the bottom of the floating cylinder is at the interface along with variation of drafts and is studied for different values as $d = 0.2$ h, 0.3 h, 0.4 h, . . ., 0.7 h, etc.

11.5.3 Results

Based on the derived expressions of the exciting force acting on the cylinder, a set of results are obtained, which shows the effect of various parameters on the exciting force. Figures (11.2) and (11.3) represent the magnitude of the exciting forces against different depth ratios in the surface wave and inner modes, respectively. From the figures, one can observe that in the surface-wave mode, the exciting forces show the same pattern, and the exciting forces reach maximum peaks for both internal and surface wave modes when the bottom of cylinder is present at the interface $\frac{h_1}{h_2} = 5/5$. The exciting force corresponding to the depth ratio $\frac{h_1}{h_2} = 5/5$ attains maximum peak value.

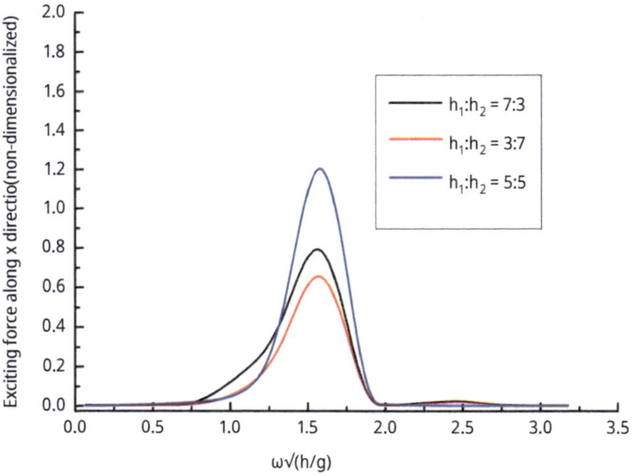

Figure 11.2: Magnitude of nondimensional exciting forces in the mode of surface wave when $y = 0.97$, $a = 0.5$ h, and $d = 0.5$ h.

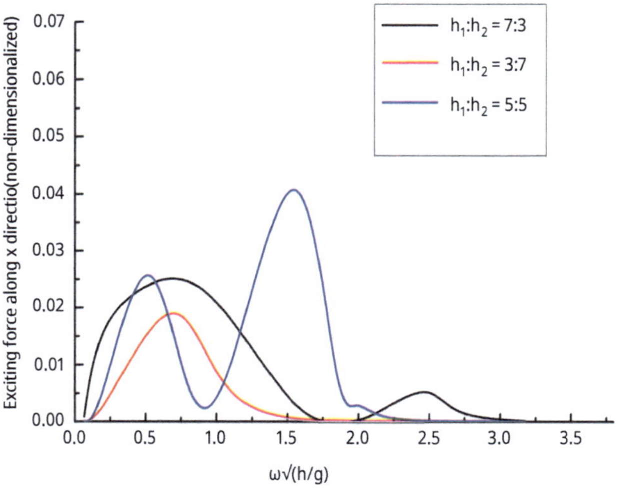

Figure 11.3: Magnitude of nondimensional exciting forces in the mode of inner wave when $y = 0.97$, $a = 0.5$ h, $d = 0.5$ h.

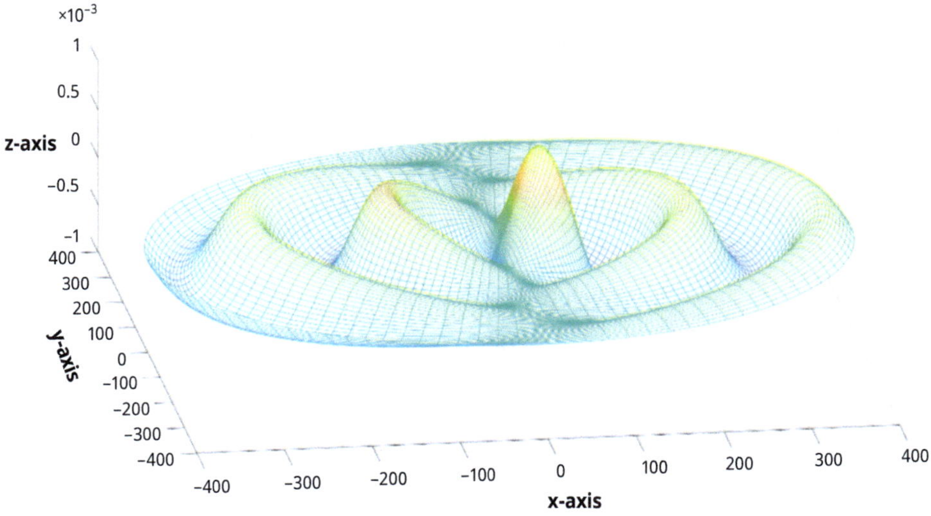

Figure 11.4: Free surface elevation by the diffracted waves of the surface wave mode: $\gamma = 0.97$, $a = 0.5$ h, $d = 0.5$ h, and $\omega\sqrt{\frac{h}{g}} = 0.3967$.

Figures 11.4 and 11.5 give the elevations at free surface and interface, respectively, by the diffracted wave of surface mode due to incident wave of surface mode, i.e., $n = 1$. We observe that both the elevations have similar shape but the elevation for the free surface is greater than that for the interface.

Figures 11.6 and 11.7 give the elevations at free surface and interface, respectively, for the diffracted wave of surface mode due to incident wave of internal mode, i.e., $n = 2$.

We see similar behavior as previous cases of Figures 11.2 and 11.3, i.e., free surface elevation greater than interface in case of diffracted wave of surface wave mode and interface elevation greater than free surface mode in the diffracted wave of internal wave mode.

Figures 11.8 and 11.9 are presented or the surface and interface elevation, respectively of diffracted waves of internal mode due to incident wave of internal mode. Here, we see as in Figures 11.4 and 11.5 that the interface elevation is higher than the free surface elevation for the diffracted wave of internal wave mode.

Now, here we discuss the elevation at free surface and interface for region 2, i.e., Ω2 by the diffracted waves of surface or internal mode due to incident mode (n=1 or 2), i.e., surface or internal mode.

The free surface and interface elevations are shown in Figures 11.10 and 11.11, respectively, by the diffracted surface mode wave caused by the incident surface wave mode, or $n = 1$. We observe that both the elevations have similar shape but the eleva-

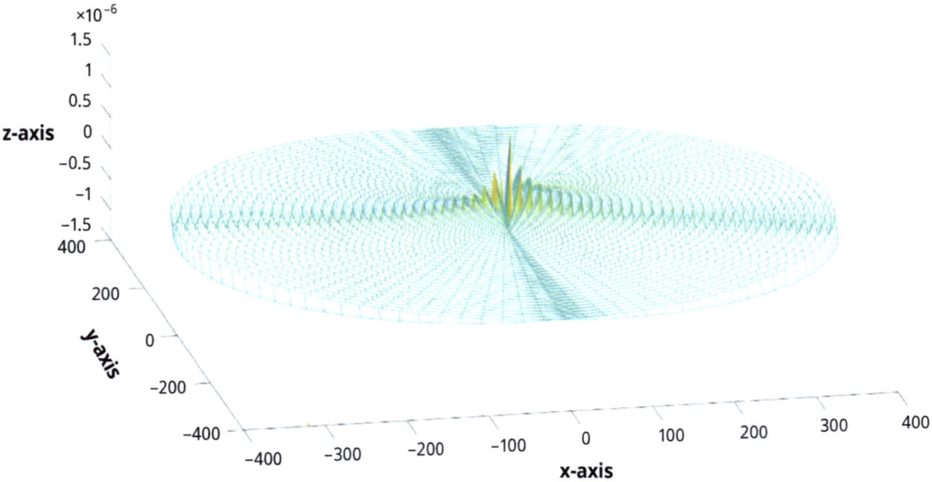

Figure 11.5: Interface elevation by the diffracted waves of internal waves mode: $\gamma = 0.97$, $a = 0.5$ h, $d = 0.5$ h, and $\omega\sqrt{\frac{h}{g}} = 0.3967$.

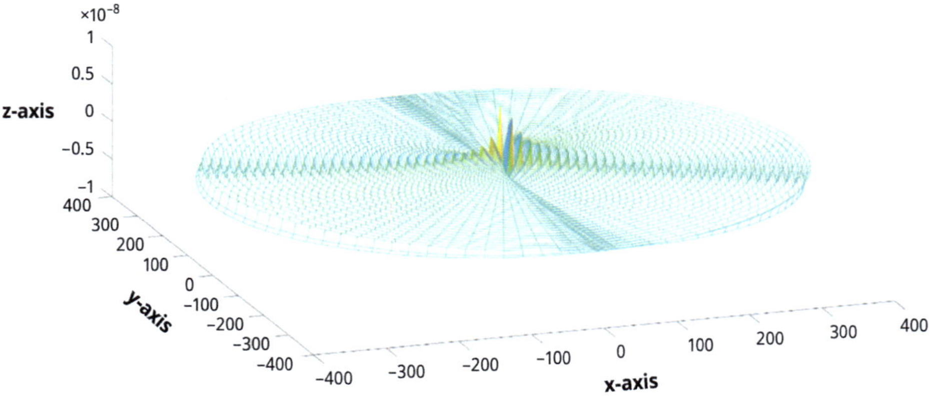

Figure 11.6: Free surface elevation formed by the diffraction of internal wave mode when taking incident wave of surface wave mode ($n = 1$) (region 1) with $\gamma = 0.97$, $a = 0.5$ h, $d = 0.5$ h, and $\omega\sqrt{\frac{h}{g}} = 0.3967$.

tion for the free surface is greater than that for the interface. All the results follow the same pattern as seen in region 1.

Figures 11.12 and 11.13 are the elevations at free surface elevation and interface, respectively, for the diffracted wave in the internal wave mode due to incident wave of surface mode. The same results can be seen as in previous cases, i.e., magnitude of

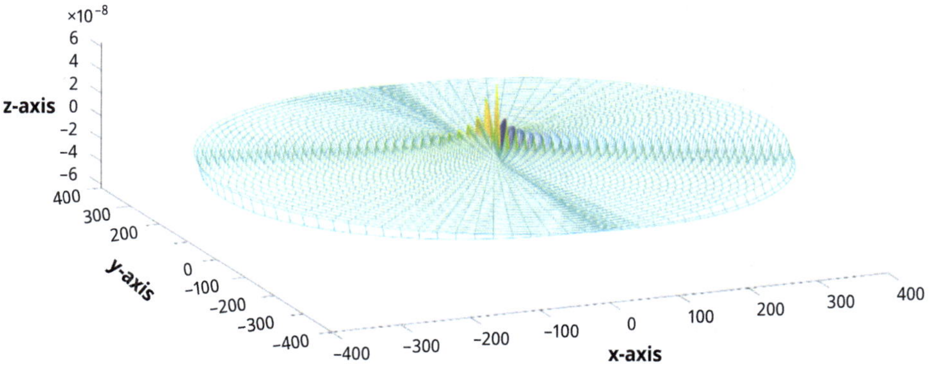

Figure 11.7: Interface elevation formed by the diffraction of internal wave mode when taking incident wave of surface wave mode ($n = 1$) (region 1) with $\gamma = 0.97$, $a = 0.5$ h, $d = 0.5$ h, and $\omega\sqrt{\frac{h}{g}} = 0.3967$.

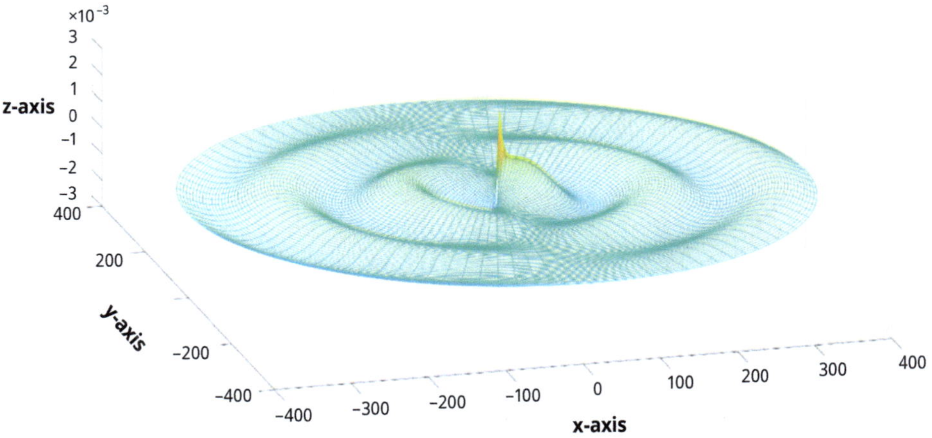

Figure 11.8: Free surface elevation formed by the diffraction of surface wave mode when taking incident wave of internal wave mode ($n = 2$) (region 1) with $\gamma = 0.97$, $a = 0.5$ h, $d = 0.5$ h, and $\omega\sqrt{\frac{h}{g}} = 0.3967$.

interface elevation is greater than surface elevation for diffracted wave of internal wave mode. When the radius of the cylinder is half of the total depth and the bottom of the cylinder is at the interface, only one peak value is observed for the surface wave mode while multiple peaks can be seen for the inner wave mode. As a result, internal and surface wave modes have a significant effect on the exciting force. Figures 11.10 and 11.11 show the 3D simulation of the free surface and interface elevation in surface and internal wave modes.

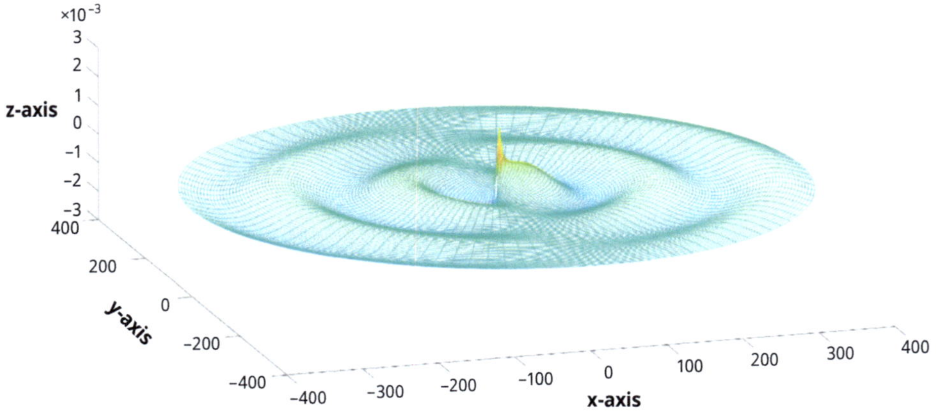

Figure 11.9: Interface elevation formed by the diffraction of surface wave mode when taking incident wave of internal wave mode ($n = 2$) (region 1) with $\gamma = 0.97$, $a = 0.5$ h, $d = 0.5$ h, and $\omega\sqrt{\frac{h}{g}} = 0.3967$.

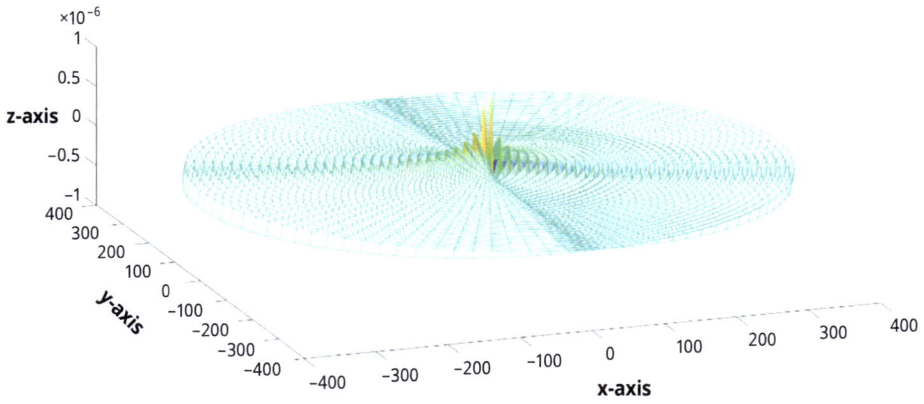

Figure 11.10: Free surface elevation formed by the diffracted wave of internal mode when incident wave of internal mode ($n = 2$) (region 1) is taken with $\gamma = 0.97$, $a = 0.5$ h, $d = 0.5$ h, and $\omega\sqrt{\frac{h}{g}} = 0.3967$.

11.6 Conclusion

In this chapter, we contemplate a floating hollow cylindrical structure immersed in the fluid of two-layer with finite impermeable depth. Linear water wave theory is applied to examine and solve the corresponding diffraction problem for diffracted potentials in identified regions. In order to solve this diffraction problem, variable separation and matched eigenfunction expansion are applied. The obtained diffracted potentials enable us to determine the exciting forces acting on the cylinder. There are a number of results presented regarding the exciting force along with several param-

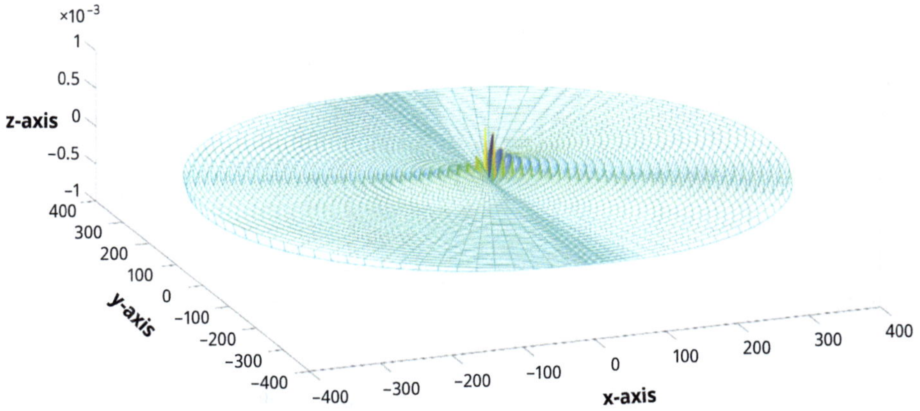

Figure 11.11: Interface elevation formed by the diffracted wave of internal mode when incident wave of internal mode ($n = 2$) (region 1) is taken with $\gamma = 0.97$, $a = 0.5$ h, $d = 0.5$ h, and $\omega\sqrt{\frac{h}{g}} = 0.3967$.

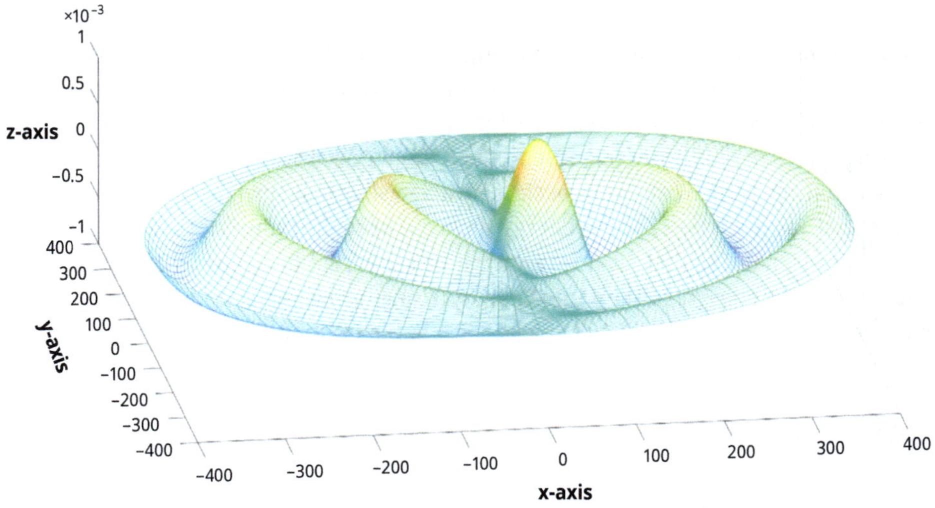

Figure 11.12: Free surface elevation formed by surface mode diffracted wave when incident wave of surface wave mode ($n = 1$) (region 2) is taken with $\gamma = 0.97$, $a = 0.5$ h, $d = 0.5$ h, and $\omega\sqrt{\frac{h}{g}} = 0.3967$.

eters of the device. With the different drafts of cylinder and different depth ratios, the exciting forces for surface wave and inner wave modes are plotted when the cylinder is located at an interface, the upper layer, or the lower layer. A remarkable effect of two-layer fluid on the exciting force is observed for a wide range of parameter values of the device. It can be seen that the exciting forces are directly proportional to the draft of the cylinder in surface and inner wave modes. For the surface and the inter-

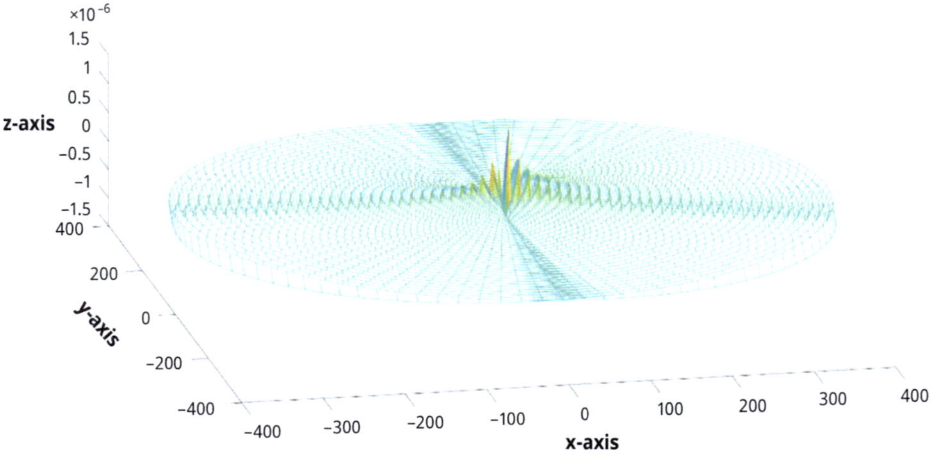

Figure 11.13: Interface elevation formed by surface mode diffracted wave when incident wave of surface wave mode ($n = 1$) (region 2) is taken with $\gamma = 0.97$, $a = 0.5$ h, $d = 0.5$ h, and $\omega\sqrt{\frac{h}{g}} = 0.3967$.

nal wave modes, oscillations in the exciting force are observed with drafts of the cylinder. It is noticed that compared to the internal wave modes, surface wave modes exhibit higher exciting forces, whereas for the internal wave modes, multiple peaks of exciting forces can be observed. Due to the hollow cylindrical structure's inclusion of the oscillating water column's basic characteristics, the proposed model can be thought of as one type of wave energy device. Through it an electrical generator is installed internally so that wave energy can be converted to electrical energy. The results of this chapter give useful information to engineers to ensure maximum energy extraction from the device by placing it in a suitable location.

References

[1] Stokes G. G. (1847). On the theory of oscillatory waves. Transactions of the Cambridge Philosophical Society, 8, 441–455.

[2] Benjamin T. B. (1967). Internal waves of permanent form in fluids of great depth. Journal of Fluid Mechanics, 29, 559–592. https://10.1017/S002211206700103X.

[3] Keulegan G. H. (1953). Characteristics of internal solitary waves. Journal of Research of the National Bureau of Standards, 51, 133–140. http://dx.doi.org/10.6028/jres.051.017.

[4] Long R. R. (1956). Solitary waves in one-and two-fluid systems. Tellus, 8, 460–471.

[5] Zhu S. P., and Mitchell L. (2009). Diffraction of ocean waves around a hollow cylindrical shell structure. Wave Motion, 46, 78–88.

[6] Sarah C. M., Meunier P., and Dizès L. S. (2021). Fluid-structure interaction of a cylinder in a stratified fluid. Journal of Fluid Mechanic, 915(A97). doi: 10.1017/jfm.2021.155s.

[7] Hassan M., Bora P., and Biswakarma M. (2019). Wave forces due to interaction of water wave with a cylindrical hollow structure in water of finite depth. International Journal of Scientific Research in Mathematical and Statistical Sciences, 6(2), 73–77.
[8] Baines P. G. (1984). A unified description of two-layer flow over topography. Journal of Fluid Mechanics, 146, 127–168.
[9] Chanda A., and Bora S. N. (2020). Investigation of water wave scattering by an elastic sea-bed of varying depth in two superposed fluids covered by an ice-sheet. Ocean Engineering, 221 108510. https://doi.org:10.1016/j.oceaneng.2020.108510
[10] Linton C. M., and McIver M. (1995). The interaction of waves with horizontal cylinders in two-layer fluids. Journal of Fluid Mechanics, 304, 213–229.
[11] Yeung R. W., and Nguyen T. (1999). Radiation and diffraction of water waves in a two layer fluid. In Proceedings of the 22nd Symposium on Naval Hydrodynamics, Washington DC USA, pp. 875–887.
[12] Cadby J. R., and Linton C. M. (2000). Three-dimensional water wave scattering in two layer-fluid. Journal of Fluid Mechanics, 423, 153–173. https://doi.org/10.1017/S0022112000002007.
[13] Linton C. M., and Jonathan R. C. (2002). Scattering of oblique waves in a two-layer fluid. Journal of Fluid Mechanics, 461, 343–364.
[14] You Y. X., Miao G. P., and Cheng J. S. (2005). The interaction of water waves with a vertical barrier in a two layer fluid. Acta Mechanica Sinica, 37(5), 529–541.
[15] Shi Q., You Y. X., Wei G., and Miao G. P. (2006). The interaction of water waves with a floating rectangular box in a two-layer fluid. Progress in Naural Science, 16(8), 992–1001.
[16] Das D., and Mandal B. N. (2006). Oblique Wave Scattering by a Circular Cylinder in Two Layer Fluid with an Ice-cover in 21st IWWWFB, Loughborough University, pp. 29–32.
[17] Kundu P. K., and Cohen I. M. (2002). Fluid Mechanics, San Diego, CA, USA: Academic Press.
[18] Barman K. K., and Bora S. N. (2021). Interaction of oblique water waves with a single chamber caisson type breakwater for a two-layer fluid flow over an elastic bottom. Ocean Engineering, 236, 109766. https://doi.org/10.1016/j.oceaneng.2021.109766.
[19] David M. O., Yang D. W., Xuliang H., and Ping C. (2014). Internal wave exciting force on a rectangular barge in a two layer fluid of finite depth. Advancement in Scientific and Engineering Research, 2(3), 48–61.
[20] Porter R. (2021). Modelling and design of a perfectly-absorbing wave energy converter. Applied Ocean Research, 113, 102724. https://doi.org/10.1016/j.apor.2021.102724.
[21] Göteman M., Engström J., Eriksson M., and Isberg J. (2015). Fast modeling of large wave energy farms using interaction distance cut-off. Energies, 8(12), 13741–13757.
[22] Saha S., and Bora S. N. (2020). Trapped flexural waves supported by a pair of identical cylinders in a two-layer fluid. SN Applied Sciences, 2, 1455. doi: 10.1007/s42452-020-03229-5.
[23] Tashi L., and Hassan M. (2024). Analysis of hydrodynamic loads by a vertical hollow cylindrical structure in a two-layer fluid of finite depth. Trends in Science, 21, 7730.
[24] Yadav D., Bhargava R., and S G. (2013). Agrawal Numerical solution of a thermal instability problem in a rotating nanofluid layer. International Journal of Heat and Mass Transfer, 63, 313–322.
[25] Champak K. N., and Hassan M. (2024). Hydrodynamic of surge radiation by a floating cylinder in presence of bottom-mounted cylindrical obstacle in a two-layer fluid system. Journal of Engineering Mathematics, 145. https://doi.org/10.1007/s10665-024-10339-1

Himanshu Upreti, Ankita Pandey, and Alok Kumar Pandey

Chapter 12 Data-driven prediction of thermal conductivity ratio in nanoparticle-enhanced 60:40 EG/water nanofluids

Abstract: This study investigates prediction of the thermal conductivity (TC) ratio of three nanofluids, i.e., Al_2O_3 + EG/water, CuO + EG/water, and ZnO + EG/water using various ML (machine learning) and NN (neural network) methods. The primary objective is to compare the performance of eleven regression techniques, including RFR (random forest regressor), LR (linear regression), decision tree regressor, K – neighbor regressor (KNN), support vector regressor, Lasso regression, ridge regression, XG Boost regressor, artificial neural network, Gaussian Naïve Bayes, and polynomial regressor, in predicting the TC ratio. The dataset, sourced from a thermophysical properties database, consists of 131 rows and 6 columns (corresponding to nanoparticle type, fluid ratio 60:40 EG/water, volume fraction, temperature, nanoparticle size, and TC ratio). The models are evaluated based on performance metrics, for example, MAE (mean absolute error), MSE (mean square error), and R^2. The results reveal that the Ridge regression consistently outperforms other models, achieving the highest R^2 score for ZnO and CuO nanofluids, while linear regression shows superior performance for Al_2O_3 nanofluid. The study demonstrates the potential of ML techniques for accurate predictions of nanofluid properties and highlights the importance of model selection in achieving optimal results. The findings provide valuable insights for future applications of ML in nanofluid research and engineering. Additionally, an explainable ML technique, SHAP, is used to interpret the output obtained from the ML models.

Keywords: nanofluid, thermal conductivity, ethylene glycol (EG), machine learning, optimization, SHapley Additive exPlanations (SHAP)

Himanshu Upreti, School of Engineering and Technology, BML Munjal University, Gurugram, Haryana, India; Center for Advanced Data and Computational Science, BML Munjal University, Gurugram, Haryana, India, https://orcid.org/0000-0003-4592-4904, e-mail: hupretidream@gmail.com
Ankita Pandey, Department of Mathematics, GLA University, Mathura, Uttar Pradesh, India, https://orcid.org/0000-0002-5242-7255, e-mail: ankitapandey0595@gmail.com
Alok Kumar Pandey, Department of Mathematics, Graphic Era (Deemed to be University), Dehradun, Uttarakhand, India, https://orcid.org/0000-0002-6460-5274, e-mails: mr.alokpandey1@gmail.com; dralokpandey@yahoo.com

https://doi.org/10.1515/9783111661674-012

12.1 Introduction

Nanofluids are colloidal suspensions of solid particles, i.e., nanoparticles in a base fluid, usually metals, oxides, or carbon nanotubes in water or oil. These fluids are beneficial in a miscellany of applications, such as solar energy systems, medicinal devices, and electronics cooling, due to their enhanced thermal conductivity (TC) and heat transfer possessions. The term "nanofluid" was originally used by Choi and Eastman [1]. They introduced that nanofluids have high TC, in comparison to other heat transfer fluids. Abdollahi et al. [2] demonstrated the impact of temperature and nanoparticles fraction on dynamic viscosity of CuO-loaded nanofluids. Afrand et al. [3] investigated the TC of Fe_3O_4 nanofluids, established a correlation, and optimized an artificial neural network (ANN) for prediction. Agarwal et al. [4] synthesized Al_2O_3 nanoparticles and discovered that the sensitivity increased with increasing volume concentrations. Ahmadi et al. [5] performed four modeling techniques and experiments to estimate the viscosity of TiO_2-water nanofluids. Ahmadi et al. [6] forecast the dynamic viscosity of Ag/H_2O nanofluid by using ANN-MLP, MARS (multivariate adaptive regression splines), and MPR (multivariate polynomial regression) algorithms. The calculation of dynamic viscosity of CuO/H_2O nanofluid from various machine learning (ML) approaches was investigated by Ahmadi et al. [7]. Alade et al. [8] found that the BSVR model is a reliable method to estimate the viscosity of nanofluids. Azmi et al. [9] established nonlinear equations to predict TC and viscosity of H_2O-based nanofluids, which was validated through experimental data. Barbés et al. [10] examined temperature and particle fraction effects, verified experimental data using models, and assessed the TC and specific heat capacities of Al_2O_3 nanofluids. Bhaumik et al. [11] predicted the viscosity of a nanofluid by physics-guided deep neural network.

ML is a branch of AI (artificial intelligence) that studies statistical algorithms that can learn from data and complete tasks without explicit instructions. Samuel [12] introduced ML in 1959 as a field that enables computers to learn without explicit programming. Ramezanizadeh et al. [13] used ANN to analyze the TC of nanofluids, examine model structures, input variables, and develop methods to improve the projected accuracy. Adun et al. [14] established an ANN model to expect the TC of hybrid nanofluid using seven input variables. Maleki et al. [15] studied several ML models, especially ANN, and then modified the presentation of ANN-based approaches by using suitable optimization methods. Liu et al. [16] improved the ANN performance with PSO and outperformed conventional neural networks by using a hybrid ANN-PSO model to predict TC in polymeric nanocomposites. Sargam et al. [17] developed a multilayer perceptron (MLP) model that predicts TC in concrete using fractional hot-deck imputation data, displaying high accuracy and practical implications for enhancing structural design and performance. The behavior of TC and dynamic viscosity of THFN (ternary hybrid nanofluid) was examined by Adun et al. [18], who discovered that temperature and volume concentration have a major impact on the fluid's thermophysical properties. Jamei et al. [19] and Rostami et al. [20] identified that ANN is the most appropriate method for predicting the TC of nanofluids. Jaliliantabar [21] discovered that ANN model is more accurate than CART (classification and regression tree) and MARS (multivariate adaptive regression spline) at predicting the TC of NEPCM (nano-enhanced phase change materials) for various types of PCM. He et al. [22] developed an ODF (optimized deep forest) model to predict the TC of nanofluids and evaluated its accuracy with other models. Several of the studies related to predicting the TC of nanofluids are mentioned in [23–31]. Qu et al. [32] investigated the thermal behavior of WO_3-CuO with the help of back-propagation algorithm. The thermophysical properties of nanofluids by ML approach are analyzed in [33–35]. Dong et al. [36] utilized CFD simulations and ANN-Multilayer Perceptron model to optimize heat transfer parameters in mixed convection, to maximize the Nusselt number. Huang et al. [37] developed a deep learning approach for predicting the TC of porous materials that uses a 3D convolutional NN. Wang et al. [38] conducted TC tests on 560 samples of scum particles to examine the impact of ice content, temperature, and particle size outlines. Meng et al. [39] utilized ML models to predict and analyze the TC of ellipsoidal particle composites, combining ML accuracy with theoretical model trend prediction. Cakiroglu et al. [40] investigated ML models to predict and analyze the TC of cement-based form.

This chapter presents a comparison of the various ML approaches to predict the TC ratio of a nanofluid. Here, the comparison is established for three different nanofluids, namely, Al_2O_3 + EG/water (60:40), CuO + EG/water (60:40), and ZnO + EG/water (60:40), referred to as nanofluid 1, nanofluid 2, and nanofluid 3. The dataset used in the study is collected from Mondejar et al. [41]:

- This study presents a comprehensive analysis of a variety of ML techniques.
- To emphasis the significance of ML techniques, the prediction performance is measured in terms of mean square error (MSE), MAE, and R^2.
- This study includes 11 well-known ML techniques: RFR (random forest regressor), LR (linear regression), DTR (decision tree regressor), K – neighbor regressor, SVR (support vector regressor), Lasso regression, ridge regression, XGB regressor, ANN, Gaussian Naïve Bayes, and polynomial regressor.
- Explainable ML technique SHAP is used to interpret the model's output.

12.2 Materials and methods

12.2.1 Dataset

The dataset has been compiled from various experimental studies on nanofluids, sourced from the thermophysical properties database of nanofluids (see Table 12.1), which aggregates data from numerous scholarly works. It consists of 131 rows and 6 columns, with no missing values. The features represented in the dataset include the

TC ratio (*k*_ratio) of the nanofluid to the base fluid, the type of nanoparticle used, the volume fraction of nanoparticles in the fluid, the temperature at which the measurement was recorded, and the size of the nanoparticles. For *k*_ratio prediction, we take three input features – volume fraction (phi), temperature (*T*), and size of nanoparticle with same fluid medium EG/water (60:40) corresponding to the particles Al_2O_3, CuO, and ZnO. Additionally, the experimental dataset is normalized by the Min-Max scalar in the range between 0 and 1.

Table 12.1: Description of the input dataset [41].

	Particle	Fluid	phi	T	Size	*k*_ratio
0	Al_2O_3	60:40 EG/W	0.01	21.911243	5.300000e-08	1.078801
1	Al_2O_3	60:40 EG/W	0.01	27.970996	5.300000e-08	1.110283
2	Al_2O_3	60:40 EG/W	0.01	36.159774	5.300000e-08	1.144509
3	Al_2O_3	60:40 EG/W	0.01	44.355310	5.300000e-08	1.196839
4	Al_2O_3	60:40 EG/W	0.01	55.113400	5.300000e-08	1.213151
5	CuO	60:40 EG/W	0.01	97.293864	2.900000e-08	1.132200
6	CuO	60:40 EG/W	0.01	91.288642	2.900000e-08	1.174891
7	CuO	60:40 EG/W	0.01	82.933551	2.900000e-08	1.222974
8	CuO	60:40 EG/W	0.01	71.184204	2.900000e-08	1.265868
9	CuO	60:40 EG/W	0.01	59.695953	2.900000e-08	1.303443
10	ZnO	60:40 EG/W	0.01	25.501685	2.900000e-08	1.053778
11	ZnO	60:40 EG/W	0.01	30.445506	2.900000e-08	1.085636
12	ZnO	60:40 EG/W	0.01	37.636517	2.900000e-08	1.137321
13	ZnO	60:40 EG/W	0.01	49.434270	2.900000e-08	1.181615
14	ZnO	60:40 EG/W	0.01	61.344382	2.900000e-08	1.204351

12.2.2 Brief description of the applied methodologies

Random forest regressor (RFR): In ML, the RF algorithm is a potent tree learning technique. It functions by organizing several decision trees during the training phase. An arbitrary subset of the dataset is applied to compute a random subset of characteristics in each subdivision when preparing the individual tree. Each tree erraticism is familiarized by this randomness, which drops the option of over fitting and improves the overall predictability. The procedure averages all of the trees' results in order to make predictions.

Linear regression (LR): It is the linear connection between a dependent variable and one or more independent factors. The model's equation contains definite coefficients that elucidate the influence of each independent variable on the dependent variable, permitting for an improved grasp of the original dynamics:

$$y = \alpha + \beta_0 x_1 + \beta_1 x_2 + \beta_2 x_3$$

Here, y is the output, α is the error term, β_0, β_1, and β_2 are parameters, and x_1, x_2, x_3 are input features.

Decision tree regressor (DTR): Regression models that use a tree-like structure to predict numerical values are called decision trees. Beginning with a root question and working its way out, it divides data according to the important characteristics. Before arriving at the leaf nodes with final predictions, each node inquiries about a feature, further segmenting the data. This method follows the path that matches the features of your data from the root to the leaf in order to get a result.

***K* – neighbor regressor:** One nonparametric technique for forecasting continuous values is KNN regression. The main idea is to average the desired values of the K – nearest neighbors in the space of features in order to estimate the target value for a new data point. Euclidean distance is commonly used to compute the distance between datasets. Choosing the number of neighbors and K, and computing the distance between the new point and all the others in the training set are the first steps. Next, determine which K points are closest to the newly added data point. Determine the forecast value for the new data point by averaging the aim values of the K adjacent neighbors (Figure 12.1).

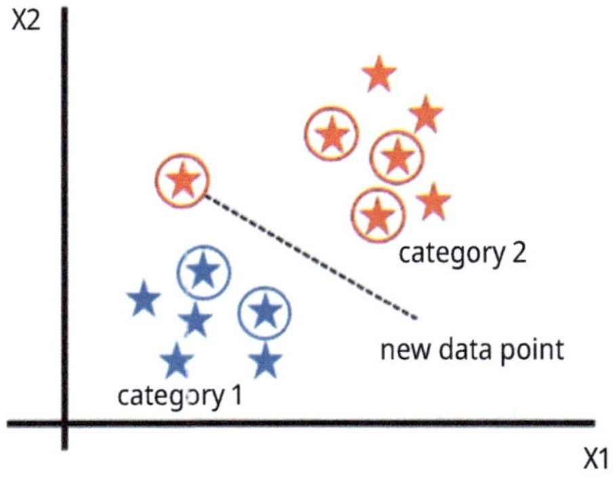

Figure 12.1: KNN representation.

Support vector regressor (SVR): A reliable ML technique for predicting continuous outcomes is support vector regression. In order to find the optimal hyperplane that best captures the data, it uses the principles of support vector machines to convert the input features into high-dimensional spaces. This approach makes support vector regression a flexible tool that may be used in a variety of domains by efficiently managing both linear and nonlinear interactions. SVR employs linear or nonlinear kernels such as the radial basis function (rbf), poly, and linear.

Lasso regression: The Lasso stands for least absolute shrinkage and selection operator. It is used for both model selection and regularization. A model is stated as lasso regression if it employs the $L1$ regularization procedure. With this shrinkage strategy, a penalization factor known as the alpha is introduced, causing the coefficients found in the linear model to decrease toward the center point as the mean. Lasso regression reduces the complexity and multicollinearity of the model by shrinking the coefficients:

$$L_{\text{lasso}}\left(\hat{\beta}\right) = \sum_{i=1}^{n}\left(y_i - x'_i\hat{\beta}\right)^2 + \lambda\sum_{j=1}^{m}\left|\hat{\beta}_j\right|$$

Here, y_i is the actual output λ is the regularization parameter, $\hat{\beta}$ represents weights, x'_i is the input, n is the number of rows, and m is the number of columns.

Ridge regression: By adding a penalty component, ridge regression places a comparable limitation on the coefficients, as lasso regression. Ridge regression, on the other hand, takes the square of the coefficients, whereas lasso regression takes the magnitude:

$$L_{\text{ridge}}\left(\hat{\beta}\right) = \sum_{i=1}^{n}\left(y_i - x'_i\hat{\beta}\right)^2 + \lambda\sum_{j=1}^{m} w_j\hat{\beta}_j^2$$

Here, y_i is the actual output λ is the regularization parameter, $\hat{\beta}$ represents weights, x'_i is the input, n is the number of rows, and m is the number of columns.

XGBoost regressor: XGBoost is an ensemble learning technique. XGBoost's goal function includes a regularization term and a loss function. Reg:linear, reg:logistics, and reg:squarederror are the most widely used loss functions. In order for the bad predictions to cancel out and the better ones to add up to the final good predictions, XGBoost relies on base learners that are consistently awful at the rest.

Artificial neural network: Artificial neurons are a component of ANNs. There are input, output, and hidden layers in an ANN. Data from the external is sent to the input layer of the neural network for the learning process. After that, the input data is converted into meaningful extracted features for the output layer by passing through one or more hidden layers. Finally, the output layer gives the response of ANNs for the input data in the form of an output. Here, we have constructed three ANN architectures, with Adam used as the optimizer and mean square error as the loss function. Their descriptions are given in Table 12.2.

Gaussian Naïve Bayes (GNB): The Bayes theorem, a probability theory principle, assists as the basis for the naïve Bayes classifier. The premise made here is that every input attribute is independent of another, which may not be true in the real-world context. Naïve Bayes takes consideration of the probability of occurrence for each class and allocates the target value to the class with the highest probability. Gaussian naïve Bayes is a form of naïve Bayes algorithm that handles continuous attributes and uses a Gaussian distribution for data features across the dataset.

Polynomial regressor: In statistics and ML, polynomial regression is a sort of regression analysis that is utilized when the relation between the independent and dependent variable is not linear and is alternatively represented as an nth-degree polyno-

Table 12.2: Descriptions of the constructed ANN models.

ANN1	ANN2	ANN3
Model.Add(dense(64, input_dim = X_train.shape[1], activation = 'relu'))	Model.Add(dense(64, input_dim = X_train.shape[1], activation = 'relu'))	Model.Add(dense(64, input_dim = X_train.shape[1], activation = 'relu'))
Model.Add(dense(32, activation = 'relu'))	Model.Add(dense(32, activation = 'relu'))	Model.Add(dense(32, activation = 'relu'))
Model.Add(dense(32, activation = 'relu'))	Model.Add(dense(32, activation = 'relu'))	Model.Add(dense(32, activation = 'relu'))
Model.Add(dense(1))	Model.Add(dense(32, activation = 'relu'))	Model.Add(dense(32, activation = 'relu'))
	Model.Add(dense(1))	Model.Add(dense(32, activation = 'relu')) Model.Add(dense(1))

mial. Nonlinear patterns in the data can be captured by polynomial regression when a curve rather than a straight line better depicts the relationship between the variables. The general expression for a second-degree polynomial regression model with three input features is

$$y = \beta_0 + \sum_{i=1}^{3} \beta_i x_i + \sum_{i \leq j}^{3} \beta_{ij} x_i x_j + \alpha$$

Here, y is the output, α is the error term, β_0, β_1, β_{ij} are parameters and x_i, x_j are input features.

The experiment was executed in Google Collaboratory with the Tensor Flow API and CPU processor. Experiment starts with the basic to advance ML techniques. Linear regression, polynomial regressor, random forest regressor, DTR, K – neighbor regressor, SVR, Lasso regression, ridge regression, XGB regressor, Gaussian naïve Bayes, and ANN.

The performances are measured in terms of MAE (mean absolute error), MSE (mean square error), and R^2. The terms MAE, MSE, R^2 are described below:

$$\text{MAE} = \frac{1}{N}\sum_{i=1}^{N} |y_i - \hat{y}|$$

$$\text{MSE} = \frac{1}{N}\sum_{i=1}^{N} (y_i - \hat{y})^2$$

$$R^2 = 1 - \frac{\sum (y_i - \hat{y})^2}{\sum (y_i - \bar{y})^2}$$

Here, y_i is the actual output, $\hat{y}$ is the predicted output, and $\bar{y}$ is the mean of y.

12.3 Results and discussion

The dataset here is divided into training and validation, in a ratio of 80:20. A glimpse of the dataset is presented in Table 12.1, and actual experimental value of the k – ratio of the nanofluids 1, 2, and 3 for the validation set is shown in Table 12.3.

The performances of multiple ML models with the predicted values of the TC ratio (k – ratio) of the three nanofluids are presented in Tables 12.4, 12.5, and 12.6. Metrics such as MAE, MSE, and the coefficient of determination (R^2) were used to asses each model's accuracy. Table 12.4 corresponds to nanofluid 1, i.e., Al_2O_3; Table 12.5 for nanofluid 2, CuO; and Table 12.6 for nanofluid 3, i.e., ZnO. It is noted from Table 12.4 that the LR model demonstrated the best predictive performance, with an R^2 score of 0.9772, followed closely by ridge regression and K-neighbor regression (n neighbor = 4), both achieving high R^2 values above 0.96. ANN 3 performed well with an R^2 of 0.9546, though it was slightly less accurate than the top performing models. Models such as Lasso regression and Gaussian naïve Bayes exhibited poor predictive capabilities, with negative R^2 score, indicating they failed to capture the underlying patterns. Polynomial regression (degree = 2) performed the worst, with an extremely high error and a nonsensical R^2 score, indicating overfitting or stability.

From Table 12.5, Ridge regression emerged as the most accurate model, achieving an R^2 score of 0.9622 and the lowest MAE of 0.0139. This was closely followed by linear regression and K-neighbors regression (n neighbors = 3), which demonstrated consistent accuracy across datasets. RF regressor, XGB regressor, and SVR (kernel = rbf) also performed well, with R^2 values above 0.90. Similar to nanofluid 1, Lasso regression, polynomial regression, and GNB performed poorly, with negative R^2 score, suggesting that these models are not suitable for this dataset. NN models showed variable performance, with ANN 1 and ANN 2 underperforming, while ANN 3 achieved moderate accuracy.

From Table 12.6, the ridge regression is again the most accurate model, achieving the highest R^2 score of 0.9847. This model demonstrated robust productive capabilities,

Table 12.3: Experimental value of k – ratios of nanofluids [41].

Nanofluid 1 (Al_2O_3)	Nanofluid 2 (CuO)	Nanofluid 3 (ZnO)
1.394323	1.409698	1.452633
1.337296	1.401879	1.288653
1.141401	1.196148	1.398315
1.444455	1.447453	1.250027
1.213151	1.303443	1.311126
1.560167	1.468128	1.143435
1.380377	1.408496	1.298204
1.345800	1.336008	1.345296
1.367086	1.368172	1.204351
–	–	1.333950

Table 12.4: Predicted k − ratio of nanofluid 1 using the applied approaches.

Model	Measuring metrics (MAE, MSE, R^2)	Predicted values
RF regressor	MAE: 0.04640908706111113 MSE: 0.002791350974703386 R^2 score: 0.7881105481603765	[1.33859228 1.32585984 1.15873405 1.35371406 1.24503393 1.48171034 1.32101179 1.37997376 1.40564769]
Linear regression	MAE: 0.014791814216562615 MSE: 0.00030008352154821307 R^2 score: 0.9772208749586887	[1.37637546 1.365207 1.13366927 1.42736615 1.1979371 1.53027321 1.38842159 1.34387307 1.37445504]
DT regressor	MAE: 0.06087570222222224 MSE: 0.004487190472482346 R^2 score: 0.6593805873461334	[1.32114541 1.29075974 1.10473806 1.32114541 1.23387404 1.48689895 1.31151204 1.40911198 1.40911198]
K − neighbor regressor		
(i) n_neighbors = 2	MAE: 0.030414172277777762 MSE: 0.0015055208478392864 R^2 score: 0.885716991495242	[1.31632873 1.36319623 1.13241538 1.47891873 1.21535629 1.53696102 1.3625631 1.27963461 1.35009311]
(ii) n_neighbors = 3	MAE: 0.024847122111111113 MSE: 0.0009044370735856183 R^2 score: 0.9313448299829605	[1.35872852 1.33862865 1.12503793 1.48174294 1.21610803 1.50581671 1.40417253 1.33277194 1.32817043]
(iii) n_neighbo·s = 4	MAE: 0.01451246919444444 MSE: 0.0004413826445812199 R^2 score: 0.9664949598028331	[1.37132438 1.31919743 1.12990565 1.43918522 1.22730696 1.50793987 1.38100741 1.34631197 1.36186171]
SVR		
(i) kernel = 'po y'	MAE: 0.07955842976878219 MSE: 0.007498187417968655 R^2 score: 0.43081796729163835	[1.26671246 1.273581 1.17940379 1.31579354 1.18738441 1.49980214 1.27850517 1.26120885 1.28164418]
(ii) kernel = 'linear'	MAE: 0.038536894902411536 MSE: 0.0022054550403184776 R^2 score: 0.8325854886092658	[1.39548314 1.40037271 1.22265782 1.43462062 1.27535152 1.50014427 1.41145997 1.36380527 1.38727822]
(iii) kernel = 'r'	MAE: 0.03564194002701355 MSE: 0.0017493335319080001 R^2 score: 0.8672093453958988	[1.39935385 1.39291747 1.19308932 1.45177225 1.27845347 1.49954356 1.41228125 1.36906359 1.38711158]
Lasso(alpha = 0.1)	MAE: 0.09661459135016842 MSE: 0.014242895870406759 R^2 score: −0.08116801718566191	[1.32108422 1.32108422 1.32108422 1.32108422 1.32108422 1.32108422 1.32108422 1.32108422 1.32108422]
Ridge(alpha = 0.1)	MAE: 0.014498603056528244 MSE: 0.00032095212459422773 R^2 score: 0.9756367542586579	[1.37511435 1.36358487 1.13789962 1.42481617 1.20038215 1.52576541 1.38653434 1.34375516 1.37356413]

Table 12.4 (continued)

Model	**Measuring metrics** (MAE, MSE, R^2)	**Predicted values**
XGB regressor (n_estimators = 100, learning_rate = 0.1, max_depth = 5, objective = 'reg:squarederror', random_state = 42)	MAE: 0.05519101129947241 MSE: 0.0037292987626424026 R^2 score: 0.7169116038349664	[1.3319201 1.290693 1.1144259 1.3319201 1.2338157 1.4841881 1.3249267 1.4041724 1.4048221]
ANN 1	MAE: 0.042041354798909504 MSE: 0.002633652747838735 R^2 score: 0.8000813075343325	[1.3160555 1.3967767 1.0787575 1.3633759 1.1644409 1.5283909 1.3683374 1.3465297 1.36344]
ANN 2	MAE: 0.027874321299696185 MSE: 0.0013287075558305805 R^2 score: 0.8991387617639074	[1.3688036 1.3940188 1.0726306 1.4284325 1.1604986 1.5577186 1.395589 1.3379114 1.3727195]
ANN 3	MAE: 0.021857032774149583 MSE: 0.0005976405200312141 R^2 score: 0.9546335364724181	[1.3630277 1.3709446 1.1366591 1.4148003 1.1821047 1.5292873 1.387801 1.332004 1.3528582]
Gaussian Naïve Bayes	MAE: 3.152331737 MSE: 13.287705718567327 R^2 score: −2.4944940363764725	[1.31719197 1.31719197 1.11285668 1.31719197 1.2490802 1.4534155 1.31719197 1.38530373 1.38530373]
Polynomial regressor(degree = 2)	MAE: 3.363890100803018 MSE: 21.76252232175656 R^2 score: −1650.9774715519486	[4.63196501 1.28014896 1.27477976 4.63425991 0.96837019 11.37948486 2.53338768 7.57555182 7.57726331]

Table 12.5: Predicted k − ratio of nanofluid 2 using the applied approaches.

Model	**Measuring metrics** (MAE, MSE, R^2)	**Predicted values**
RF regressor	MAE: 0.017803266265555714 MSE: 0.0005308562037208256 R^2 score: 0.9139233444803416	[1.40274096 1.40543178 1.2067362 1.46758876 1.33224007 1.45055413 1.40568493 1.28354957 1.35081689]
Linear regression	MAE: 0.014725130377399396 MSE: 0.000256810217938703 R^2 score: 0.9583590348789383	[1.40108961 1.43270408 1.21077552 1.46055148 1.28581052 1.47622559 1.42228119 1.3237136 1.35461504]
DT Regressor	MAE: 0.030627633111111125 MSE: 0.0011164856784282378 R^2 score: 0.8189653761958489	[1.3738603 1.38417477 1.17489083 1.49702396 1.31762055 1.44866332 1.3813651 1.28269831 1.33097576]

Table 12.5 (continued)

Model	Measuring metrics (MAE, MSE, R^2)	Predicted values
K – neighbor regressor		
(i) n_neighbors = 2	MAE: 0.02654174077777781 MSE: 0.0010064690776420077 R^2 score: 0.8368042202763077	[1.3776127 1.37685714 1.20061 1.49371982 1.29174426 1.44668084 1.41403452 1.27660528 1.40112526]
(ii) n_neighbors = 3	MAE: 0.011477810259259251 MSE: 0.00024513416153805473 R^2 score: 0.9602522705189898	[1.40129624 1.40408127 1.19203694 1.48198973 1.31235703 1.47172301 1.40314471 1.30939178 1.37774209]
(iii) n_neighbors = 4	MAE: 0.016835339861111163 MSE: 0.00045116736047769824 R^2 score: 0.9268446385342132	[1.4010802 1.3914566 1.20767116 1.49693154 1.30515955 1.45390027 1.42496245 1.31113242 1.35398115]
SVR		
(i) kernel = 'poly'	MAE: 0.031137671282200612 MSE: 0.0014471179054402806 R^2 score: 0.7653544056378409	[1.39133547 1.37725268 1.26951368 1.40255621 1.36149117 1.49241252 1.38297209 1.33939581 1.37591626]
(ii) kernel = 'linear'	MAE: 0.02186441475359493 MSE: 0.0009062195034988669 R^2 score: 0.8530593718579018	[1.38919395 1.39085021 1.27190047 1.42106444 1.30447363 1.44475401 1.39290799 1.35536605 1.37192867]
(iii) kernel = 'rbf'	MAE: 0.015917438269534154 MSE: 0.000471833322879136 R^2 score: 0.9234937180511472	[1.39894952 1.38496566 1.25002834 1.43966474 1.30470045 1.45032997 1.39870826 1.35728143 1.37198337]
Lasso (alpha = 0.1)	MAE: 0.0701702670707071 MSE: 0.006816627978723084 R^2 score: −0.10529468096558547	[1.34556412 1.34556412 1.34556412 1.34556412 1.34556412 1.34556412 1.34556412 1.34556412 1.34556412]
Ridge (alpha = 0.1)	MAE: 0.013915809388284763 MSE: 0.00023252139914396822 R^2 score: 0.9622973900751663	[1.39978865 1.43041711 1.21390855 1.45779833 1.28700407 1.47330355 1.42035572 1.32440929 1.35455605]
XGB regressor ($n_{estimators}$=100, $learning_{rate}$ 0.1, max_{depth}=5, objective = 'reg: squarederror', $random_{state}$ = 42)	MAE: 0.01883988219656711 MSE: 0.0005513832252826413 R^2 score: 0.9105949528152605	[1.408478 1.4432487 1.1926963 1.4778502 1.3205189 1.4530886 1.4472635 1.3201143 1.3618286]
ANN 1	MAE: 0.09684698701034884 MSE: 0.01265256653189733 R^2 score: −1.051573671310869	[1.3373812 1.2029157 1.2964438 1.3738636 1.1627716 1.5980574 1.2803148 1.3568879 1.3613743]
ANN 2	MAE: 0.07729163052954778 MSE: 0.00756440602897845 R^2 score: −0.22654452826103633	[1.3566041 1.2587098 1.2872436 1.4178442 1.2049168 1.5876237 1.311419 1.3766738 1.3910658]

Table 12.5 (continued)

Model	Measuring metrics (MAE, MSE, R^2)	Predicted values
ANN 3	MAE: 0.03764139646261255 MSE: 0.0016765665958012191 R^2 score: 0.7281500257300537	[1.3416876 1.3551539 1.2245033 1.4267156 1.2657963 1.4796351 1.366974 1.3044099 1.3154999]
Gaussian Naïve Bayes	MAE: 3.8717224931296292 MSE: 17.2414833240217 R^2 score: −6.1987636559059665	[1.36796368 1.36796368 1.26317999 1.47274737 1.36796368 1.36796368 1.36796368 1.26317999 1.31557183]
Polynomial regressor (degree = 2)	MAE: 3.91551250477662 MSE: 23.73585036404083 R^2 score: −3847.693112996294	[5.61454755 2.35127635 2.34930454 5.6156584 1.38935374 10.66847148 3.75953537 7.91507853 7.91580987]

Table 12.6: Predicted k − ratio of nanofluid 3 using the applied approaches.

Model	Measuring metrics (MAE, MSE, R^2)	Predicted values
RF regressor	MAE: 0.01693282357799968 MSE: 0.00044739432983084476 R^2 score: 0.9383021308253107	[1.44748173 1.29977752 1.42883823 1.26570494 1.27540203 1.14696367 1.29797327 1.33249908 1.24175422 1.35111747]
Linear regression	MAE: 0.010045450366344456 MSE: 0.0001509114178962538 R^2 score: 0.9791885763910994	[1.46671374 1.28015686 1.41731991 1.24547927 1.3141856 1.14205261 1.28841413 1.35207672 1.19632237 1.35923277]
DT regressor	MAE: 0.039755447700000016 MSE: 0.0019337901534982591 R^2 score: 0.7333208671934932	[1.49188147 1.24170522 1.46888081 1.29507994 1.26494823 1.14923992 1.24170522 1.33535105 1.23469149 1.38092091]
K – neighbor regressor		
(i) $n\_neighbors = 2$	MAE: 0.017763793799999993 MSE: 0.0008582480468554926 R^2 score: 0.881643391112392	[1.48038114 1.29154406 1.48038114 1.26084247 1.30374532 1.13536926 1.30073254 1.34424433 1.21166006 1.36173166]
(ii) $n\_neighbors = 3$	MAE: 0.021261550300000008 MSE: 0.0008109776588844928 R^2 score: 0.888162209117918	[1.44955937 1.30614639 1.44722773 1.23396991 1.30848204 1.16578118 1.28105677 1.34962547 1.22146581 1.39744804]

Table 12.6 (continued)

Model	Measuring metrics (MAE, MSE, R^2)	Predicted values
(iii) *n*_neighbors = 4	MAE: 0.026090585750000006 MSE: 0.000823653129789537 R^2 score: 0.886414199602637‌1	[1.43239976 1.32296765 1.43239976 1.21348528 1.32659176 1.18300875 1.28122659 1.32264541 1.20613571 1.37322924]
SVR		
(i) kernel = 'poly'	MAE: 0.068715045607512 MSE: 0.00543075227729096 R^2 score: 0.25107266412802165	[1.40844786 1.2082345 1.32139418 1.17823047 1.21560443 1.15581757 1.20880057 1.24894454 1.16708847 1.25104256]
(ii) kernel = 'linear'	MAE: 0.03332474717721297 MSE: 0.0016603297353994465 R^2 score: 0.7710323981078333	[1.37022995 1.25997647 1.3476974 1.25845536 1.30064945 1.19697588 1.26029666 1.30363651 1.21113702 1.32119913]
(iii) kernel = 'rbf'	MAE: 0.03189235611960597 MSE: 0.0015833159173394423 R^2 score: 0.7816529807895718	[1.38289253 1.25590039 1.36796493 1.24153504 1.29372374 1.22087489 1.25958189 1.31301896 1.21032605 1.32807799]
Lasso (alpha = 0.1)	MAE: 0.08429365963783772 MSE: 0.010241872390977538 R^2 score: −0.4124043617658504	[1.24791358 1.24791358 1.24791358 1.24791358 1.24791358 1.24791358 1.24791358 1.24791358 1.24791358 1.24791358]
Ridge (alpha = 0.1)	MAE: 0.008497344842122324 MSE: 0.0001110675574483961 R^2 score: 0.9846832398801425	[1.46149137 1.27946552 1.41329717 1.24576861 1.3126677 1.14471544 1.28724523 1.34950009 1.19732088 1.3566208]
XGB regressor (*n*_estimators = 100, learning_rate = 0.1, max_depth = 5, objective = 'reg:squarederror', *random_state* = 42)	MAE: 0.04134001277557982 MSE: 0.002440846560435134 R^2 score: 0.6633953053938887	[1.4827409 1.2447512 1.4607079 1.2486442 1.264374 1.1427766 1.2457676 1.2999144 1.2364962 1.4321903]
ANN 1	MAE: 0.013039131389776592 MSE: 0.00020197331838268446 R^2 score: 0.9721468902409554	[1.4663044 1.2727053 1.4082036 1.2368065 1.3014687 1.1280189 1.2820994 1.320283 1.2066103 1.343162]

Table 12.6 (continued)

Model	**Measuring metrics** (MAE, MSE, R^2)	**Predicted values**
ANN 2	MAE: 0.019723771445684823 MSE: 0.0005228523105623976 R^2 score: 0.9278961057308	[1.4692019 1.2833371 1.4042195 1.2104785 1.3141537 1.1655707 1.2768092 1.3168335 1.1717172 1.3561937]
ANN 3	MAE: 0.010554384145410189 MSE: 0.00018609965731874854 R^2 score: 0.9743359458421212	[1.4845494 1.2946357 1.4026921 1.2609283 1.3053886 1.1488698 1.3021213 1.3604925 1.2009275 1.3526074]
Gaussian Naïve Bayes	MAE: 4.3979632077 MSE: 22.508578634897397 R^2 score: −5.600756197917125	[1.37018616 1.37018616 1.37018616 1.17547367 1.32150804 1.12679555 1.37018616 1.37018616 1.2241518 1.32150804]
Polynomial regressor (degree = 2)	MAE: 247074.38838969398 MSE: 286879906404.8041 R^2 score: −39562144072998.016	[8.48443442e + 00 4.50196117e + 05 8.48380266e + 00 1.56922261e + 06 8.48248125e + 00 4.51297057e + 05 1.56807129e + 00 4.14901226e + 00 9.34123057e–01 8.48305880e + 00]

with low MAE and MSE values. Linear regression and ANN 3 also performed exceptionally well, achieving an R^2 score of 0.9792 and 0.9743, respectively, closely rivaling ridge regression. RF regressor and K-neighbor regressor also performed well, with R^2 score in the range of 0.88–0.94. As with the other nanofluids, Lasso regressor, Gaussian naïve Bayes, and polynomial regression were among the poorest performers.

Figures 12.2–12.4 compare the performances of the top three ML models in predicting the properties of three nanofluids: Al_2O_3, CuO, and ZnO. Each model is evaluated based on three metrics: MAE, MSE, and R^2 score. For Al_2O_3 nanofluid (Figure 12.2), linear regression, ridge regression (alpha = 0.1), and KNN ($n = 4$) models show the best outcomes, out of the 11 implemented ML techniques. The R^2 score of linear regression, ridge regression (alpha = 0.1), and KNN ($n = 4$) models are 0.9772, 0.9756, and 0.9665, respectively. In terms of MSE, linear regression obtained the lowest value of 0.000300, compared to Ridge regression (alpha = 0.1) and KNN ($n = 4$). For the CuO nanofluid (Figure 12.3), the Ridge regression outperforms other models, with the lowest MAE (0.0139) and MSE (0.000233), and highest R^2 score of 0.9623, followed closely by KNN ($n = 3$). Linear regression has the highest error but still achieves a respectable R^2 score. For ZnO nanofluid (Figure 12.4), again the ridge regression achieved the best

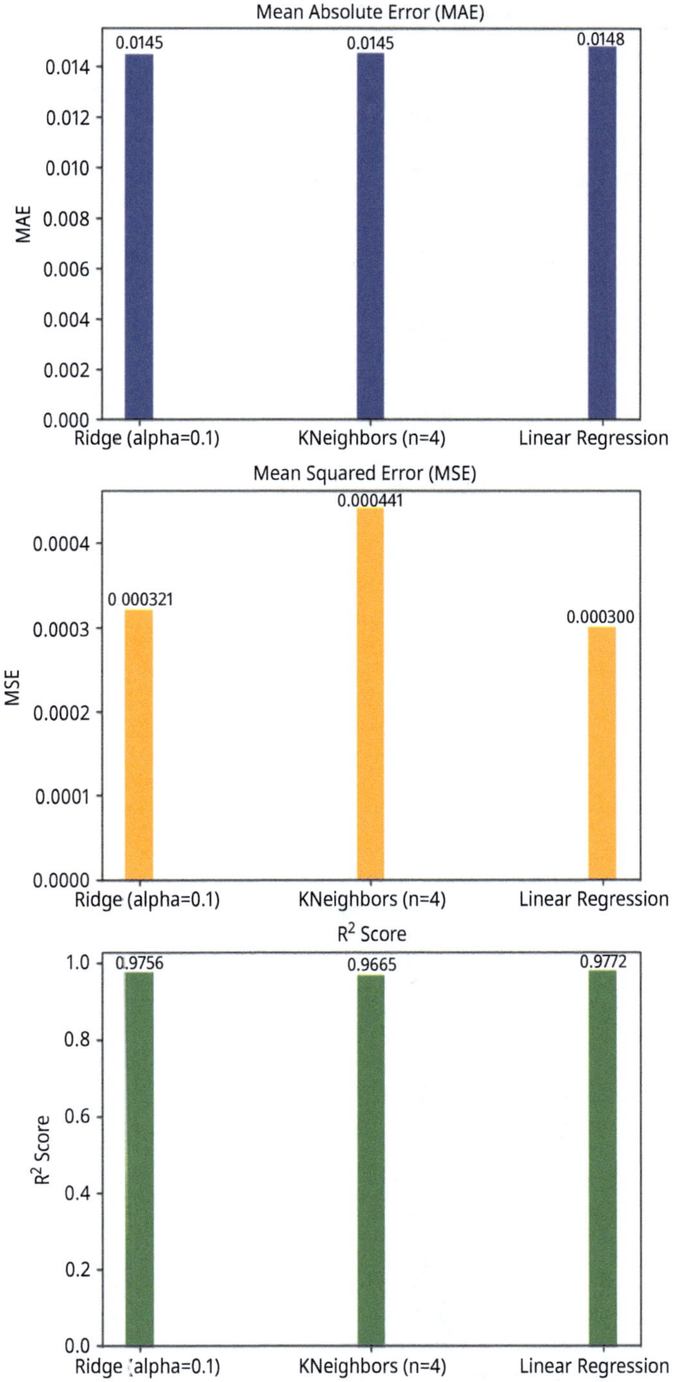

Figure 12.2: MAE, MSE, and R^2 of the best three ML models on Al_2O_3 nanofluid.

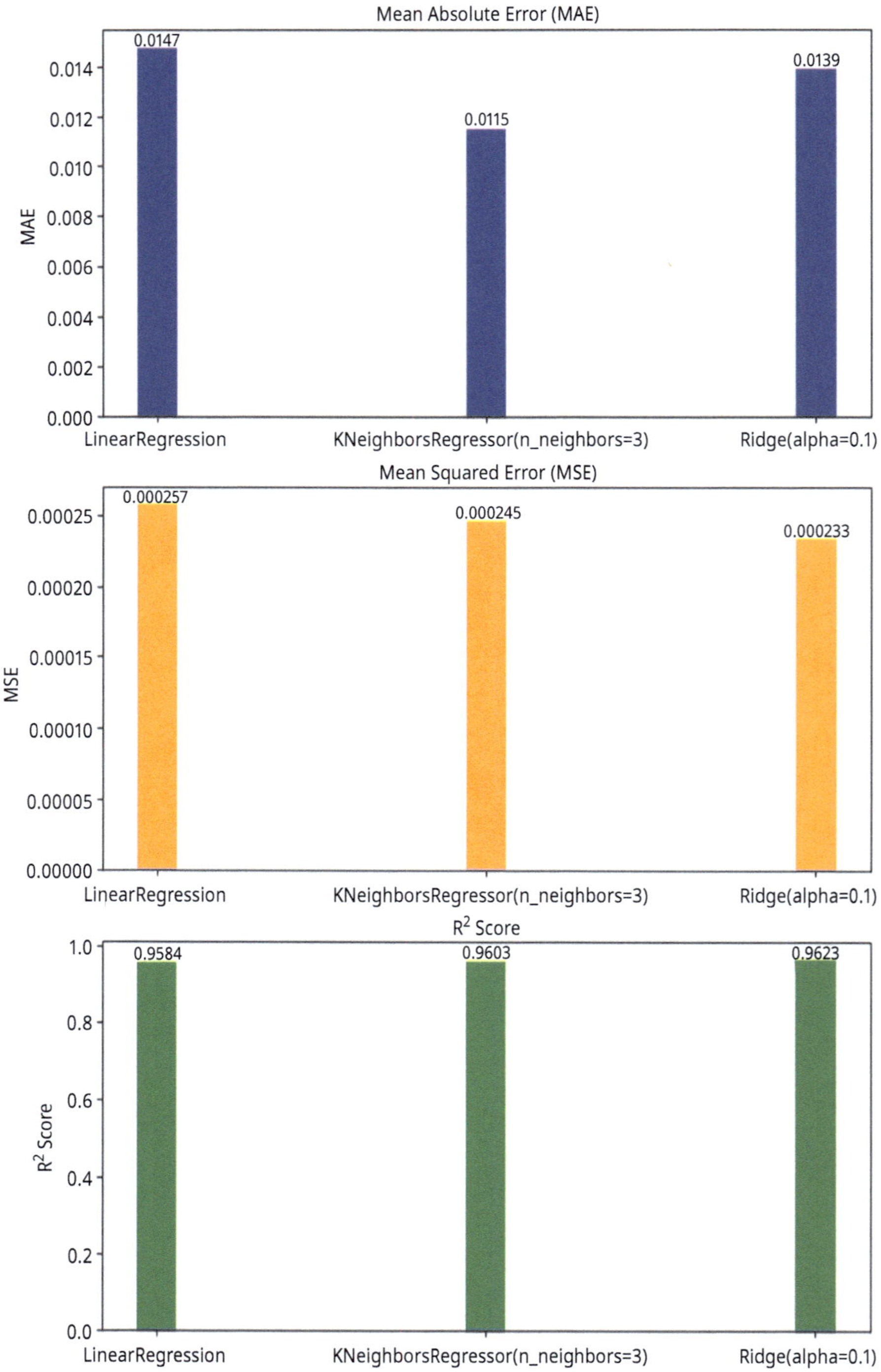

Figure 12.3: MAE, MSE, and R^2 of best three ML models on the CuO nanofluid.

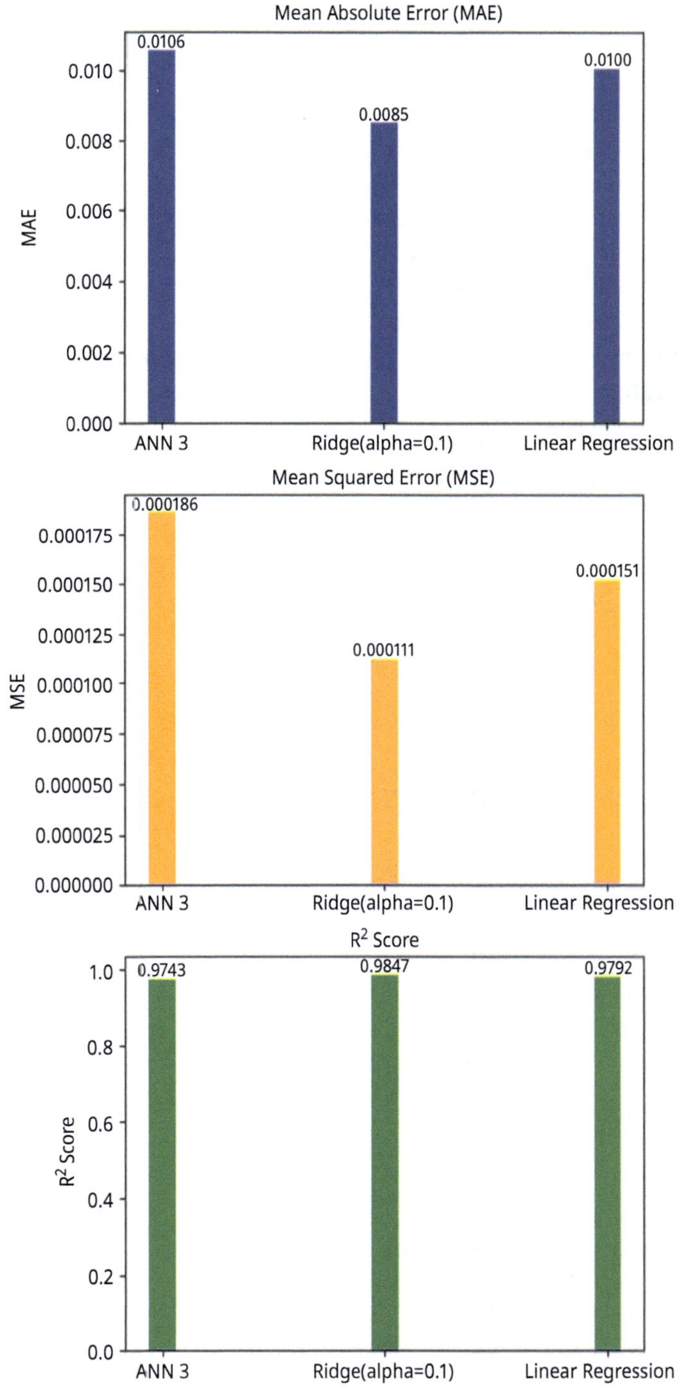

Figure 12.4: MAE, MSE, and R^2 of the best three ML models on the ZnO nanofluid.

performance, with the lowest MAE (0.0085) and MSE (0.000111), and the highest R^2 score of 0.9847. The R^2 score is above 0.95 in all the three best models, highlighting excellent predictive capability.

Additionally, an explainable ML technique, SHapley (SHAP), is implemented to interpret the output obtained by the ML models. We used SHAP on the best performers linear regression on Al_2O_3 and Ridge (alpha = 0.1) on ZnO and CuO, to measure the contribution of each input (volume fraction, temperature, and size) to the final output TC ratio. Figure 12.5 shows that for Al_2O_3 nanofluid, the highest performance is given by the linear regression model, and feature 1, i.e., volume fraction (phi), out of the three features, has a high impact on the model's output. Figure 12.6 shows that for the CuO nanofluid, the highest performance is given by the Ridge regression (alpha = 0.1), and feature 2, i.e., temperature (T) has a high impact on the model's output. Similarly, Figure 12.7 shows that for ZnO nanofluid, the highest performance is given by again the Ridge regression (alpha = 0.1), and feature 2, i.e., temperature (T) has a high impact on the model's output.

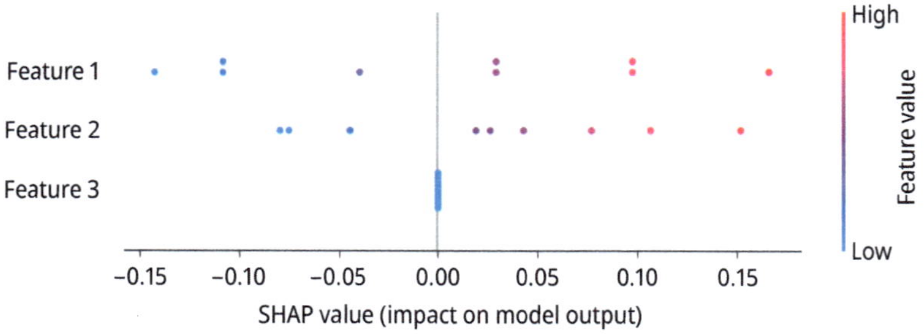

Figure 12.5: SHAP-based results' interpretation for Al_2O_3.

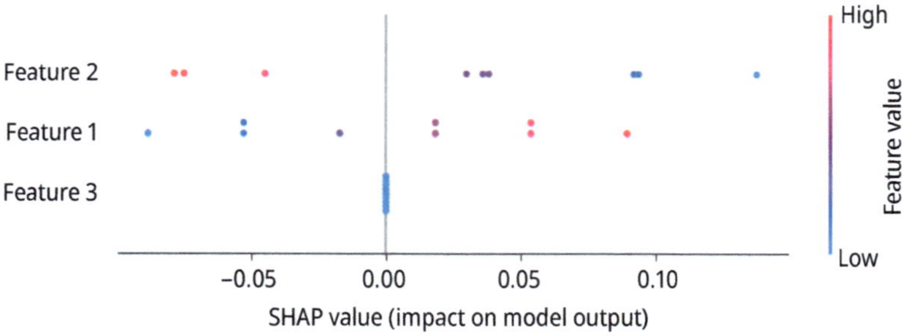

Figure 12.6: SHAP-based results' interpretation for CuO.

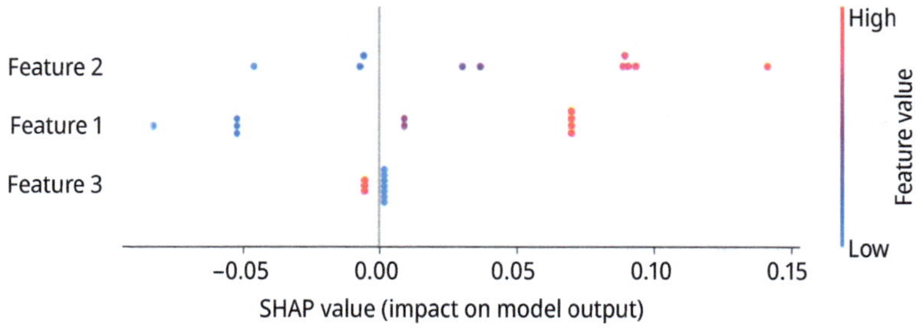

Figure 12.7: SHAP-based results' interpretation for ZnO.

12.4 Conclusions

The main findings of this work are:

- Ridge regression consistently outperformed other models across all three nanofluids, achieving the highest R^2 score and lowest error metrics on CuO and ZnO.
- Linear regression outperformed other models on Al_2O_3 nanofluids, achieving the highest R^2 score and lowest error metrics.
- Linear regression and K – neighbors' regression (n neighbor = 3) were also highly effective, with comparable performance to ridge regression.
- Lasso regression, Gaussian Naïve Bayes, and polynomial regression (degree = 2) failed to provide reliable predictions, often producing negative R^2 scores or extremely high errors.
- Tree-based models like random forest and decision trees show moderate performance and their accuracy was generally lower than ridge and linear regression.
- NN demonstrated variable performance, with some ANN architectures underperforming, while others, such as ANN 3, achieved string results.

References

[1] Choi SU, Eastman JA. Enhancing thermal conductivity of fluids with nanoparticles. Argonne National Lab.(ANL), Argonne, IL (United States); 1995.

[2] Abdollahi A, Karimi DMH, Karimipour A, Safaei MR. Experimental study to obtain the viscosity of CuO-loaded nanofluid: effects of nanoparticles' mass fraction, temperature and base fluid's types to develop a correlation. Meccanica 2018,53,3739–57.

[3] Afrand M, Toghraie D, Sina N. Experimental study on thermal conductivity of water-based Fe3O4 nanofluid: Development of a new correlation and modeled by artificial neural network. International Communications in Heat and Mass Transfer 2016,75,262–69.

[4] Agarwal R, Verma K, Agrawal NK, Singh R. Sensitivity of thermal conductivity for Al2O3 nanofluids. Experimental Thermal and Fluid Science 2017,80,19–26.

[5] Ahmadi MH, Baghban A, Ghazvini M, Hadipoor M, Ghasempour R, Nazemzadegan MR An insight into the prediction of TiO2/water nanofluid viscosity through intelligence schemes. Journal of Thermal Analysis and Calorimetry 2020,139, 2381–94.

[6] Ahmadi MH, Mohseni-Gharyehsafa B, Farzaneh-Gord M, Jilte RD, Kumar R, Chau KW. Applicability of connectionist methods to predict dynamic viscosity of silver/water nanofluid by using ANN MLP, MARS and MPR algorithms. Engineering Applications of Computational Fluid Mechanics 2019,13, 220–28.

[7] Ahmadi MH, Mohseni-Gharyehsafa B, Ghazvini M, Goodarzi M, Jilte RD, Kumar R. Comparing various machine learning approaches in modeling the dynamic viscosity of CuO/water nanofluid. Journal of Thermal Analysis and Calorimetry, 2020,139, 2585–99.

[8] Alade IO, Rahman MAA, Hassan A, Saleh TA. Modeling the viscosity of nanofluids using artificial neural network and Bayesian support vector regression. Journal of Applied Physics, 2020,128, 085306.

[9] Azmi WH, Sharma KV, Mamat R, Alias ABS, Misnon II. Correlations for thermal conductivity and viscosity of water based nanofluids. Materials Science and Engineering, 2012, 36, 012029.

[10] Barbes B, Paramo R, Blanco E, Pastoriza-Gallego MJ, Pineiro MM, Legido JL, Casanova C. Thermal conductivity and specific heat capacity measurements of Al_2O_3 nanofluids. Journal of Thermal Analysis and Calorimetry, 2013,111,1615–25.

[11] Bhaumik B, Chaturvedi S, Changdar S, De S. A unique physics-aided deep learning model for predicting viscosity of nanofluids International Journal for Computational Methods in Engineering Science & Mechanics, 2023,24(3),167–81.

[12] Samuel AL. Some studies in machine learning using the game of checkers. IBM Journal of Research and Development 1959,3(3),210–29.

[13] Ramezanizadeh M, Alhuyi Nazari M, Ahmadi MH, Lorenzini G, Pop I. A review on the applications of intelligence methods in predicting thermal conductivity of nanofluids. Journal of Thermal Analysis and Calorimetry 2019,138(1),827–43.

[14] Adun H, Wole-Osho I, Okonkwo EC, Bamisile O, Dagbasi M, Abbasoglu S. A neural network-based predictive model for the thermal conductivity of hybrid nanofluids. International Communications in Heat and Mass Transfer 2020,119,104930.

[15] Maleki A, Haghighi A, Mahariq I. Machine learning-based approaches for modeling thermophysical properties of hybrid nanofluids: A comprehensive review. Journal of Molecular Liquids 2021,322,114843.

[16] Liu B, Vu-Bac N, Rabczuk T. A stochastic multiscale method for the prediction of the thermal conductivity of Polymer nanocomposites through hybrid machine learning algorithms. *Composite Structures* 2021,273,114269.

[17] Sargam Y, Wang K, Cho IH. Machine learning based prediction model for thermal conductivity of concrete. Journal of Building Engineering 2021,34,101956.

[18] Adun H, Kavaz D, Dagbasi M, Umar H, Wole-Osho I. An experimental investigation of thermal conductivity and dynamic viscosity of Al2O3-ZnO-Fe3O4 ternary hybrid nanofluid and development of machine learning model. Powder Technology, 2021,394,1121–40.

[19] Jamei M, Olumegbon IA, Karbasi M, Ahmadianfar I, Asadi A, Mosharaf-Dehkordi M. On the thermal conductivity assessment of oil-based hybrid nanofluids using extended Kalman filter integrated with feed-forward neural network. International Journal of Heat and Mass Transfer 2021,172,121159.

[20] Rostami S, Kalbasi R, Sina N, Goldanlou AS. Forecasting the thermal conductivity of a nanofluid using artificial neural networks. Journal of Thermal Analysis and Calorimetry 2021 145,2095–104.

[21] Jaliliantabar, F. Thermal conductivity prediction of nano enhanced phase change materials: a comparative machine learning approach. Journal of Energy Storage 2022,46,103633.

[22] He D, Li C, Chen Y, Li X. Prediction of thermal conductivity of hybrid nanofluids based on deep forest model. Heat Transfer Research 2022;53,55–71.

[23] Kocak E, Aylı E, Turkoglu H. A comparative study of multiple regression and machine learning techniques for prediction of nanofluid heat transfer. Journal of Thermal Science and Engineering Applications 2022,14,061002.
[24] Boobalan C, Kannaiyan SK. A correlation to predict the thermal conductivity of MXene-silicone oil-based nano-fluids and data driven modeling using artificial neural networks. International Journal of Energy Research 2022,46, 21538–47.
[25] Oh Y, Guo Z. Applicability of machine learning techniques in predicting specific heat capacity of complex nanofluids. Heat Transfer Research 2024,55, 39–60.
[26] Saha A, Basu A, Banerjee S. Enhancing thermal management systems: a machine learning and metaheuristic approach for predicting thermophysical properties of nanofluids. Engineering Research Express 2024,6,045537.
[27] Ganga S, Uddin Z, Asthana R, Hassan H, Bhardwaj A. Modelling of Viscosity and Thermal Conductivity of Water-Based Nanofluids using Machine-Learning Techniques. The International Journal of Mathematical, Engineering and Management Sciences 2023 8(5),817.
[28] Qin G, Wei Y, Yu L, Xu J, Ojih J, Rodriguez AD, Wang H, Qin Z, Hu M. Predicting lattice thermal conductivity from fundamental material properties using machine learning techniques. Journal of Materials Chemistry A 2023 11,5801–10.
[29] Sahin F, Genc O, Gökcek M, Çolak AB. An experimental and new study on thermal conductivity and zeta potential of Fe3O4/water nanofluid: Machine learning modeling and proposing a new correlation. Powder Technology 2023,420,118388.
[30] Gao Y, Shigidi IM, Ali MA, Homod RZ, Safaei MR. Thermophysical properties prediction of carbon-based nano-enhanced phase change material's using various machine learning methods. Journal of the Taiwan Institute of Chemical Engineers 2023,148,104662.
[31] Dehury P, Chaudhari S, Banerjee T, Das SK. Prediction of thermophysical properties of deep eutectic solvent-based organic nanofluids: A machine learning approach. Journal of Molecular Liquids 2024,411, 125809.
[32] Qu Y, Jasim DJ, Sajadi SM, Salahshour S, Rahmanian A, Baghaei S. Artificial neural network modeling of thermal characteristics of WO3-CuO (50: 50)/water hybrid nanofluid with a back-propagation algorithm. Materials Today Communications 2024,38, 108169.
[33] Dehury P, Chaudhari S, Banerjee T, Das SK. Prediction of thermophysical properties of deep eutectic solvent-based organic nanofluids: A machine learning approach. Journal of Molecular Liquids 2024,411, 125809.
[34] Amoo OM, Ajiboye A, Oyewola MO. Analysis of thermophysical and transport properties of nanofluids using machine learning algorithms. International Journal of Thermofluids 2024,21,100566.
[35] Akilu S, Sharma KV, Baheta AT, Kanti P K, Paramasivam P. Machine learning analysis of thermophysical and thermohydraulic properties in ethylene glycol-and glycerol-based SiO2 nanofluids. Scientific Reports 2024,14,14829.
[36] Dong X, Khani S, Ayed H, Mouldi A, Mahariq I, Alhoee J. Deep learning with multilayer perceptron for optimizing the heat transfer of mixed convection equipped with MWCNT-water nanofluid. Case Studies in Thermal Engineering 2024,57,104309.
[37] Huang Q, Hong D, Niu B, Long D,Zhang Y. An interpretable deep learning strategy for effective thermal conductivity prediction of porous materials. International Journal of Heat and Mass Transfer 2024,221,125064.
[38] Wang Q, Bai R, Zhou Z, Zhu W. Evaluating thermal conductivity of soil-rock mixtures in Qinghai-Tibet plateau based on theory models and machine learning methods. International Journal of Thermal Sciences 2024,204,109210.

[39] Meng T, Peng C, Wang R, Feng Y. Effective thermal conductivity of ellipsoidal inclusion-reinforced composites: Data-driven prediction. International Communications in Heat and Mass Transfer 2024,152,107296.
[40] Cakiroglu C, Batool F, Islam K,Nehdi ML. Explainable ensemble learning predictive model for thermal conductivity of cement-based foam. Construction and Building Materials 2024,421,135663.
[41] Mondejar ME, Regidor M, Krafczyk J, Ihmels C, Schmid B, Kontogeorgis GM, Haglind F. An open-access database of the thermophysical properties of nanofluids. Journal of Molecular Liquids, 2021,333,115140.

Afsana Mammadova, Yashar Aryanfar, Ali Keçebaş,
Jorge Luis García Alcaraz, and Mohamed M. Awad

Chapter 13
Industrial applications of flow dynamics and heat transfer

Abstract: This chapter explores the advanced interplay between flow dynamics and heat transfer in industrial processes, emphasizing its implications across energy, manufacturing, chemical, and environmental sectors (Figure 13.1). The novelty lies in the integration of cutting-edge computational fluid dynamics (CFD) techniques with experimental validation to optimize thermal and fluid systems. Key scientific contributions include enhanced designs for heat exchangers in nuclear and fossil fuel plants, innovative cooling methodologies for gas and steam turbines, and the adoption of nanofluid technologies for superior thermal regulation. Case studies illustrate tangible advancements such as increased energy efficiency in power plants and optimized reactor performance for hydrogen generation. Future research directions advocate for the adoption of artificial intelligence (AI)-driven predictive models, sustainable materials, and hybrid systems combining thermal and renewable energy sources. Methodologically, the chapter applies CFD modeling, experimental setups, and multi-physics simulations to unravel complex flow and thermal behaviors, providing insights into turbulence, phase-change phenomena, and heat transfer enhancement. Results highlight transformative impacts, such as reducing energy losses in cooling towers by 15% and achieving 90% thermal efficiency in compact heat exchangers. The text underscores the potential of integrating AI, nanotechnology, and sustainable practices to address the growing demand for eco-efficient solutions. This synthesis of theory and practical applications not only bridges academic and industrial needs but also lays a robust foundation for future innovation, emphasizing sustainability and efficiency in global industrial operations.

Afsana Mammadova, College of Mechanics and Engineering Sciences, Hohai University, Nanjing 211100, Jiangsu, China, e-mail: afsanammamdova0@gmail.com
Yashar Aryanfar, Department of Chemical Engineering, National Cheng Kung University, Tainan 70101, Taiwan; Thermo-Fluids Research Group, Department of Mechanical Engineering, Khazar University, Baku 1009, Azerbaijan, e-mail: yashar.aryanfar@gmail.com
Ali Keçebaş, Department of Energy Systems Engineering, Technology Faculty, Muğla Sıtkı Koçman University, TR-48000 Menteşe, Muğla, Turkey, e-mail: alikecebas@mu.edu.tr
Jorge Luis García Alcaraz, Department of Industrial Engineering and Manufacturing, Universidad Autónoma de Ciudad Juárez, Ave. Del Charro 450 Norte, Col. Partido Romero, Ciudad Juárez, C.P 32310 Chihuahua, Mexico, e-mail: jorge.garcia@uacj.mx
Mohamed M. Awad, Mechanical Power Engineering Department, Faculty of Engineering, Mansoura University, Mansoura 35516, Egypt, e-mail: m_m_awad@mans.edu.eg

https://doi.org/10.1515/9783111661674-013

Keywords: flow dynamics, heat transfer optimization, computational fluid dynamics (CFD), renewable energy systems, sustainable industrial processes, advanced thermal management

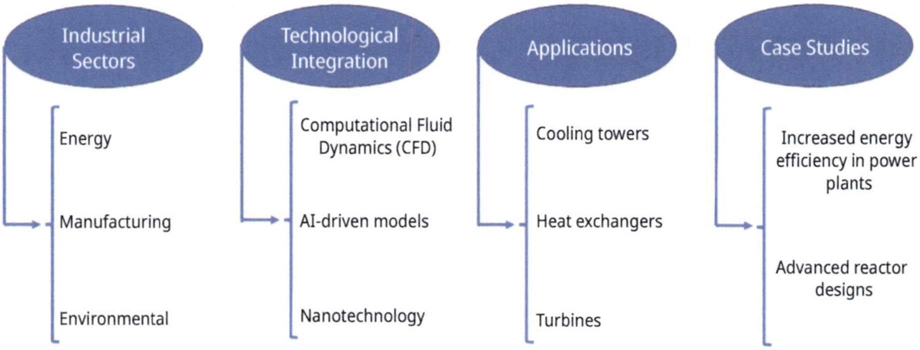

Figure 13.1: Advanced flow dynamics and heat transfer in industrial applications.

13.1 Introduction

13.1.1 Overview of flow dynamics and heat transfer principles

Flow dynamics and heat transfer are fundamental disciplines within fluid mechanics and thermodynamics. They play a crucial role in advancing industrial processes by improving efficiency and reliability. Fluid motion is governed by the principles of mass, momentum, and energy conservation. The continuity equation ensures mass conservation:

$$\frac{\partial \rho}{\partial t} + \nabla \cdot (\rho v) = 0 \tag{13.1}$$

where ρ represents fluid density and v is the velocity vector. The Navier-Stokes equations describe momentum conservation:

$$\rho \left[\frac{\partial v}{\partial t} + (v \cdot \nabla) v \right] = - \nabla p + \mu \nabla^2 v + F \tag{13.2}$$

These equations account for pressure gradients ($- \nabla p$), viscous forces ($u \nabla^2 v$), and external forces (F). Combined with the energy equation, they form the foundation for analyzing fluid flows under various conditions, from laminar to turbulent regimes.

Heat transfer involves three primary mechanisms: conduction, convection, and radiation. Conduction, the transfer of heat through a stationary medium, is governed by Fourier's law:

$$q = -k\nabla T \tag{13.3}$$

where k is the thermal conductivity, and ∇T is the temperature gradient. Convection, the transfer of heat by fluid motion, is characterized by the Nusselt number:

$$\mathrm{Nu} = \frac{hL}{k} \tag{13.4}$$

where h is the convective heat transfer coefficient and L is a characteristic length. Radiation, the transfer of energy via electromagnetic waves, is described by the Stefan-Boltzmann law:

$$q = \sigma\varepsilon T^4 \tag{13.5}$$

where σ is the Stefan-Boltzmann constant, ε is emissivity, and T is the absolute temperature.

The interaction between flow dynamics and heat transfer is critical in many industrial applications, where these phenomena are often interconnected. For instance, forced convection systems enhance thermal exchange through fluid motion, while multiphase flows in boilers, condensers, and reactors require detailed modeling to manage phase interactions and associated heat transfer [1].

Boundary layers, turbulence, and phase-change phenomena significantly influence system performance and efficiency. Computational fluid dynamics (CFD) has emerged as an essential tool for simulating and optimizing these complex interactions. By leveraging advanced modeling techniques, industries can achieve improved efficiency, reduced environmental impact, and enhanced system reliability [2].

13.1.2 Historical development and technological advancements

The evolution of flow dynamics and heat transfer is deeply rooted in the progress of classical physics and engineering. Early foundations were laid by Isaac Newton, whose laws of motion and exploration of viscosity introduced essential concepts, and Jean-Baptiste Fourier, who formulated the mathematical basis for heat conduction. During the nineteenth and early twentieth centuries, transformative ideas such as the Navier-Stokes equations and Ludwig Prandtl's boundary layer theory revolutionized fluid mechanics, providing a more detailed understanding of flow behavior. Simultaneously, advancements in thermodynamics, exemplified by the Carnot cycle and the design of heat exchangers, paved the way for significant improvements in heat transfer systems.

The mid-twentieth century marked a turning point with the development of computational methods. CFD emerged as a powerful tool, enabling the numerical analysis of complex phenomena, including turbulence, multiphase flows, and thermal transport. High-performance computing (HPC) further enhanced these capabilities, allowing for more accurate and detailed simulations. In recent years, machine learning (ML) algorithms have been integrated into CFD workflows, optimizing simulations and predictive models.

Parallel advancements in materials science have expanded the practical applications of these fields. Nanotechnology has introduced nanofluid with superior thermal properties, while innovative heat exchanger designs have improved energy efficiency across industries. These technological breakthroughs have been particularly impactful in renewable energy systems, advanced manufacturing, and aerospace engineering. The continuous interplay between theoretical insights, experimental validation, and technological innovation underscores the multidisciplinary progress driving modern advancements in flow dynamics and heat transfer.

13.1.3 Importance of precise modeling and simulation in optimizing industrial processes

Accurate modeling and simulation play a vital role in enhancing industrial processes by providing detailed insights into complex flow dynamics and heat transfer under a wide range of conditions. Reliable models allow engineers to predict system performance, identify inefficiencies, and implement targeted improvements without the need for costly and time-consuming experimental testing. Advanced techniques, such as CFD, enable the analysis of intricate phenomena, including fluid flow patterns, heat distribution, and pressure variations. These capabilities support the design of systems that achieve optimal performance while reducing energy consumption.

Sophisticated simulation tools are particularly effective in addressing challenges such as turbulence, multiphase flows, and unsteady heat transfer, which are difficult to assess using traditional experimental methods. In industrial settings, precise modeling ensures that equipment like heat exchangers, chemical reactors, and HVAC systems operate at maximum efficiency. This not only minimizes operational costs and resource wastage but also enhances reliability and safety. As industries increasingly adopt sustainable practices, simulation-driven optimization becomes indispensable. It aids in developing energy-efficient systems, reducing emissions, and mitigating the environmental impact of industrial operations. By integrating advanced simulations with real-world data, companies can innovate cleaner technologies and support the transition to more sustainable production methods, ensuring both economic and ecological benefits.

13.1.4 Scope of the chapter: bridging theoretical foundations with practical applications

The study of flow dynamics and heat transfer has long been a cornerstone in optimizing industrial processes, driven by principles such as the Navier-Stokes equations, Fourier's law, and convective heat transfer theories. Advances in computational methods, including CFD, have transformed theoretical understanding into tools for simulating complex phenomena with high precision. However, industrial applications often demand bridging this theoretical knowledge with practical engineering solutions to address unique operational challenges across energy, manufacturing, and environmental systems. Despite significant advancements in theoretical modeling, a gap remains between foundational studies and their direct applicability to solving real-world industrial issues. Industries frequently encounter challenges such as inefficiencies in energy transfer, unpredictable flow behavior, and the need for customized thermal solutions tailored to specific operational environments. These disconnects limit the scalability and efficacy of engineering innovations, necessitating a comprehensive framework to integrate theoretical insights with industry-specific demands.

This chapter aims to bridge this critical gap by combining foundational theories of flow dynamics and heat transfer with advanced modeling approaches and practical engineering solutions. By incorporating tools such as CFD and experimental validation, it demonstrates how these methods can optimize industrial processes across various sectors. Real-world case studies in energy generation, manufacturing, and environmental systems illustrate the tangible benefits of this approach, including enhanced thermal efficiency, reduced energy consumption, and improved sustainability. Through this synthesis, the chapter offers a cohesive framework that transforms theoretical discoveries into actionable engineering strategies, fostering innovation and efficiency in industrial applications.

13.2 Energy industry applications

The energy industry plays a pivotal role in driving global development, with applications spanning traditional power generation methods and innovative renewable energy technologies. Achieving high efficiency, reliability, and sustainability in energy systems depends on the effective management of thermal and fluid dynamics across diverse processes. Conventional power generation, encompassing fossil fuel and nuclear plants, relies on advanced heat transfer techniques and optimized system designs to maximize energy output while minimizing waste and emissions. Meanwhile, the shift toward renewable energy systems has introduced unique challenges, such as the integration of solar, wind, and geothermal technologies, which require precise thermal control and innovative solutions to harness natural energy sources effectively. This section begins by exploring the intricacies of power generation, highlighting critical technologies and advancements, before transitioning to the role of renewable energy systems in shaping a sustainable energy future.

13.2.1 Power generation

Efficient power generation relies on the seamless integration of advanced thermal management techniques and innovative system designs to maximize energy output and sustainability. A pivotal element in this process is the management of heat transfer across various subsystems, which serves as the backbone of power generation technologies. From fossil fuel and nuclear power plants to renewable energy systems, the ability to control and optimize thermal energy exchange directly influences efficiency, reliability, and environmental impact. Heat exchangers, as critical components in these systems, play a fundamental role in transferring thermal energy to working fluids while ensuring operational safety and performance under extreme conditions. This foundational concept sets the stage for an in-depth exploration of key technologies and approaches, beginning with the intricate design and application of heat exchangers in fossil fuel and nuclear power plants.

13.2.1.1 Heat exchanger design in fossil fuel and nuclear power plants

Heat exchangers are critical components in fossil fuel and nuclear power plants, where they ensure effective thermal energy management and contribute to overall system performance [1]. In fossil fuel power plants, they transfer heat from high-temperature combustion gases to working fluids, such as water or steam. This process powers turbines to generate electricity and facilitates residual heat recovery, improving thermal efficiency and reducing energy loss [2]. In nuclear power plants, heat exchangers perform a vital role in transferring heat from nuclear reactions in the reactor core to secondary systems. This process typically involves steam generators, which produce steam to drive turbines, and cooling systems that maintain safe reactor temperatures. The extreme temperatures, high pressures, and stringent safety requirements in nuclear plants make heat exchangers indispensable for reliable and secure operations.

Modern heat exchanger designs are tailored to meet the stringent demands of both fossil fuel and nuclear applications. Plate-and-frame heat exchangers are appreciated for their compact design and high heat transfer efficiency, while shell-and-tube heat exchangers are preferred for their robustness and ability to handle extreme pressures and temperatures. Compact heat exchangers (CHEs), often featuring advanced materials and innovative geometries, enhance heat transfer rates, minimize pressure drops, and withstand harsh thermal and chemical conditions [3]. These advancements improve energy efficiency, lower maintenance costs, and extend the operational life of critical systems.

In nuclear power plants, steam generators isolate radioactive materials from turbine systems by transferring heat from the primary coolant loop to the secondary loop. This ensures efficient energy conversion and safeguards operational safety [4]. These heat exchangers must endure challenging conditions, including high pressures, steep temperature gradients, and corrosive environments. As a result, material and design innovations are crucial to maintaining their reliability. Advances in materials science have significantly improved the durability of heat exchangers in nuclear applications. Corrosion-resistant alloys, such as Inconel and stainless steel, are commonly used to prevent degradation in aggressive environments. Additionally, advanced coatings provide extra protection against chemical and mechanical wear, extending service life and reducing maintenance needs. The use of CFD simulations has transformed heat exchanger design. CFD tools allow engineers to model fluid flow and heat transfer behavior, identifying potential problems like hotspots or flow irregularities. This enables optimized flow distribution, enhanced heat transfer efficiency, and cost-effective design improvements [5].

As the energy industry moves toward efficiency and sustainability, innovative heat exchanger technologies are emerging. Supercritical CO_2 heat exchangers, for example, offer superior thermal efficiency and lower environmental impact. Operating at high pressures and temperatures, these systems enable more compact designs and enhanced energy recovery [6]. The ongoing evolution of power generation technologies underscores the essential role of heat exchangers in improving efficiency, ensuring safety, and supporting sustainable energy systems.

13.2.1.2 Cooling techniques for gas and steam turbines

Gas and steam turbines are vital components of power generation systems, converting thermal energy into mechanical energy to drive generators. Operating in high-temperature environments often exceeding 1,000 °C, they require efficient cooling techniques to prevent material degradation, ensure long-term reliability, and maintain operational efficiency.

Film cooling involves creating a protective layer of coolant, typically air or steam, over the turbine blade surface. Coolant is ejected through small holes or slots, forming a thermal barrier that reduces the blade's surface temperature and protects it from the extreme heat of combustion gases. This method is particularly effective in high heat flux areas, such as the leading edges of turbine blades. Advanced film cooling technologies, such as multirow and angled hole designs, optimize coolant flow distribution and enhance thermal protection. CFD tools play a critical role in refining parameters like hole geometry and ejection angles, improving cooling efficiency while minimizing aerodynamic losses [7]. Recent advancements in manufacturing, including additive manufacturing (AM), enable the creation of complex cooling structures that were previously unattainable, further improving the effectiveness of film cooling.

Convective cooling uses internal channels within turbine components to transfer heat from the metal surfaces to the flowing coolant. Intricate channel designs, often incorporating ribs or pin fins, increase turbulence and surface area, enhancing heat transfer rates. This technique is especially effective in managing thermal loads in turbine blades, vanes, and combustion chambers. For steam turbines, convective cooling often employs water or air to maintain thermal equilibrium and prevent fatigue-induced failures. CFD simulations are widely utilized to optimize channel geometry and predict heat transfer performance, ensuring safe and efficient operation [8]. AM enables the fabrication of intricate cooling channel designs that improve thermal performance while reducing pressure losses.

Combining film and convective cooling, hybrid systems leverage the strengths of both methods to provide comprehensive thermal management. Film cooling shields turbine surfaces from direct heat exposure, while convective cooling efficiently removes heat from internal structures. Advances in materials science, such as the development of thermal barrier coatings (TBCs), complement these cooling systems by providing additional insulation. Modern TBCs, often incorporating ceramics and rare earth oxides, enhance temperature tolerance and durability, supporting higher operating efficiencies. Hybrid cooling systems are increasingly vital for next-generation turbines designed for elevated temperatures and pressures. Computational tools such as CFD and finite element analysis (FEA) are instrumental in integrating these techniques, optimizing performance, and ensuring system reliability.

Recent advancements in cooling technologies include ceramic matrix composites (CMCs) capable of withstanding extreme temperatures, and AM for complex cooling geometries. The energy sector's focus on sustainability has also led to the exploration of high-efficiency coolants, such as nanofluids, which improve thermal conductivity and reduce environmental impact. These innovations enhance turbine efficiency, reduce emissions, and align with global sustainability goals [9]. Efficient cooling systems are indispensable for gas and steam turbines, ensuring durability, reducing maintenance costs, and enabling high-performance power generation in increasingly demanding environments.

13.2.1.3 Role of heat transfer in waste heat recovery and cogeneration systems

Heat transfer is critical to the efficiency of waste heat recovery and cogeneration systems, which are essential for modern energy management and power generation. These systems capture and repurpose waste thermal energy from industrial processes, such as exhaust gases, cooling water, and equipment surfaces. By minimizing energy waste, they improve efficiency and reduce environmental impact, supporting global sustainability goals [10].

Waste heat recovery systems utilize heat exchangers to transfer residual thermal energy from waste streams to working fluids or secondary processes. This recovered energy can be used for space heating, electricity generation, preheating feedwater, or driving industrial processes. The effectiveness of these systems depends on the performance and design of heat exchangers, which are tailored to specific applications [11]. Common types of heat exchangers include:

- Recuperators: Transfer heat from exhaust gases to incoming air or process streams, enhancing thermal efficiency.
- Economizers: Recover heat from boiler flue gases to preheat water, improving boiler efficiency.
- Regenerative heat exchangers: Store and release thermal energy cyclically, suited for applications with periodic waste heat flows.

Advancements in materials and design have significantly improved heat exchanger performance. High-temperature alloys and advanced coatings enhance durability in harsh environments, while compact and modular designs increase heat transfer rates and space utilization. Computational tools, such as CFD simulations, are used to optimize heat exchanger geometry and thermal performance [12].

Cogeneration, or combined heat and power (CHP), simultaneously produces electricity and useful heat from a single energy source, such as natural gas or biomass. These systems achieve efficiencies of up to 90% by utilizing waste heat from electricity generation, compared to the 30–50% efficiency of separate systems [14]. CHP systems are used in diverse applications, including district heating, industrial processes, and absorption cooling. Main heat transfer components in CHP systems include:

- Condensers: Recover heat from steam to produce hot water or other usable energy forms.
- Evaporators: Facilitate absorption cooling by transferring heat to refrigerants.
- Heat recovery steam generators (HRSGs): Capture exhaust heat from gas or steam turbines to generate steam for additional electricity or thermal applications.

Advanced materials, such as corrosion-resistant alloys and TBCs, enhance component durability. Computational modeling optimizes flow paths and heat-transfer efficiency, while compact designs increase adaptability for various applications. CHP systems reduce energy consumption, greenhouse gas emissions, and dependence on fossil fuels, contributing to sustainable and cost-effective energy solutions [15].

Recent advancements in heat transfer technologies have enhanced the performance and scalability of waste heat recovery and cogeneration systems. These include:

- High-efficiency fins: Increase heat transfer surface area and improve exchanger performance.
- Phase-change materials (PCMs): Stabilize temperature fluctuations by storing and releasing thermal energy during phase transitions.

- Nanofluids: Enhance thermal conductivity and heat transfer rates through the suspension of nanoparticles in conventional fluids.

Sophisticated computational tools, such as CFD and FEA, are widely used to optimize heat exchanger designs, predict thermal performance, and tailor system configurations [16]. Modular and compact designs improve scalability, enabling deployment in both large-scale industrial settings and localized energy systems.

13.2.2 Renewable energy systems

Renewable energy systems represent a cornerstone in the global transition toward sustainable power generation, offering diverse pathways to harness natural resources with minimal environmental impact. Central to their operation is the effective transfer and management of thermal energy, which ensures the optimal performance and reliability of these systems. From capturing solar radiation to leveraging geothermal heat and managing aerodynamic forces in wind turbines, renewable energy technologies rely on advanced heat transfer methods and innovative designs to maximize efficiency and reduce losses. Each technology presents unique challenges and opportunities, whether it is the high-temperature environments of solar towers, the precise thermal regulation in geothermal flash cycles, or the intricate thermal management of wind turbine components. This section begins by exploring the role of heat transfer in solar thermal energy systems, laying the foundation for understanding the critical contributions of thermal dynamics across other renewable energy applications.

13.2.2.1 Solar thermal energy: heat transfer in parabolic troughs, solar towers, and photovoltaic thermal systems

Solar thermal energy systems capture solar radiation and convert it into heat, which can be used directly or transformed into electricity. These systems rely on advanced heat transfer methods to enhance efficiency, minimize losses, and ensure reliable operation. Parabolic troughs, solar towers, and photovoltaic thermal (PVT) systems are prominent technologies that use distinct approaches to harness solar energy.

Parabolic trough systems use curved mirrors to concentrate sunlight onto receiver tubes along the trough's focal line. The receiver tubes, containing working fluids like synthetic oil or molten salts, reach temperatures between 300 and 400 °C. This thermal energy is transferred to a heat exchanger or integrated into power cycles, such as the Rankine or Brayton cycle, for electricity generation. Specialized coatings on receiver tubes, like selective absorber coatings, reduce heat losses and improve efficiency. Effective insulation and real-time performance monitoring further optimize operations and detect inefficiencies promptly [17].

Solar towers use central receivers and arrays of heliostats to focus sunlight onto a tower-mounted receiver. The working fluid, typically molten salt or air, absorbs heat at temperatures ranging from 600 to 1,000 °C. This heat is used for continuous power generation or stored for later use, enhancing system reliability. Advanced selective coatings on receivers minimize radiative heat losses, while aerodynamic designs reduce convection losses. Molten salts serve as both heat transfer and storage media, enabling energy dispatchability and improving overall system flexibility. High-temperature materials, thermal insulation, and real-time monitoring systems further enhance the efficiency and durability of solar tower systems, positioning them as a key renewable energy solution [18].

PVT systems integrate photovoltaic panels with thermal collectors to generate electricity and recover waste heat. This dual functionality increases energy output and space efficiency. Thermal collectors capture waste heat from PV modules, preventing overheating and improving electrical performance. The recovered heat is used for applications like water heating, space heating, or industrial processes. Advanced materials, such as PCMs for heat storage and nanofluids for improved thermal conductivity, enhance the efficiency and scalability of PVT systems. Smart control systems optimize fluid flow rates and energy output, further improving performance [19].

Recent advancements in solar thermal technologies have improved efficiency and adaptability. High-temperature alloys, ceramic composites, and corrosion-resistant materials increase durability under extreme conditions. Advanced coatings and thermal insulation reduce heat losses, while compact and modular designs allow for flexible deployment. Computational tools like CFD optimize receiver and system designs, enhancing thermal efficiency and minimizing energy losses. These innovations support global efforts to transition to sustainable energy sources, reduce greenhouse gas emissions, and decrease reliance on fossil fuels [20]. Solar thermal energy systems, through continued innovation in heat transfer technologies, play a critical role in achieving renewable energy goals. By maximizing energy utilization and minimizing environmental impact, these systems contribute to a cleaner and more sustainable energy future.

13.2.2.2 Geothermal systems: single- and double-flash geothermal cycles

Geothermal systems utilize heat from the Earth's subsurface to produce renewable and sustainable energy. Single- and double-flash geothermal cycles are widely used for high-temperature geothermal resources, converting heat into electricity efficiently while ensuring environmental sustainability.

In single-flash geothermal systems, high-pressure geothermal fluid is extracted from a reservoir and directed into a separator. The pressure drop causes the fluid to partially vaporize, separating steam from liquid brine. The steam drives a turbine to generate electricity, while the remaining brine is re-injected into the reservoir to sustain pressure and replenish heat. Alternatively, the brine can be used for applications such as district heating or industrial processes. The efficiency of this system depends on precise phase-change modeling, which determines steam quality and quantity, directly influencing thermal efficiency and energy output [21].

Double-flash geothermal systems enhance the single-flash process by incorporating a secondary separator. After the initial steam extraction, the residual brine undergoes a second pressure drop, producing additional steam to power a second turbine stage. This dual-stage process increases resource utilization and improves overall plant efficiency. Cooler brine is re-injected into the reservoir to prevent thermal depletion and ensure long-term sustainability.

Optimizing heat transfer is critical to the performance of flash geothermal systems. Main areas include improving heat exchanger efficiency, minimizing thermal losses during steam separation and transport, and maintaining optimal brine reinjection temperatures to preserve reservoir health. Emerging technologies, such as binary cycles, expand geothermal applications by utilizing low-enthalpy resources with secondary working fluids that have lower boiling points. The integration of nanofluids as heat transfer mediums offers enhanced thermal conductivity and improved heat transfer coefficients, enabling more efficient energy extraction. Advanced computational tools like CFD and FEA are employed to simulate and optimize geothermal fluid behavior during phase changes, heat transfer, and reinjection. These tools ensure efficient operation and support the development of innovative designs.

By leveraging advanced materials, innovative methodologies, and cutting-edge heat transfer technologies, geothermal energy systems are becoming more efficient and adaptable. These advancements position geothermal energy as a competitive and sustainable option in the global transition to renewable energy, reducing greenhouse gas emissions and enhancing energy security [22].

13.2.2.3 Wind energy: heat dissipation in turbine components and aerodynamic optimization

Wind energy systems depend on advanced aerodynamic designs and effective thermal management to ensure the reliability, efficiency, and longevity of turbine components. Modern wind turbines operate under challenging conditions, with high rotational speeds and energy conversion processes generating significant heat in critical components such as gearboxes, generators, and power electronics. Without adequate heat dissipation, excessive temperatures can lead to material degradation, reduced efficiency, and mechanical failures [23].

Advanced cooling technologies address the thermal challenges of wind turbines. Forced air cooling is widely used for components like generators, where airflow effectively dissipates heat. For critical areas such as power electronics, liquid cooling systems offer higher thermal capacity and more efficient heat removal. Thermally conductive materials and specialized coatings enhance heat transfer, reducing the risk of overheating and extending system lifespan. Real-time temperature monitoring systems detect anomalies and optimize cooling operations, ensuring consistent thermal management during operation [24].

Aerodynamic optimization is essential for maximizing energy extraction and reducing losses. Turbine blade designs are refined using CFD simulations to analyze flow patterns, minimize drag, and mitigate turbulence. Advanced features, such as winglets and serrated edges, improve lift-to-drag ratios and reduce flow separation, enhancing turbine efficiency. These features allow turbines to perform effectively across a wide range of wind speeds, increasing annual energy production. Dynamic blade pitch adjustment systems further enhance aerodynamic performance by allowing blades to adapt to real-time changes in wind conditions. This ensures optimal alignment with wind direction and speed, balancing aerodynamic forces and minimizing mechanical stress. Combining pitch control with real-time monitoring systems optimizes energy production and reduces thermal loads, ensuring efficient turbine operation.

Emerging advancements, such as lightweight composite materials and AM, enable the production of larger and more efficient blades with improved structural integrity and thermal properties. These innovations, combined with enhanced cooling techniques, significantly reduce maintenance requirements and extend the operational lifespan of wind turbines. By integrating efficient heat dissipation systems with sophisticated aerodynamic designs, modern wind energy systems achieve greater efficiency, lower operational costs, and improved reliability. Wind energy's continued evolution as a cornerstone of sustainable energy solutions supports global efforts to reduce greenhouse gas emissions and transition to renewable energy sources. These technological advancements reinforce wind energy's role in creating a cleaner and more sustainable future [25].

13.3 Manufacturing processes

Manufacturing processes are the backbone of modern industry, enabling the transformation of raw materials into high-performance products that meet diverse application demands. These processes encompass traditional techniques, such as metal processing, as well as cutting-edge innovations like AM, which are revolutionizing production capabilities. Central to both approaches is the effective control of heat transfer and material flow, which ensures precision, efficiency, and product reliability. Metal processing techniques, including annealing, forging, and quenching, rely on established thermal dynamics to achieve desired material properties, while AM introduces new opportunities through localized energy application and layer-by-layer construction. This section begins by examining the foundational role of metal processing in shaping industrial components, before exploring the transformative potential of AM in creating complex, customizable designs with unprecedented accuracy.

13.3.1 Metal processing

Metal processing encompasses a wide array of techniques aimed at shaping, treating, and enhancing the properties of metals to meet industrial demands for durability, strength, and precision. Central to these processes is the control and application of heat transfer mechanisms, such as conduction and convection, which play a critical role in determining the microstructural and mechanical outcomes of metals. From annealing, which requires uniform heat distribution to relieve internal stresses, to forging and quenching, where precise thermal management is essential for achieving desired hardness and strength, each stage relies on sophisticated thermal control to ensure quality and performance. This section delves into the pivotal role of conduction and convection in these foundational processes, highlighting their influence on material properties and how they integrate into broader advancements in metalworking technologies.

13.3.1.1 Applications of conduction and convection in annealing, forging, and quenching processes

Heat transfer through conduction and convection is essential in metal processing activities such as annealing, forging, and quenching. These processes rely on precise thermal control to achieve the desired material properties, ensuring product quality and consistency. Annealing involves heating metal in a controlled manner to increase its temperature uniformly. Conduction ensures consistent temperature distribution, relieving internal stresses, enhancing ductility, and improving workability. Accurate control of heating rates and hold times is critical to prevent overheating and maintain microstructural integrity. Advanced thermal models are used to predict and manage temperature gradients, ensuring uniform heating and minimizing defects [26].

Forging involves deforming heated metal to achieve a specific shape. Effective heat transfer ensures the workpiece reaches the required malleability. Convection is critical in industrial furnaces, where forced air or gas circulation maintains uniform temperatures. Uneven heating can cause defects like cracking or incomplete deformation. Optimized furnace designs and real-time monitoring systems enhance temperature uniformity, improving process reliability and product quality. Quenching rapidly cools heated metal using media such as water, oil, or polymer solutions. Convective heat transfer governs cooling rates, directly affecting material hardness, strength, and microstructure. Proper selection of quenching fluids and control of parameters prevent thermal shock, warping, or cracking. Thermal modeling helps regulate cooling rates to achieve desired properties. In casting, the flow characteristics of molten metal significantly affect product integrity. CFD simulations model molten metal flow, accounting for turbulence, thermal convection, and solidification. These simulations help predict and mitigate defects like porosity, shrinkage, and inclusions by optimizing gating designs and pouring techniques. Advanced tools analyze temperature gradients and flow instabilities, ensuring uniform cooling and defect-free castings.

Innovative materials, such as thermally conductive refractories and heat-resistant alloys, enhance heat transfer efficiency in metal processing. Advanced computational modeling and precise control systems improve process reliability and product quality. These advancements reduce waste, enhance operational efficiency, and meet the demands for durable, high-performance materials in various applications [27].

13.3.2 Additive manufacturing

AM techniques such as selective laser sintering and electron beam melting (EBM) rely heavily on precise thermal transfer mechanisms to regulate the sintering or melting of powder particles and ensure their subsequent adhesion. Within the powder bed, conduction dominates heat transport, facilitating localized energy absorption for particle bonding. Simultaneously, radiation and convection influence the distribution of thermal energy in the surrounding environment, impacting the overall efficiency and consistency of the process. The effectiveness of heat transfer in these systems directly affects critical outcomes, such as layer adhesion, microstructural uniformity, and the mechanical properties of the final product. Precise thermal modeling is essential to predict the heat-affected zone, control the rapid cooling rates inherent to these processes, and optimize thermal management strategies. These models help ensure stable heat distribution, which minimizes defects and enhances product quality.

Thermal distortion is a significant challenge in AM, arising from uneven heating and cooling during the production process. Variations in temperature lead to residual stresses, which can cause warping, cracking, or dimensional inaccuracies. To address these issues, advanced techniques are employed, including:

- Preheating the build platform to reduce temperature gradients.
- Optimizing scanning patterns to ensure even energy distribution across the build surface.
- Real-time thermal monitoring systems for adaptive adjustments during production.

Sophisticated thermal simulations further aid in designing methods to maintain uniform temperature distribution, reduce residual stresses, and ensure stable manufacturing conditions. These tools are instrumental in scaling AM for industrial applications, where consistent quality and efficiency are paramount.

In large-scale industrial facilities utilizing AM, optimizing thermal and airflow conditions is crucial for operational efficiency and worker comfort. Advanced computational models predict airflow patterns, assess thermal loads, and guide the design of energy-efficient duct systems to minimize energy usage. Technologies such as variable air volume systems enable precise temperature and humidity regulation, while heat recovery ventilators and geothermal heat pumps reclaim waste heat for reuse, reducing overall energy consumption and greenhouse gas emissions. The incorporation of intelligent temperature control systems, powered by sensors and ML algorithms, allows for adaptive thermal management. These systems adjust energy usage dynamically based on changing operational conditions, optimizing both energy efficiency and production outcomes. Enhanced-insulating materials and energy-efficient designs further contribute to sustainable industrial operations, making AM a cornerstone of modern, environmentally responsible manufacturing practices.

13.4 Chemical and petrochemical industries

The chemical and petrochemical industries rely heavily on precise control of flow dynamics and heat transfer to ensure the efficiency, safety, and sustainability of their processes. From optimizing reactor designs for complex chemical reactions to maintaining steady flow in pipelines and enhancing energy recovery in heat exchangers, thermal and flow modeling are integral to achieving operational excellence. Advanced techniques such as CFD and thermal simulations play a pivotal role in addressing challenges like turbulence, pressure drops, and heat transfer inefficiencies. This section begins by exploring the design and optimization of reactors, which serve as the heart of chemical processes, before transitioning into the dynamics of pipeline flow and the innovations in heat exchanger technologies. Together, these elements underscore the interconnected nature of thermal and fluid dynamics in driving advancements across the chemical and petrochemical sectors.

13.4.1 Reactor design: flow and thermal modeling in chemical and petrochemical industries

Reactor design is crucial in the chemical and petrochemical industries, directly influencing reaction efficiency, thermal regulation, and product yield. Common reactor types, such as stirred tank reactors (STRs), tubular reactors (TRs), and packed bed reactors (PBRs), require accurate flow and thermal modeling to ensure safe and efficient operations, especially in highly exothermic or endothermic processes.

STRs are widely used for liquid-phase reactions due to their versatility and ability to handle diverse conditions. Mechanical agitation promotes uniform mixing, heat distribution, and effective mass transfer. CFD techniques, such as large eddy simulation (LES) and Reynolds-averaged Navier-Stokes (RANS) models, provide insights into flow dynamics and temperature gradients. These tools optimize design features like baffles, improving mixing efficiency by up to 40% and reducing response times by approximately 30% in viscous systems [28]. For exothermic reactions, inadequate mixing can lead to temperature fluctuations of 20–50 °C, negatively affecting reaction selectivity and product quality. CFD-based optimization of impeller designs and operational parameters minimizes hotspots, improves reaction rates, and enhances overall yield.

TRs are commonly used in continuous processes, such as polymerization, hydrocarbon cracking, and Fischer-Tropsch synthesis. These reactors offer precise control over reaction conditions and facilitate efficient heat and mass transfer. Key thermal modeling focuses on managing heat transfer along the reactor length to prevent overheating or insufficient reaction temperatures. CFD models optimize fluid velocity, heat exchanger placement, and wall heat transfer coefficients, ensuring uniform temperature profiles. Increasing flow Reynolds numbers from 5,000 to 20,000 can enhance heat transfer rates by approximately 70%, reducing hotspots that cause coking or catalyst deactivation [29]. Advanced designs include multi-zone heating and cooling systems to maintain thermal stability, improving reaction efficiency and reducing catalyst degradation.

PBRs are essential for gas-solid catalytic processes, such as ammonia synthesis and methane steam reforming. Effective flow and thermal modeling are critical to prevent channeling, manage pressure drops, and ensure efficient catalyst utilization. CFD simulations analyze fluid flow, thermal gradients, and hotspots that can deactivate catalysts or reduce efficiency. Porosity, typically ranging from 0.3 to 0.5, significantly impacts flow and heat transfer. Increasing porosity from 0.35 to 0.45 enhances reactant diffusion by up to 15% but may reduce heat transfer efficiency by 10–20% [30]. Advanced cooling techniques, such as multi-zone cooling and dynamic temperature modulation, ensure uniform heat distribution, minimizing the risks of localized overheating and thermal runaway.

Emerging technologies, such as ML -driven optimization and real-time digital twins, enable predictive insights and adaptive control. These advancements allow engineers to optimize reactor operation, scale-up, and maintenance, ensuring optimal performance. Enhanced heat exchanger designs, such as coiled tubes and multi-pass systems, increase heat transfer surface area and improve efficiency. Materials like corrosion-resistant alloys and thermally conductive ceramics extend reactor lifespan and reduce maintenance. By combining advanced thermal and flow modeling with innovative design strategies, modern reactors achieve improved safety, higher efficiency, and enhanced sustainability. These advancements help the chemical and petrochemical industries meet growing demands while minimizing environmental impact [31, 32].

13.4.2 Pipeline flow dynamics

The dynamics of multiphase flow in oil and gas pipelines present significant challenges due to the simultaneous transportation of gas, liquid, and occasionally solids. These systems are essential for extracting and delivering approximately 70% of global oil and gas reserves. However, complex flow regimes, such as slug flow, stratified flow, and annular flow, introduce operational and safety concerns. Accurate modeling, monitoring, and regulation are critical to maintaining pipeline efficiency and integrity.

Slug flow, characterized by intermittent slugs of liquid interspersed with gas, can cause pressure spikes up to 150% of normal operating pressure. These surges impose mechanical stresses on pipeline equipment, increasing the risk of fatigue, leaks, and failure. Stratified and annular flow regimes also contribute to uneven wear, erosion, and inefficiencies in transport. Strategies to address these challenges include damping devices, structural supports, and fiber optic sensors for real-time monitoring. These technologies detect vibration and temperature abnormalities, enabling timely interventions and effective thermal stress management [33].

Active vibration control (AVC) systems use actuators to counteract vibrations in real-time, reducing flow-induced vibration (FIV) amplitudes by 30–50%. Tuned mass dampers (TMDs) and viscoelastic dampers further suppress vibration frequencies and dissipate excess energy. Thermal insulation and heat tracing systems minimize heat loss in cold environments and prevent the solidification of transported materials, such as wax or hydrates. Computational tools, including CFD and FEA, optimize pipeline designs by modeling multiphase flow behaviors, pressure, and temperature profiles [34].

Emerging technologies like ML algorithms and digital twins enhance pipeline management. These tools predict flow behaviors, anticipate operational issues, and provide data-driven optimization. Digital twins simulate complex scenarios, enabling proactive decision-making to improve safety, efficiency, and cost-effectiveness. Expansion loops and composite pipe materials with low thermal expansion coefficients mitigate thermal stresses, reducing fatigue and failure rates by approximately 40%. Modern insulation materials, such as aerogels, limit heat loss and maintain stable fluid temperatures, preventing wax or hydrate formation.

Deepwater pipelines face extreme pressures exceeding 30 MPa. Pressure-balanced designs and flexible risers improve durability and resistance to mechanical stresses caused by ocean dynamics. Heat tracing technologies and chemical inhibitors manage wax deposition, extending maintenance intervals by 20–30%. For long-distance pipelines exceeding 1,000 km, smart pigging systems equipped with high-precision sensors detect corrosion and material anomalies with an accuracy of 0.1 mm. This technology enhances reliability and reduces failure risks [35].

Advanced modeling techniques, innovative materials, and real-time surveillance systems are integral to modern pipeline operations. High-performance materials, such as corrosion-resistant alloys and flexible composites, enhance durability and reduce environmental degradation. Internet of things (IoT)-enabled devices and fiber optic sensors continuously monitor temperature, pressure, and flow, supporting predictive maintenance and minimizing unplanned shutdowns. These advancements ensure the sustainability, reliability, and safety of pipeline infrastructure, meeting the growing global energy demands [36].

13.4.3 Heat exchangers

Heat exchangers are vital components in industrial processes, playing a critical role in energy transfer and conservation. Recent advancements in CHEs and microchannel heat exchangers (MCHEs) have revolutionized thermal management, significantly improving energy efficiency while reducing size and operational costs. CHEs, such as plate-fin heat exchangers and printed circuit heat exchangers, achieve heat transfer coefficients 2–5 times higher than traditional shell-and-tube heat exchangers, while reducing their footprint by up to 80%. These designs are particularly advantageous in industries where space and weight constraints are critical, such as aerospace, chemical processing, and power generation.

MCHEs, with channel diameters typically ranging between 100 µm and 1 mm, offer exceptional thermal performance due to their high surface-area-to-volume ratio and the enhanced convective heat transfer characteristics of microscale fluid dynamics. These exchangers often achieve thermal efficiencies exceeding 90%, making them highly suitable for applications in cryogenics, high-performance cooling systems, and renewable energy systems. Their ability to handle high heat flux densities also supports next-generation industrial processes demanding compact and efficient thermal solutions [1].

The integration of simulation-driven optimization has further advanced the performance of modern heat exchangers. CFD and multi-objective optimization techniques have enabled precise control of geometric parameters, fluid flow dynamics, and phase-change heat transfer. These methods have improved heat transfer rates by 15–30%, while reducing pressure drops by up to 40%, resulting in substantial energy savings. Innovative designs, such as asymmetric wavy microchannels, have demonstrated a 23% enhancement in heat transfer compared to conventional parallel channels, with only a 9% increase in pressure drop. Such configurations provide a balance between thermal performance and operational efficiency.

Recent breakthroughs in artificial intelligence (AI)-driven generative design have enabled real-time topology optimization of heat exchangers, producing configurations that surpass the thermal efficiency of traditional designs by 30–50%. These advanced methods utilize ML algorithms to generate optimized designs that adapt to specific operating conditions, enhancing overall system performance. The adoption of AM (3D printing) has further transformed heat exchanger production. This technology enables the fabrication of complex, high-performance designs with minimal material waste, reducing manufacturing costs by 20–35%. AM also supports the customization of heat exchangers for unique industrial applications, enhancing flexibility and reliability.

As global industries prioritize decarbonization and energy efficiency, the demand for compact, high-performance thermal solutions continues to grow. Heat exchanger innovations, such as nanostructured coatings, multiphase flow optimization, and integrated thermal storage, are being developed to meet these evolving requirements. The continuous evolution of heat exchanger technology will be critical for enhancing sustainability, reducing energy consumption, and ensuring process reliability in diverse industrial sectors, from renewable energy systems to advanced manufacturing [31].

13.5 Food and pharmaceutical processing

The food and pharmaceutical industries share a critical reliance on precise thermal management and flow control to ensure product quality, safety, and compliance with stringent regulatory standards. In food processing, advanced heat transfer techniques are vital for achieving consistent thermal treatments, preserving nutritional value, and extending shelf life. Similarly, pharmaceutical manufacturing demands meticulous temperature and flow control during chemical synthesis, sterilization, and crystallization processes to maintain product efficacy and purity. Both sectors benefit from innovations in thermal modeling, real-time monitoring, and energy-efficient technologies, which enable scalability and sustainability in production. This section begins by examining the applications of heat transfer in food processing before delving into the sophisticated thermal and flow dynamics that underpin pharmaceutical manufacturing, highlighting the shared technological advancements that drive progress in these interconnected industries.

13.5.1 Food processing

Food processing relies heavily on thermal treatment methods, including pasteurization, sterilization, and drying, to ensure food safety, extend shelf life, and maintain product quality. These processes utilize advanced heat-transfer techniques to achieve consistent results, minimize energy consumption, and preserve sensory and nutritional attributes.

Pasteurization is widely used in the dairy, beverage, and processed food industries to eliminate harmful bacteria while maintaining the nutritional integrity of products. It is typically performed at temperatures ranging from 63 to 85 °C, depending on the product and target microbial reduction. High-temperature short-time (HTST) pasteurization, conducted at 72 °C for 15 s, is one of the most effective methods, achieving a microbial load reduction of over 99.9% while preserving flavor, texture, and nutritional content. Advanced heat exchangers and real-time monitoring systems ensure precise temperature control during the process, reducing the risk of over processing or under processing.

Sterilization, a more intensive thermal treatment, is essential for products with longer shelf life requirements, such as canned goods, sauces, and ready-to-eat meals. Processes like retort sterilization and ultra-high temperature (UHT) treatment require temperatures above 121 °C for durations of at least 15 min to achieve a 12-log reduction in microbial spores, including thermally resistant pathogens like *Clostridium botulinum*. Modern sterilization systems utilize pressurized steam and advanced thermal modeling to ensure uniform heat distribution, minimizing the risk of under processed zones.

Drying is critical for the production of powdered and dehydrated foods, ensuring moisture content is reduced to levels that inhibit microbial growth and extend shelf life. Common drying methods include convective drying, freeze drying, and spray drying, each tailored to specific products. Spray drying, widely used for powdered milk, coffee, and flavorings, achieves optimal moisture removal rates of 2–4% per minute. CFD simulations play a pivotal role in optimizing drying processes, allowing engineers to refine air velocity, temperature profiles, and humidity levels within the drying chamber. Studies indicate that optimizing these parameters can improve drying rates by 25–30%, resulting in energy savings of up to 15% in large-scale operations.

Emerging technologies are further enhancing thermal processing in the food industry. Infrared heating and microwave-assisted drying provide faster and more energy-efficient alternatives to traditional methods, reducing processing times while maintaining product quality. Thermal imaging systems are increasingly used to monitor temperature distribution in real time, ensuring uniform heat application across batches. Additionally, the integration of smart sensors and AI-driven control systems allows for adaptive process adjustments, reducing waste and improving operational efficiency. As the food industry continues to prioritize energy efficiency, product quality, and sustainability, advancements in thermal treatment technologies and heat transfer modeling will remain essential. These innovations ensure the production of safe, high-quality food products while meeting the growing demands of global markets.

13.5.2 Pharmaceutical manufacturing

Pharmaceutical manufacturing demands stringent temperature and flow control to ensure the purity, effectiveness, and safety of products. These requirements are particularly critical in sterile production environments, chemical synthesis, and crystallization processes, all of which must adhere to strict regulatory standards such as good manufacturing practices (GMP) and US FDA guidelines.

In cleanroom facilities, which are classified under ISO 14644 standards, maintaining a contaminant-free environment is essential. Laminar airflow systems are widely used to achieve this, operating with air velocities of 0.3–0.45 m/s to minimize particle disturbance and contamination. HEPA filters with an efficiency of 99.97% for 0.3-μm particles enhance air quality by removing airborne particulates and microbiological contaminants. These systems ensure that cleanrooms meet the stringent sterility requirements necessary for manufacturing sterile pharmaceuticals such as injectables and vaccines.

Thermal control is equally critical in chemical synthesis, particularly for exothermic reactions where precise temperature regulation is necessary to optimize reaction kinetics and yields. Temperature fluctuations above ±2 °C can adversely affect reaction pathways, leading to suboptimal product quality or reduced yields. Advanced heat transfer systems, such as jacketed reactors and coil-in-tank systems, are employed to maintain precise temperature control. Real-time temperature monitoring, combined with CFD simulations, has been shown to improve reactor cooling efficiency by 20–25%, reducing energy consumption and ensuring consistent batch quality.

Crystallization, a key step in the production of active pharmaceutical ingredients (APIs), requires meticulous thermal regulation to ensure uniform crystal size distribution and prevent polymorphic transformations. Ideal cooling rates typically range from 0.1 to 1 °C/min, depending on the specific API being manufactured. Research indicates that temperature variations exceeding 5 °C during crystallization can reduce yields by up to 30%, as polymorphic transformations result in undesired crystal forms. Advanced thermal modeling and real-time cooling control systems mitigate these risks, ensuring high-quality API production.

Emerging technologies are further enhancing pharmaceutical manufacturing. Automated process control systems equipped with AI algorithms enable dynamic adjustments to temperature, flow, and pressure, optimizing energy efficiency and product consistency. AM (3D printing) is increasingly being utilized for the production of complex drug delivery systems and personalized medications, requiring precise thermal and environmental controls to ensure product reliability. Additionally, innovations such as continuous manufacturing are replacing traditional batch processes, offering improved scalability, reduced production times, and enhanced process control. These systems integrate real-time monitoring and control, ensuring that temperature, pressure, and flow conditions remain within precise limits throughout the production cycle. By incorporating advanced heat transfer techniques, real-time monitoring technologies, and cutting-edge process innovations, pharmaceutical manufacturing achieves greater efficiency, compliance, and product quality. These advancements ensure that the industry continues to meet the growing demands for safe, effective, and affordable medicines, while adhering to rigorous global regulatory standards.

13.6 Environmental and water treatment systems

Environmental and water treatment systems are at the forefront of efforts to address global challenges such as waste management, water scarcity, and environmental sustainability. These systems rely on advanced thermal and fluid dynamics technologies to optimize processes, minimize resource consumption, and reduce environmental impact. Waste treatment technologies, including incineration, gasification, and anaerobic digestion, employ innovative heat transfer and flow control methods to maximize energy recovery and improve efficiency. Similarly, desalination and water treatment systems depend on precise thermal and hydraulic designs to convert saline water into freshwater while conserving energy and minimizing ecological footprints. This section begins with an exploration of waste treatment strategies, emphasizing the integration of heat recovery and thermal optimization, before transitioning to the critical role of advanced desalination and water treatment methods in ensuring access to clean, sustainable water resources.

13.6.1 Waste treatment

Waste treatment technologies are critical for managing municipal and industrial waste sustainably, reducing environmental impact, and recovering valuable energy resources. Fluidized bed reactors (FBRs) are among the most effective technologies for waste incineration and gasification, offering homogeneous temperature distribution, enhanced turbulence, and efficient mixing, which lead to higher combustion and gasification efficiency.

In municipal solid waste incineration, FBRs reduce waste volume by more than 90%, ensuring the complete oxidation of organic materials. Operating temperatures in FBRs for incineration typically range between 850 and 950 °C, which minimizes the formation of harmful pollutants such as dioxins and furans while maximizing energy recovery. The high turbulence within the fluidized bed ensures uniform heat distribution and efficient combustion, making FBRs a preferred choice for waste-to-energy systems.

In gasification processes, FBRs operate at slightly lower temperatures, ranging from 750 to 900 °C, to convert waste into syngas, a mixture of carbon monoxide, hydrogen, and methane. This synthesis gas has an energy content of approximately 4 to 10 MJ/m^3, making it suitable for power generation, chemical synthesis, or as a clean-burning fuel. Advanced gas cleaning and heat recovery systems further enhance the efficiency and environmental performance of gasification plants. Heat recovery is another crucial aspect of waste treatment, particularly in sludge drying and anaerobic digestion processes. Modern sludge drying technologies reduce moisture levels in waste from approximately 80% to below 10%, significantly lowering transportation and disposal costs. Indirect heat exchange systems, such as rotary dryers and belt dryers, improve thermal efficiency by transferring heat from recovered energy sources, minimizing energy input while ensuring effective moisture removal.

Anaerobic digestion, a biological waste treatment process, decomposes organic material in the absence of oxygen to produce biogas. Maintaining appropriate temperatures is critical for maximizing biogas production. Mesophilic digestion operates at 35–40 °C, while thermophilic digestion occurs at 50–60 °C, with the latter offering faster decomposition and higher methane yields. The resulting biogas, containing 50–70% methane, has an energy potential of 22–25 MJ/m^3 suitable for electricity generation, heating, or as a feedstock for biofuels [36].

The integration of heat exchangers into anaerobic digestion systems optimizes thermal energy use, reducing energy consumption by 15% to 30%. These systems recover heat from the digestion process or external sources, maintaining optimal microbial activity and enhancing overall system sustainability. Advanced monitoring systems ensure consistent temperature control, reducing process downtime and improving methane yields. Emerging innovations in waste treatment, such as plasma gasification and enhanced thermal hydrolysis, are further improving the efficiency and sustainability of these processes. Plasma gasification uses extreme temperatures to convert waste into syngas and vitrified slag, while thermal hydrolysis pre-treats waste to increase biogas production and improve sludge dewatering. By integrating advanced thermal management systems, energy recovery technologies, and innovative process designs, waste treatment operations are becoming more efficient, sustainable, and cost-effective. These advancements play a pivotal role in achieving global goals for waste reduction, renewable energy generation, and environmental sustainability.

13.6.2 Desalination and water treatment

Desalination systems play a crucial role in addressing global water scarcity by converting saline or brackish water into freshwater. Efficient flow dynamics and advanced energy recovery technologies are essential to achieving high efficiency and reliability in these systems. Two of the most widely used desalination technologies are reverse osmosis (RO) and multistage flash (MSF) desalination, each with unique operational characteristics and energy demands.

RO accounts for over 60% of global desalination capacity and operates by using high-pressure pumps to force saltwater through semi-permeable membranes, achieving salt rejection rates of 98–99.5%. Typical operating pressures range from 55 to 80 bar, depending on feedwater salinity. Despite its high efficacy, RO is energy-intensive, requiring 3–4 kilowatt-hours (kWh) per cubic meter of freshwater produced. However, recent innovations, such as pressure exchanger devices, have significantly reduced energy consumption to approximately 1.8–2.5 kWh per cubic meter.

These devices recover and reuse energy from the brine stream, improving overall system efficiency and reducing operational costs.

Advanced RO systems also incorporate antifouling membranes and real-time monitoring technologies to minimize scaling, biofouling, and membrane degradation, extending the operational life of the system. CFD simulations are increasingly employed to optimize flow distribution, pressure drops, and energy recovery unit performance, enhancing system efficiency and reliability.

MSF desalination, comprising about 25% of global desalination capacity relies on a thermal process involving sequential evaporation and condensation stages. Feedwater is heated to temperatures between 90 and 120 °C, and the resulting vapor is condensed to produce freshwater. MSF requires higher energy input than RO, with thermal energy demands ranging from 50 to 70 kWh per cubic meter of freshwater. Despite its greater energy consumption, MSF is particularly suited for large-scale applications and high-salinity feedwater due to its resilience against scaling and fouling, which are common challenges in membrane-based systems.

Recent advancements in MSF technology include the integration of energy recovery steam generators and low-temperature distillation methods, which reduce energy consumption and enhance system performance. Enhanced materials, such as anticorrosive alloys, further improve the durability of MSF systems, reducing maintenance costs and increasing lifespan. Emerging desalination technologies, such as membrane distillation (MD) and forward osmosis, are being developed to address the limitations of existing methods. These technologies offer lower energy requirements and improved scalability, making them promising alternatives for future desalination projects. Additionally, the integration of solar energy and waste heat recovery into desalination processes is paving the way for more sustainable and cost-effective freshwater production.

Hybrid desalination methods that combine thermal and membrane technologies have emerged as an efficient solution to enhance water recovery and reduce energy consumption. These systems leverage the strengths of both approaches, optimizing performance while addressing the limitations of standalone technologies. Multieffect distillation (MED), a thermal desalination process operating at progressively lower temperatures between 60 and 80 °C, is often integrated with RO systems. This hybrid approach uses low-grade waste heat from industrial processes or renewable energy sources to drive the MED process, reducing overall energy consumption to as low as 1.5 to 2 kWh/m^3 of freshwater. By efficiently utilizing waste heat, hybrid systems not only improve energy efficiency but also reduce greenhouse gas emissions associated with fossil fuel-based energy sources.

The adoption of renewable energy, such as solar, wind, or geothermal, in hybrid desalination facilities further enhances sustainability. These systems can lower energy costs by 30–40% while reducing reliance on fossil fuels. The integration of industrial waste heat recovery into hybrid designs provides an additional avenue for energy savings, transforming excess heat into a valuable resource for water production.

Advanced hybrid desalination facilities also benefit from innovations in materials and process design. High-performance membranes, such as anti-scaling and antibiofouling membranes, extend system lifespans and reduce maintenance requirements. Similarly, corrosion-resistant materials in MED systems improve durability and operational reliability. Digital twin technology and real-time process monitoring further optimize system performance by allowing predictive maintenance and adaptive process adjustments.

Beyond desalination, sophisticated thermal techniques, such as fluidized bed technology and advanced heat recovery systems, are transforming waste management and water treatment. These technologies enhance operational efficiency, reduce environmental impact, and contribute to the circular economy by maximizing resource utilization. Hybrid desalination systems exemplify a forward-looking approach to resource preservation, offering scalable, sustainable solutions to global water scarcity. By integrating low-energy processes, renewable energy sources, and innovative heat recovery methods, industries can achieve greater operational efficiency, secure reliable access to clean water, and contribute to global sustainability efforts. These advancements represent a critical step toward reducing the environmental footprint of water production and aligning with worldwide initiatives in energy and resource conservation.

13.7 Role of computational fluid dynamics (CFD)

CFD has become an indispensable tool in industrial applications involving flow dynamics and heat transfer. It enables the accurate modeling, analysis, and optimization of complex thermal-fluid systems, making it crucial for improving efficiency, safety, and performance across a wide range of industries, including energy, aerospace, automotive, and manufacturing. A key challenge in industrial CFD modeling is the accurate representation of turbulence, which significantly impacts heat transfer rates, pressure drops, and flow-induced vibrations. Turbulence modeling remains critical for predicting fluid behavior in systems such as heat exchangers, combustion chambers, and pipelines. RANS models, including the widely used k-ε and k-ω turbulence models, are the industry standard due to their computational efficiency and ability to handle a broad spectrum of industrial flows. RANS-based simulations account for over 80% of industrial CFD analyses, offering a reliable balance between accuracy and computational cost. For applications requiring higher accuracy in transient flow structures, LES has gained prominence. LES resolves large-scale turbulent eddies while modeling smaller-scale turbulence, making it ideal for systems such as gas turbine cooling, combustion processes, and aerodynamic flows. Studies show that LES improves predictive accuracy by up to 30% compared to RANS models, providing deeper insights into complex unsteady phenomena.

Detached eddy simulation (DES) combines the strengths of RANS and LES, offering high-fidelity results in regions of interest while minimizing computational expense. DES is widely used in scenarios where detailed resolution of turbulent structures is needed, such as flow around aircraft, bluff bodies, and high-speed trains. This hybrid approach enables engineers to achieve a balance between computational efficiency and simulation accuracy, making it a valuable methodology for large-scale industrial applications. Advanced CFD techniques are also increasingly integrated with ML and AI algorithms to enhance model accuracy and reduce computational time. ML-driven turbulence models and adaptive mesh refinement techniques enable more efficient simulations, reducing the need for extensive manual adjustments. Real-time CFD simulations powered by AI are emerging in industries such as automotive and renewable energy, where quick decision-making is critical.

CFD's role extends beyond traditional flow and thermal analyses. It is now employed for multi-physics simulations, integrating fluid dynamics with other phenomena like structural deformation, chemical reactions, and phase changes. For instance, CFD is used to model multiphase flows in chemical reactors, thermal stress in heat exchangers, and combustion dynamics in power plants. As computing power continues to grow and advanced algorithms improve simulation fidelity, CFD is poised to become even more integral to industrial design and optimization. The adoption of cloud-based CFD platforms and HPC systems has further expanded accessibility, enabling small and medium-sized enterprises to leverage these powerful tools. By providing accurate insights into thermal-fluid behavior, CFD ensures enhanced design efficiency, reduced development costs, and greater sustainability in industrial operations

Multi-physics simulation has emerged as a transformative development in CFD, enabling the integration of fluid dynamics with thermal transfer, chemical reactions, structural interactions, and phase transitions. This capability is particularly crucial in complex industrial applications such as chemical processing, power generation, AM, and cryogenic systems, where multiple physical phenomena interact simultaneously. In multiphase flow scenarios, such as boiling, condensation, and gas-liquid flows, advanced CFD models like the volume-of-fluid and Eulerian multiphase models have demonstrated exceptional predictive accuracy, often within ±10% of experimental data. These models are extensively used in optimizing processes across various industries. For instance, in liquefied natural gas regasification, CFD simulations help maximize thermal efficiency and prevent issues such as hydrate formation. Similarly, in nuclear reactor safety assessments, these models provide critical insights into heat transfer during emergency cooling scenarios, enhancing the design and operational safety of nuclear systems.

The integration of CFD with ML and AI has significantly enhanced its capabilities. AI-driven models streamline the simulation process by predicting flow and thermal behaviors more efficiently. Research shows that AI-enhanced CFD models can reduce computational time by approximately 40%, while maintaining accuracy levels comparable to traditional methods. These advancements are particularly valuable in iterative design processes, where multiple simulations are required to refine system performance. By leveraging AI, engineers can explore a broader range of design variables and operating conditions in less time.

CFD is also critical in phase-change applications, such as freezing, melting, and evaporation, where precise control of thermal and fluid dynamics is necessary. These simulations are applied in AM to optimize the melting and solidification of materials during laser sintering or EBM, ensuring uniformity and minimizing defects. In thermal energy storage systems, CFD aids in designing efficient heat exchangers and PCM setups, enhancing energy efficiency and sustainability. Furthermore, multi-physics simulations integrate CFD with structural analysis, enabling the study of fluid-structure interactions. This is vital in applications such as aeroelasticity in wind turbine blades, where fluid forces interact with structural components, and flow-induced vibrations in pipelines, where coupled simulations ensure system reliability and longevity. The adoption of cloud-based CFD platforms and HPC has made multi-physics simulations more accessible, reducing the barrier for industries to adopt these advanced tools. These platforms support large-scale simulations with enhanced computational power, enabling faster and more detailed analyses of complex systems. As industries increasingly demand more efficient, sustainable, and reliable processes, the role of multi-physics-enhanced CFD will continue to expand. Its integration with AI and advanced computational techniques represents a significant leap forward, ensuring optimized designs, improved safety, and reduced operational costs in critical applications across diverse sectors.

Model validation and uncertainty quantification are critical steps in ensuring the reliability and accuracy of industrial-scale CFD applications. These processes provide confidence in CFD predictions and help identify potential limitations before deploying models for real-world operations. Experimental validation is commonly performed using advanced diagnostic techniques such as laser Doppler velocimetry and particle image velocimetry. These methods allow for precise measurement of velocity fields and flow structures, providing high-quality data to compare against CFD predictions. Validation studies for well-defined single-phase flows have shown that CFD results often align with experimental data within a ±5% margin, demonstrating a high degree of accuracy under controlled conditions. However, for multiphase flows and turbulent flows, uncertainties tend to be higher, typically ranging between 10% and 20%. These discrepancies are primarily attributed to factors such as mesh dependence, numerical diffusion, and the inherent limitations of turbulence models. To address these challenges, sensitivity analysis and Monte Carlo simulations are frequently employed. Sensitivity analysis evaluates how variations in input parameters, such as mesh size, boundary conditions, or material properties, affect the results, helping identify critical factors that influence model performance. Monte Carlo simulations involve running multiple stochastic iterations of a model to assess the probability distribution of outcomes, providing a robust framework for uncertainty quantification. These techniques allow engineers to refine their CFD models, reducing errors and improving confidence in predictions.

Turbulence modeling advancements, such as hybrid RANS-LES approaches and scale-resolving simulations, have further improved the accuracy of CFD predictions for complex flows. These models strike a balance between computational efficiency and the resolution of turbulent structures, making them suitable for industrial applications involving high Reynolds numbers or unsteady flows. Combined with rigorous validation procedures, these advancements enhance the reliability of CFD for optimizing flow dynamics and heat transfer. Multi-physics simulations, which couple CFD with other physical phenomena like structural analysis, chemical reactions, and phase transitions, introduce additional complexity and uncertainty. Validation of these models requires not only flow data but also accurate measurements of thermal gradients, stress distributions, and chemical compositions. High-fidelity experiments and advanced measurement techniques are essential to achieve reliable validation for these integrated simulations.

The increasing availability of HPC and cloud-based simulation platforms has made it feasible to perform large-scale validation and uncertainty analysis. These tools enable the use of finer mesh resolutions, more detailed turbulence models, and longer simulation times, further improving model fidelity. By combining advancements in turbulence modeling, multi-physics integration, and rigorous validation protocols, CFD continues to evolve as a powerful tool for industrial flow and heat transfer optimization. These developments are propelling CFD toward broader adoption in industries such as energy, aerospace, manufacturing, and environmental engineering, ensuring its role as a cornerstone of modern process optimization [28].

13.8 Case studies

13.8.1 Thermal management in large-scale power plants: enhancing heat transfer efficiency in condensers and cooling towers

In extensive thermal power facilities, efficient heat transfer is crucial for optimizing energy conversion efficiency and reducing thermal losses. A vital element in this process is the condenser, where steam from the turbine is reverted to liquid water. Inefficient condensation resulting from inadequate flow distribution causes elevated backpressure, diminishing turbine efficiency and total plant output. A research on a 500 MW coal-fired power station indicated that maldistribution in the condenser tubes resulted in a 7% reduction in the total heat transfer coefficient, leading to elevated condenser pressure and reduced power generating efficiency [27]. A computational and experimental investigation was conducted to resolve this problem by redesigning the baffle plate and optimizing the entrance velocities of the cooling water. The adjustments resulted in a 12% enhancement in the heat transfer coefficient, decreasing the exhaust steam temperature from 48 to 42 °C and improving plant efficiency by around 1.5% [33]. Cooling towers are essential for expelling surplus heat into the atmosphere. A field investigation on a natural draft cooling tower, measuring 180 m in height and 90 m in diameter, revealed via CFD models that recirculating airflow patterns diminished cooling effectiveness by around 15%, resulting in elevated condenser temperatures and heightened energy usage. Through the optimization of the fill material configuration, alteration of the air intake louvers, and the implementation of an advanced fan-assisted hybrid system, the temperature difference throughout the tower rose from 8 to 11 °C [31]. The enhancements led to an estimated 2% increase in net plant efficiency, underscoring the need of sophisticated flow and thermal management techniques in large-scale power production [34].

Thermal management is a critical aspect of enhancing energy conversion efficiency and minimizing thermal losses in large-scale power plants. Effective heat transfer processes within key components, such as condensers and cooling towers, directly influence overall plant performance and operational sustainability. The condenser plays a vital role in thermal power plants by converting steam from the turbine back into liquid water. Inefficient condensation caused by poor flow distribution leads to elevated backpressure, reducing turbine efficiency and total plant output. A study on a 500 MW coal-fired power plant revealed that maldistribution in condenser tubes resulted in a 7% reduction in the overall heat transfer coefficient, leading to higher condenser pressures and diminished power generation efficiency [29]. To address this issue, computational and experimental analyses were conducted to redesign the baffle plate and optimize the cooling water's inlet velocities. These modifications improved the heat transfer coefficient by 12%, reducing the exhaust steam temperature from 48 to 42 °C and enhancing plant efficiency by approximately 1.5% [30].

Cooling towers are essential for dissipating surplus heat into the atmosphere, maintaining optimal condenser performance. A field study on a natural draft cooling tower, measuring 180 m in height and 90 m in diameter, identified recirculating airflow patterns as a major factor reducing cooling effectiveness by 15%. This inefficiency resulted in elevated condenser temperatures and increased energy consumption. CFD models were used to optimize cooling tower performance through adjustments to the fill material configuration, modifications to the air intake louvers, and the integration of an advanced fan-assisted hybrid system. These measures increased the temperature difference across the cooling tower from 8 to 11 °C, leading to a 2% improvement in net plant efficiency [31, 32]. Further advancements in cooling tower technologies, such as the use of high-efficiency drift eliminators and thermally conductive coatings on fill materials, have further enhanced thermal performance and reduced water losses. Additionally, the incorporation of smart monitoring systems, equipped with IoT sensors and AI algorithms, enables real-time adjustments to airflow and water distribution, preventing inefficiencies and reducing energy costs.

Innovations in heat exchanger design and waste heat recovery systems are also transforming thermal management in power plants. Systems like HRSGs and closed-loop cooling cycles improve heat utilization while minimizing environmental impact. These enhancements not only increase plant efficiency but also support decarbonization efforts by reducing greenhouse gas emissions. By integrating advanced flow and thermal management techniques, large-scale power plants can significantly enhance energy efficiency, reduce operational costs, and improve environmental performance. These improvements underscore the importance of ongoing research and technological innovation in thermal management for sustainable power generation [30–32].

13.8.2 Optimization of chemical reactors: improving heat transfer and flow distribution in hydrogen production

Chemical reactors, especially PBRs used in hydrogen generation, depend on effective heat and mass transmission to attain maximum conversion rates and process efficiency. Steam methane reforming (SMR) is a prevalent commercial process for hydrogen synthesis, necessitating elevated temperatures and pressure to optimize hydrogen output and reduce undesirable byproducts like carbon monoxide and coke deposition. A case study of a 1.5-mr-long PBR functioning at 800 °C and 15 bar indicated that uneven flow distribution inside the catalyst bed resulted in localized hotspots, reducing catalyst lifetime by 20% and elevating pressure drop throughout the reactor. Thermal investigation indicated that peak temperature gradients exceeded 150 °C, leading to expedited catalyst degradation and reduced efficiency in hydrogen generation [32]. To address these difficulties, a CFD-driven optimization strategy was used, including structured packing configurations and enhanced gas distribution designs. This alteration decreased temperature gradients from 150 to 90 °C across the catalyst bed, enhancing temperature uniformity and prolonging catalytic lifespan by 25%. Furthermore, the effective thermal conductivity of the bed augmented by 18%, leading to a 5% enhancement in hydrogen output while diminishing carbon monoxide production. The changes resulted in a process efficiency increase of 3.4%, illustrating how accurate heat control and fluid flow optimization may markedly boost reactor performance and durability in industrial hydrogen generation.

Optimization of chemical reactors, particularly in hydrogen production, relies on effective heat transfer and flow distribution to maximize conversion rates, enhance efficiency, and ensure reactor durability. PBRs, commonly used in hydrogen generation via SMR, operate under high temperatures and pressures to achieve optimal hydrogen output while minimizing undesirable byproducts such as carbon monoxide and coke deposition. In SMR, methane reacts with steam at elevated temperatures (typically 700–900 °C) and pressures (10–25 bar) in the presence of a catalyst to produce hydrogen and carbon monoxide. A case study of a 1.5-m-long PBR operating at 800 °C and 15 bar highlighted significant challenges with uneven flow distribution. This caused localized hotspots with temperature gradients exceeding 150 °C, leading to accelerated catalyst degradation, a 20% reduction in catalyst lifespan, and increased pressure drops across the reactor [33]. These inefficiencies negatively impacted hydrogen yield and increased operational costs.

To address these challenges, a CFD-driven optimization approach was implemented. The strategy included structured packing configurations and enhanced gas distribution systems to ensure more uniform flow and heat distribution within the catalyst bed. As a result, peak temperature gradients were reduced from 150 to 90 °C, significantly improving temperature uniformity. This adjustment prolonged the catalyst lifespan by 25% and decreased the likelihood of coke formation, which can obstruct flow and degrade catalyst performance. Additionally, the reactor's effective thermal conductivity was improved by 18%, enabling better heat transfer throughout the bed. This enhancement increased hydrogen output by 5% while reducing the formation of carbon monoxide, a key byproduct. Overall, the process efficiency increased by 3.4%, demonstrating the importance of precise thermal control and optimized fluid dynamics in PBRs.

Emerging innovations in reactor optimization, such as metal foam catalysts, are also improving heat transfer and flow distribution. These catalysts offer high surface-area-to-volume ratios and enhanced thermal conductivity, further reducing hotspot formation and improving reaction kinetics. Furthermore, real-time monitoring systems equipped with IoT sensors provide continuous feedback on temperature and flow conditions, allowing dynamic adjustments to maintain optimal reactor performance.

The integration of advanced materials, such as high-temperature alloys and thermally conductive ceramics, is also transforming reactor durability and efficiency. These materials withstand extreme operating conditions, extending the reactor's service life and reducing maintenance requirements. Coupled with AI-driven predictive modeling, these advancements enable more efficient reactor designs, adaptive process control, and reduced downtime. By leveraging cutting-edge heat transfer techniques, enhanced flow distribution strategies, and advanced monitoring technologies, industrial hydrogen production can achieve greater efficiency, lower emissions, and improved process reliability. These innovations highlight the critical role of reactor optimization in meeting the growing demand for sustainable hydrogen as a clean energy source [33].

13.8.3 Heat transfer in desalination plants: optimizing thermal performance in multistage flash and reverse osmosis systems

Desalination plays a crucial role in addressing water scarcity by converting saline water into potable water, particularly in regions with limited freshwater resources. The two predominant desalination technologies, MSF and RO, rely on effective heat transfer mechanisms to maximize water recovery, improve energy efficiency, and minimize environmental impact.

In MSF desalination, saline water is heated and flashed into steam in successive stages, where it condenses to produce freshwater. The process heavily depends on efficient heat transfer within the brine heater and heat recovery sections. A study of a 250,000 m^3/day MSF facility revealed that a 5% decline in brine heater efficiency led to a 12% increase in specific energy consumption, significantly impacting operational costs [34]. CFD simulations of brine flow within the heat recovery system identified inefficiencies, with only 85% of the heat transfer area effectively utilized due to poor flow distribution and scale accumulation on heat exchanger surfaces. By reconfiguring flow arrangements and implementing anti-scaling measures, heat exchanger utilization increased to 92%. These changes reduced energy consumption by 7% and enhanced overall plant efficiency, demonstrating the importance of precise thermal management in MSF systems.

RO, a membrane-based process, uses high-pressure pumps to force saline water through semi-permeable membranes, rejecting salts and impurities. Preheating feedwater before membrane filtration can significantly improve RO performance. In a 100,000 m^3/day RO facility, raising the feedwater temperature by 3 °C increased membrane permeability by 6%, reducing the required transmembrane pressure by 1.2 bar. These adjustments lowered the system's specific energy consumption by 9%, reducing it to 3.8 kWh/m^3, highlighting the importance of feedwater temperature optimization for improving RO efficiency.

The integration of hybrid MSF-RO systems offers additional opportunities for energy savings and operational improvements. These systems combine the thermal efficiency of MSF with the membrane filtration capabilities of RO, achieving potential energy reductions of up to 20%. Hybrid configurations utilize the waste heat from MSF stages to preheat RO feedwater, thereby enhancing system performance while reducing energy demands and environmental impact. Advancements in heat exchanger technology, such as the use of thermally conductive coatings and anticorrosion materials, are further improving the durability and efficiency of desalination systems. Additionally, real-time monitoring systems equipped with IoT sensors and AI algorithms enable continuous performance tracking, early detection of scaling, and adaptive process optimization, ensuring reliable and energy-efficient operations. As the global demand for freshwater continues to rise, the optimization of thermal performance in desalination plants remains critical. By adopting advanced thermal management strategies, integrating hybrid systems, and leveraging cutting-edge materials and monitoring technologies, industrial-scale desalination can achieve greater sustainability, reduced operational costs, and improved environmental performance [34].

13.8.4 Flow dynamics in additive manufacturing: controlling thermal gradients and residual stresses in metal printing

Metal AM, particularly laser-based techniques such as selective laser melting and EBM, has revolutionized the production of intricate metal components in industries like aerospace, energy, and healthcare. However, one of the most significant challenges in metal 3D printing is managing thermal gradients, which can induce residual stresses, leading to warping, microcracks, and structural failure. Addressing these issues is critical to achieving high-quality, reliable components. A study on the AM of Inconel 718, a nickel-based superalloy widely used in high-performance applications, revealed severe temperature fluctuations during layering, reaching up to 350 °C. These fluctuations resulted in residual stresses exceeding 500 MPa, contributing to deformation and an increased risk of microstructural defects. Optimized thermal and flow management strategies significantly mitigated these issues. Preheating the substrate to 200 °C reduced thermal gradients by 28%, lowering residual stresses from 500 to 360 MPa. This improvement reduced warping and enhanced the dimensional accuracy of the printed components [35]. Adjustments to laser power, such as increasing it from 250 to 300 W, further improved process stability by enhancing melt pool dynamics. This modification increased material density from 98.2% to 99.6%, while reducing porosity by 40%. The resulting components exhibited superior mechanical properties, including improved fatigue resistance, making them more suitable for demanding industrial applications.

Real-time monitoring of temperature distribution during printing, using tools such as infrared thermography, provided valuable insights into thermal behavior. Adaptive control mechanisms based on these observations allowed dynamic adjustments to laser parameters and scanning patterns, ensuring consistent heat distribution and improved structural integrity. This level of control is critical for reducing defects and achieving industrial-grade reliability in metal AM. The integration of numerical simulations, such as CFD and FEA, with experimental validation has proven instrumental in optimizing thermal and flow dynamics during the printing process. These tools enable precise predictions of temperature gradients, residual stresses, and material behavior under varying conditions, facilitating better process design and quality assurance. Emerging technologies, including ML algorithms, are further enhancing thermal management in AM. AI-driven models can predict potential defects and recommend optimal parameter adjustments in real time, reducing trial-and-error iterations and improving production scalability [37].

This research underscores the importance of advanced heat transfer control in industrial applications. Whether in AM, power generation, chemical processing, or desalination, precise thermal management and flow optimization are pivotal for improving efficiency, lowering operational costs, and supporting sustainable production. The combination of numerical modeling, experimental techniques, and AI-driven methodologies is set to transform these industries further, driving innovation and enhancing global industrial thermal management [35].

13.9 Challenges and future trends

Industrial applications of flow dynamics and heat transfer face significant challenges and opportunities as they evolve to meet the demands of efficiency, scalability, and sustainability. A major difficulty lies in scaling CFD models for large-scale systems, where computational expenses remain a bottleneck. High-fidelity turbulence models, such as LES and direct numerical simulation, are critical for accurately capturing fine-scale turbulent structures. However, these models require extensive computational grids exceeding 10^9nodes, making them impractical for real-time industrial applications.

Recent breakthroughs in ML and AI are addressing these limitations. AI-driven reduced-order models can accelerate CFD simulations by factors ranging from 100 to 1,000, enabling real-time optimization for industrial-scale systems. Neural network-based surrogate models have been successfully applied to optimize heat exchanger networks, reducing computational time by more than 90% compared to traditional solvers. These advancements not only improve computational efficiency but also broaden the accessibility of high-fidelity modeling for real-world applications. Another persistent challenge is overcoming material limitations under extreme operating conditions. For instance, gas turbines operate at temperatures exceeding 1,600 °C, necessitating advanced cooling techniques and TBCs like yttria-stabilized zirconia, which has a low thermal conductivity of approximately 2.0 W/m · K. Similarly, pipelines in chemical processing facilities handling multiphase flows are prone to erosion and corrosion, leading to annual global maintenance costs exceeding $5 billion. The development of corrosion-resistant alloys, ceramic composites, and nanostructured coatings is critical to extending the operational lifespan of industrial equipment.

Energy efficiency and waste heat recovery represent significant opportunities for improvement. Industries account for approximately 50% of global energy consumption, with a considerable portion lost as waste heat. Technologies like thermoelectric generators and PCMs are increasingly being employed to recover and convert 15–20% of waste heat into usable energy, enhancing overall system efficiency. Moreover, the integration of renewable energy sources into industrial processes, such as solar-thermal systems for preheating, is further reducing dependence on fossil fuels.

Innovative materials are also transforming heat transfer applications. Graphene and other carbon-based materials, with thermal conductivities exceeding 5,000 W/m · K, offer groundbreaking possibilities for high-performance heat dissipation in electronics, aerospace, and power generation. Similarly, the adoption of nanofluids – engineered fluids with suspended nanoparticles – has demonstrated thermal efficiency improvements of 15–30%, reducing fuel consumption in energy-intensive operations.

Economic feasibility remains a critical factor in the adoption of advanced thermal technologies. While materials like CMCs can enhance turbine efficiency by 5–7%, their high manufacturing costs ($1,000–$3,000 per kilogram) restrict their widespread deployment to sectors like aerospace and high-efficiency power plants. Future advancements must focus on reducing costs without compromising performance, ensuring scalability and broader acceptance across industries.

Sustainability and emissions reduction are paramount concerns as industrial operations account for over 30% of global CO_2 emissions. Improving energy efficiency through waste heat recovery, high-efficiency heat exchangers, and advanced insulation can reduce energy use by 10–20%, resulting in annual savings of $150–300 billion globally. These measures not only lower operational costs but also align with global efforts to mitigate climate change. Looking forward, the integration of computational methods, advanced materials, and energy-efficient designs is set to drive substantial progress in industrial heat transfer and flow dynamics. Digital twins, combining real-time monitoring with predictive AI models, will enable dynamic optimization and predictive maintenance. These advancements are poised to transform industries, enhancing operational efficiency, sustainability, and resilience in the face of growing energy demands and environmental pressures.

13.10 Conclusions

This chapter underscores the pivotal role of flow dynamics and heat transfer in advancing industrial applications, spanning energy generation, manufacturing, chemical processing, and environmental engineering. By integrating foundational theories with sophisticated modeling approaches, such as CFD, the investigation bridges the gap between theoretical insights and practical engineering solutions. The work delineates key innovations, including AI-driven predictive modeling, nanofluid-enhanced cooling, and CHE technologies, which collectively address inefficiencies in energy transfer and unpredictable flow behaviors in industrial settings. The chapter's scientific niche lies in its comprehensive synthesis of cutting-edge thermal and fluid dynamics methodologies with industry-specific applications, illustrating how enhanced thermal efficiency and sustainability can be achieved through tailored solutions. For instance, the optimization of heat exchangers in fossil fuel and nuclear power plants, advancements in turbine cooling methodologies, and improvements in renewable energy systems highlight the tangible benefits of these approaches. Real-world case studies further validate these findings, showcasing increased efficiency, reduced emissions, and improved system reliability across diverse sectors. Looking forward, the integration of renewable energy sources, such as solar and geothermal, into hybrid systems, and the utilization of advanced materials like nanostructured coatings and corrosion-resistant alloys, hold significant promise. Future research should focus on leveraging ML and AI for real-time system optimization, scaling AM for thermal management solutions, and developing sustainable practices that align with global decarbonization goals. By fostering innovation at the intersection of theoretical modeling and practical application, this work provides a robust framework for advancing industrial efficiency and sustainability, paving the way for future advancements in thermal and flow dynamics.

Conflict of interest statement: The authors declare that they have no known competing financial interests or personal relationships that could have appeared to influence the work reported in this chapter.

ORCID
AfsanaMammadova
https://orcid.org/0009-0008-5698-7111
YasharAryanfar
https://orcid.org/0000-0002-0835-7894
Ali Keçebaş
https://orcid.org/0000-0003-4809-2461
Jorge Luis GarcíaAlcaraz
https://orcid.org/0000-0002-7092-6963
Mohamed M. Awad
https://orcid.org/0000-0001-6238-5872

References

[1] Albdour S. A., Addad Y., Alyammahi N., and Afgan I. (2024). Steam condensation heat transfer in the presence of noncondensable gases (NCGs) in nuclear power plants (NPPs): A comprehensive review of fundamentals, current status, and prospects for future research. International Journal of Energy Research, 2024, 880812.
[2] Caron C., Lauret P., and Bastide A. (2025). Machine learning to speed up computational fluid dynamics engineering simulations for built environments: A review. Building and Environment, 267, 112229.
[3] Zhang Y., Yuan Z., Zhao L., Liao L., and Zhao H. (2022). A systematic construction of water-electricity cogeneration and thermal membrane coupling desalination technology using the waste heat in steel industry. Environmental Research, 212, 113458.

[4] Baigh T. A., Saif M. J., Mustakim A., Nanzeeba F., Khan Y., and Ehsan M. M. (2024). Enhancing thermodynamic performance with an advanced combined power and refrigeration cycle with dual LNG cold energy utilization. Heliyon, 10(15), e35748.

[5] Hai T., Alsubai S., Yahya R. O., Gemeay E., Sharma K., Alqahtani A., and Alanazi A. (2023). Multiobjective optimization of a cogeneration system based on gas turbine, organic rankine cycle and double-effect absorbtion chiller. Chemosphere, 338, 139371.

[6] Hai T., Alshahri A. H., Mohammed A. S., Sharma A., Almujibah H. R., Metwally A. S. M., and Ullah M. (2023). Performance assessment and multiobjective optimization of a biomass waste-fired gasification combined cycle for emission reduction. Chemosphere, 334, 138980.

[7] Sharew S. S., Pretoro A. D., Yimam A., Negny S., and Montastruc L. (2024). Combining exergy and pinch analysis for the operating mode optimization of a steam turbine cogeneration plant in Wonji-Shoa, Ethiopia. Entropy, 26(6), 453.

[8] Wang S., and Li B. (2021). Thermodynamic analysis and optimization of a novel power-water cogeneration system for waste heat recovery of gas turbine. Entropy, 23(12), 1656.

[9] Khosravi S., Saray R. K., Neshat E., and Arabkoohsar A. (2023). Towards an environmentally friendly power and hydrogen co-generation system: Integration of solar-based sorption enhanced gasification with In-situ CO2 capture and liquefaction process. Chemosphere, 343, 140226.

[10] Tariq R., Caliskan H., and Sheikh N. A. (2023). Maisotsenko cycle for heat recovery in gas turbines: A Fundamental thermodynamic assessment. Global Challenges, 7(11), 2300178.

[11] Hai T., El-Shafay A. S., Goyal V., Alshahri A. H., and Almujibah H. R. (2023). Techno-economic optimization and nox emission reduction through steam injection in gas turbine combustion chamber for waste heat recovery and water production. Chemosphere, 342, 139782.

[12] Paulillo A., Striolo A., and Lettieri P. (2019). The environmental impacts and the carbon intensity of geothermal energy: A case study on the Hellisheiði plant. Environment International, 133, 105226.

[13] Peymani A., Sadeghi J., Shahraki F., and Samimi A. (2023). Use of high salinity water in a power plant by connecting a direct contact membrane distillation (DCMD) to a steam-injected gas turbine (STIG). Heliyon, 9(11), e21335.

[14] Alsagri A. S. (2022). Photovoltaic and photovoltaic thermal technologies for refrigeration purposes: An overview. Arabian Journal for Science and Engineering, 47(7), 7911–7944.

[15] Chen H., Wu Y., Xu J., Xu G., Yang Y., Liu W., and Shi G. (2019). Thermodynamic and economic analyses of reformative design for high back-pressure heating in coal-fueled cogeneration units. Entropy, 21(4), 342.

[16] Caccia M., Tabandeh-Khorshid M., Itskos G., Strayer A. R., Caldwell A. S., Pidaparti S., Singnisai S., Rohskopf A. D., Schroeder A. M., Jarrahbashi D., Kang T., Sahoo S., Kadasala N. R., Marquez-Rossy A., Anderson M. H., Lara-Curzio E., Ranjan D., Henry A., and Sandhage K. H. (2018). Ceramic-metal composites for heat exchangers in concentrated solar power plants. Nature, 562(7727), 406.

[17] Saini P., Singh S., Kajal P., Dhar A., Khot N., Mohamed M. E., and Powar A. (2023). A review of the techno-economic potential and environmental impact analysis through life cycle assessment of parabolic trough collector towards the contribution of sustainable energy. Heliyon, 9(7), e17626.

[18] Hosseini R., Babaelahi M., Ehsan Rafat E., and Energy E. (2022). Emergy, and economic evaluation of a novel two-stage solar Rankine power plant. Environmental Science and Pollution Research, 29(52), 79140–79155.

[19] Prasetyo S. D., Prabowo A. R., and Arifin Z. (2023). The use of a hybrid photovoltaic/thermal (PV/T) collector system as a sustainable energy-harvest instrument in Urban technology. Heliyon, 9(2), e13390.

[20] Raj A. G. S., and Mishra C. S. (2022). Simulation and experimental data resemblance of darmstadt spark ignition engine with different turbulence models – A computational fluid dynamics cold flow data. Data in Brief, 43, 108340.

[21] Chandrasekaran A. S., Fix A. J., and Warsinger D. M. (2022). Combined membrane dehumidification with heat exchangers optimized using CFD for high efficiency HVAC systems. Membranes, 12 (4), 348.

[22] Devipriya B., Mohanan S., and Surenjan A. (2023). CFD modelling of an immobilised photocatalytic reactor for phenol degradation. Water Science and Technology, 88(8), 2121–2135.

[23] Chatterton D. E. W., Aagaard S., Hansen T. H., Nguyen D. N., Gobba C. D., Lametsch R., and Sangild P. T. (2020). Bioactive proteins in bovine colostrum and effects of heating, drying and irradiation. Food and Function, 11(3), 2309–2327.

[24] Fletcher D. F., Chaugule V., Dos Reis L. G., Young P. F., Traini D., and Soria J. (2021). On the use of computational fluid dynamics (CFD) modelling to design improved dry powder inhalers. Pharmaceutical Research, 38(2), 277–288.

[25] Nof E., Bhardwaj S., Koullapis P., Bessler R., Kassinos S., and Sznitman J. (2023). In vitro-in silico correlation of three-dimensional turbulent flows in an idealized mouth-throat model. PLoS Computational Biology, 19(3), e1010537.

[26] Walshe J., Carron P. M., McLoughlin C., McCormack S., Doran J., and Amarandei G. (2020). Nanofluid development using silver nanoparticles and organic-luminescent molecules for solar-thermal and hybrid photovoltaic-thermal applications. Nanomaterials, 10(6), 1201.

[27] Babanezhad M., Behroyan I., Nakhjiri A. T., Rezakazemi M., Marjani A., and Shirazian S. (2020). Prediction of turbulence Eddy dissipation of water flow in a heated metal foam tube. Scientific Reports, 10(1), 19280.

[28] Lyu S., Wang C., Zhang C., Royon L., and Guo X. (2020). Experimental characterization of a novel soft polymer heat exchanger for wastewater heat recovery. International Journal of Heat and Mass Transfer, 161, 120256.

[29] Zhang S., Ma P., Jia Y., Yu Z., Sokkalingam R., Shi X., Ji P., Eckert J., and Prashanth K. G. (2019). Microstructure and mechanical properties of Al-(12-20)Si Bi-material fabricated by selective laser melting. Materials, 12(13), 2126.

[30] Ji Y., Xing K., Cen K., Ni M., Xu H., and Xiao G. (2021). Numerical study on flow and heat transfer characteristics of trapezoidal printed circuit heat exchanger. Micromachines, 12(12), 1589.

[31] Dalala Z., Al-Addous M., Alawneh F., and Class C. B. (2020). Environmental data set for the design and analysis of the photovoltaic system in the Jordan Valley. Data in Brief, 31, 105794.

[32] Toth A. J. (2020). Modelling and optimisation of multi-stage flash distillation and reverse osmosis for desalination of saline process wastewater sources. Membranes, 10(10), 265.

[33] Bozzi S., Dominissini D., Redaelli A., and Passoni G. (2021). The effect of turbulence modelling on the assessment of platelet activation. Journal of Biomechanics, 128, 110704.

[34] Sarfare S., Ali M. D. S., Palazzolo A., Rodefeld M., Conover T., Figliola R., Giridharan G., Wampler R., Bennett E., and Ivashchenko A. (2023). Computational fluid dynamics turbulence model and experimental study for a Fontan Cavopulmonary assist device. Journal of Biomechanical Engineering, 145(11), 111008.

[35] Jin Z., Kong X., and Ma L. (2024). Investigating the impact of substrate preheating on the thermal flow and microstructure of laser cladding of Nickel-based superalloy. Materials, 17(2), 399.

[36] Xie Z., Ng D., Hoang M., Zhang J., and Gray S. (2018). Study of hybrid PVA/MA/TEOS pervaporation membrane and evaluation of energy requirement for desalination by pervaporation. International Journal of Environmental Research and Public Health, 15(9), 1913.

[37] Chen Y., Feng X., Shi X., Cai W., Li B., and Zhao Y. (2023). Evaluation of computational fluid dynamics models for predicting pediatric upper airway airflow characteristics. Medical and Biological Engineering and Computing, 61(1), 259–270.

P. Balakrishnan

Chapter 14
Optimization techniques in flow dynamics and heat transfer

Abstract: This chapter comprehensively explores optimization methodologies in flow dynamics and heat transfer, addressing critical challenges in energy efficiency, system performance, and sustainability. It examines both classical approaches (gradient-based, Lagrangian, and variational methods) and modern techniques (genetic algorithms, particle swarm optimization, and simulated annealing). Classical methods provide systematic frameworks for problems with smooth, differentiable functions, while modern approaches effectively handle complex, multi-objective scenarios. Applications demonstrate significant performance improvements: tube bundle configurations optimized with GAs showed heat flux improvements of 2,708.27–3,641.25 W/m^2 with pressure drops of 380.32–1,117.74 Pa; plate-fin heat exchangers optimized via NSGA-II achieved oil pressure drops of 13.63 kPa with heat transfer rates of 9.79 kW; and integrated AI solutions reduced computational demands by 30% while improving accuracy by 20–25%. Despite these advancements, challenges persist in computational scalability, handling nonlinearity, and balancing competing objectives. Hybrid methodologies combining traditional techniques with AI offer promising solutions, with implementations demonstrating improvements in thermal and structural performance by 20.44% and 9.25%, respectively, while significantly reducing computational costs. This chapter establishes a foundation for addressing future challenges in flow dynamics and heat transfer optimization, emphasizing the integration of established theories with emerging technologies to drive sustainable engineering innovation.

Keywords: flow dynamics optimization, heat transfer enhancement, gradient-based methods, Lagrangian optimization techniques, variational methods, genetic algorithms (GAs), particle swarm optimization (PSO), simulated annealing (SA), artificial intelligence in optimization, multi-objective optimization, thermal system design

---

P. Balakrishnan, Department of Mechanical Engineering, SRM Institute of Science and Technology, Kattankulathur 603203, Tamil Nadu, e-mail: balakrip@srmist.edu.in

https://doi.org/10.1515/9783111661674-014

14.1 Introduction

14.1.1 Importance of optimization in flow dynamics and heat transfer

Optimization in flow dynamics and heat transfer is an integral focus in engineering, aimed at improving the performance, efficiency, and sustainability of various systems. Using computational tools, numerical simulations, and advanced algorithms, optimization techniques enable precise control over fluid behavior and heat exchange processes. These methods facilitate energy savings, minimize operational costs, and enhance overall system reliability. The significance of this optimization extends across multiple fields, including energy systems, mechanical engineering, and environmental applications. For instance, optimizing heat transfer in energy systems can reduce thermal losses, improve energy efficiency, and lower carbon footprints. Similarly, advancements in flow dynamics optimization contribute to more efficient hydrodynamics, enhanced cooling mechanisms in industrial processes, and improved aerodynamic designs. Mathematical modeling and iterative computational methods are at the core of these advancements. By simulating real-world scenarios and testing various parameters, engineers can identify optimal configurations and operating conditions. This approach not only enhances performance but also supports sustainable practices by reducing energy consumption and environmental impact.

Optimization in flow dynamics and heat transfer provides significant advantages, but it also comes with challenges. Managing the inherent nonlinearity and complexity of dynamic systems demands sophisticated approaches. The use of advanced computational techniques, combined with optimization methods, is crucial for tackling these challenges. However, balancing model accuracy with computational efficiency remains a key concern, as high-precision models often require substantial resources. Furthermore, implementing these methods in practical settings requires thorough validation through case studies and simulations to confirm their reliability and applicability. The use of machine learning and data-driven strategies has a crucial role to overcome such barriers and support more efficient and stable solutions. Enduring research progress in optimization models and algorithms continually promotes this domain's evolution, unlocking prospects for increased performance across a range of applications. The comprehensive optimization of flow and heat transfer is illustrated in Figure 14.1.

14.1.2 Scope and objectives of the chapter

The scope of this chapter is to explore the application of optimization techniques in enhancing flow dynamics and heat transfer systems, focusing on their critical role in improving engineering performance, energy efficiency, and industrial sustainability.

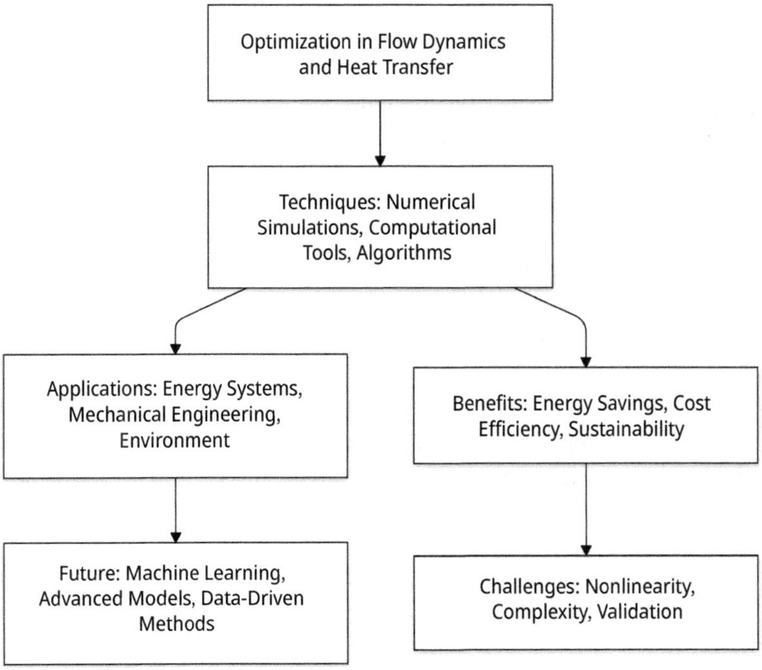

Figure 14.1: Optimization framework in flow dynamics and heat transfer.

Utilizing advancements in computational tools, algorithmic strategies, and artificial intelligence, the chapter aims to provide a comprehensive understanding of how these techniques address complex, multi-objective challenges. Integrating classical methods such as gradient-based, Lagrangian, and variational approaches with modern innovations like genetic algorithms, particle swarm optimization, and simulated annealing, the chapter bridges foundational theories with cutting-edge solutions.

The objective of this chapter is to present a detailed exploration of optimization methodologies that combines theoretical foundations with practical applications. It highlights the transformative potential of AI-driven solutions, emphasizing their ability to enhance predictive accuracy and streamline computational processes. The chapter also addresses real-world challenges, such as system complexity, computational cost, and nonlinearity, while exploring hybrid approaches that integrate traditional and modern methods to overcome these barriers. By providing comparative insights into various techniques, their advantages, limitations, and specific applications, this chapter serves as a valuable resource for engineers, researchers, and practitioners. It lays the groundwork for designing efficient, sustainable systems and inspires future advancements in optimization techniques to address global engineering challenges effectively.

14.2 Fundamentals of flow dynamics and heat transfer

14.2.1 Basic principles of fluid mechanics

A fluid is any substance, such as a liquid, gas, or other material, that can flow and undergo continuous deformation when subjected to an external force or shear stress. As illustrated in Figure 14.2, is a material that experience sun remitting deformation $\Delta\alpha$ under the application of shear stress τ, where the force F acts tangentially to the surface A, causing the fluid layers to move relative to each other. Whereas, fluid mechanics is a fundamental branch of physics that explores the properties and dynamics of fluids, encompassing both liquids and gases, whether in motion or at rest. This field is integral to various engineering and scientific domains, particularly in sectors such as energy, aerospace, and environmental science, where understanding fluid behavior is essential for innovation and problem-solving.

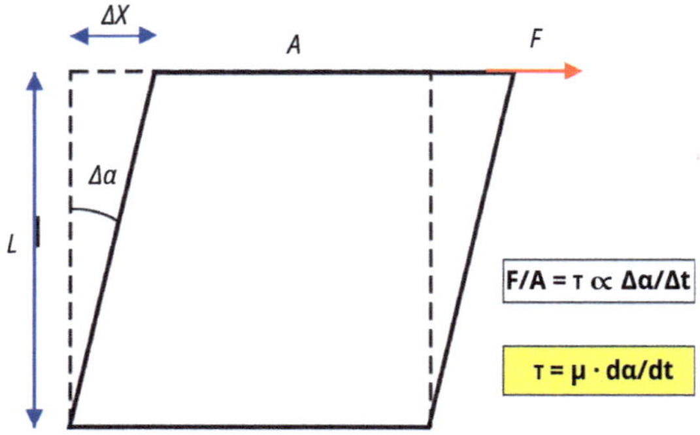

Figure 14.2: Continuous deformation of a fluid under shear stress.

Fluid mechanics can be categorized into three major areas:
1. *Fluid statics*, which deals with fluids at rest.
2. *Fluid kinematics*, which studies the motion of fluids without considering the forces causing the motion.
3. *Fluid dynamics*, which focuses on the effects of forces on fluid motion.

The parameters like density ρ, which is the mass per unit volume, and viscosity μ, which measures a fluids resistance to flow, are critical in understanding fluid behavior. Fluids with high viscosity, such as honey, resist flow, while low-viscosity fluids like water flow more easily.

The continuity equation, a fundamental concept, ensures the conservation of mass in fluid flow systems. It is represented mathematically as,

$$\frac{\partial \rho}{\partial t} + \nabla \cdot (\rho \vec{v}) = 0$$

This equation states that the mass of fluid entering the control volume $\rho \vec{v}_{in}$ is equal to the mass of fluid leaving the control volume $\rho \vec{v}_{out}$, ensuring the conservation of mass within the system. In Figure 14.3, dashed boundary represents the control volume, where the rate of change of mass within the system, combined with the net flux across its boundary, equals zero.

Figure 14.3: Illustration of the continuity equation.

Another critical principle is Newton's Second Law, which governs fluid motion under applied forces. This is mathematically represented by the Navier-Stokes equation,

$$\rho\left(\frac{\partial \vec{v}}{\partial t} + (\vec{v} \cdot \nabla)\vec{v}\right) = -\nabla p + \mu \nabla^2 \vec{v} + \vec{f}$$

Here, p is pressure, μ represents dynamic viscosity, and $\vec{f}$ accounts for external forces such as gravity.

In the context of energy conservation, the energy equation is essential, combining the principles of heat and work transfer,

$$\frac{\partial}{\partial t}\left(\frac{1}{2}\rho|\vec{v}|^2 + \rho e\right) + \nabla \cdot \left(\vec{v}\left[p + \frac{1}{2}\rho|\vec{v}|^2 + \rho e\right]\right) = \nabla \cdot (k \nabla T)$$

This equation is crucial for analyzing systems involving heat transfer, such as in power plants or cooling systems.

Fluid flow behavior is also categorized based on the Reynolds number (Re), which identifies whether the flow is laminar or turbulent.

$$Re = \frac{\rho v L}{\mu}$$

For pipe flows with $Re < 2000$ are laminar (smooth flow), while $Re > 4{,}000$ are turbulent (chaotic flow). Another important parameter is the Mach number M, which measures the compressibility of a fluid, defined as:

$$M = \frac{v}{c}$$

Here, v is the fluid velocity, and c is the speed of sound in the fluid.

14.2.2 Basics of heat transfer mechanisms

Heat transfer is a fundamental concept in thermal sciences, describing the movement of thermal energy due to temperature differences. This process is critical for applications across engineering disciplines, including energy systems, electronics cooling, and environmental engineering. Heat transfer occurs through three fundamental mechanisms: conduction, convection, and radiation. Below is an overview of each mechanism with their governing equations.

14.2.2.1 Modes of heat transfer

Heat transfer takes place by three main mechanisms: conduction, convection, and radiation. Conduction is the direct transfer of heat between molecules in physical contact, such as when a metal spoon warms up in a hot liquid. Convection is the transfer of heat by the physical movement of fluids, such as warm air rising and cold air falling, creating circulation currents. Radiation is the transfer of energy by electromagnetic waves, such as the warmth received from sunlight. These processes often work together to facilitate heat exchange in various environments. Figure 14.4 illustrates the three primary modes of heat transfer.

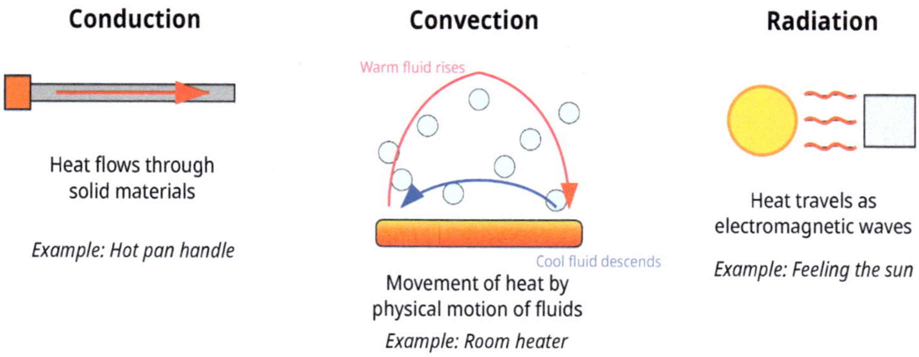

Figure 14.4: Illustration of heat transfer modes.

Conduction is the passage of heat through a solid object or immobile fluid due to temperature differences. It is described by Fourier's Law of heat conduction,

$$q = -k\nabla T$$

where q is heat flux (W/m^2), k is thermal conductivity (W/m · K), and ∇T is temperature gradient (K/m)

For 1D steady-state conduction, Fourier's law can be simplified,

$$q = -k\frac{dT}{dx}$$

Heat transfer from a solid surface to a moving fluid occurs during convection. It combines the fluid's bulk motion with conduction at the boundary layer. Newton's law of cooling describes the rate of heat transmission,

$$q = hA(T_s - T_\infty)$$

where q is heat transfer rate (W), h is convective heat transfer coefficient (W/m^2 · K), A is surface area (m^2), T_s is surface temperature (K), and T_∞ is fluid temperature far from the surface (K).

There are two main types of convective heat transfer mechanisms: forced convection and natural convection. When buoyant forces from density differences within the fluid, usually brought on by temperature changes, drive fluid motion, natural convection takes place. On the other hand, forced convection is created by outside forces that actively move the fluid to improve heat transmission, like fans or pumps.

The transport of heat by electromagnetic waves without the use of a medium is known as radiation. The Stefan-Boltzmann law is the source of the governing equation,

$$q = \sigma\epsilon A(T_s^4 - T_\infty^4)$$

where σ is Stefan-Boltzmann constant (5.67×10^{-8} W/m^2 · K^4) and ϵ is emissivity of the surface.

Each of the three heat transfer modes – radiation, convection, and conduction – can take place concurrently in real-world systems and contribute to the system's overall thermal dynamics. For example, as heat moves from one fluid to another in a heat exchanger, conduction takes place via the walls. Heat transfer is improved via convection, which occurs between the fluid and the surface. Furthermore, radiation can be important in high-temperature settings because it permits heat to be released from hot surfaces and absorbed by adjacent cooler ones.

Hence, combined modes of heat transfer

$$q_{\text{total}} = q_{\text{cond}} + q_{\text{conv}} + q_{\text{rad}}$$

14.2.2.2 Dimensionless numbers in heat transfer

As previously mentioned, the Reynolds Number, *Re* is a critical dimensionless parameter in fluid mechanics, generally used in heat transfer. In addition, the Nusselt number *Nu* and Prandtl Number, *Pr* are also widely used in heat transfer. The Nusselt number characterizes the efficiency of convective heat transfer, while the Prandtl number defines the connection between thermal diffusivity and momentum diffusivity, providing insights into the fluid's thermal and flow properties.

Nusselt number, *Nu*

$$Nu = \frac{hL}{k}$$

The Nusselt number indicates the effectiveness of convection relative to conduction. A higher *Nu* suggests that convection dominates heat transfer.

Prandtl number, *Pr*

$$Pr = \frac{c_p \mu}{k}$$

The Prandtl number characterizes the comparative thickness of the thermal boundary layer to the velocity boundary layer,

For $Pr < 1$ (e.g., liquid metals), thermal diffusivity dominates, and heat diffuses quickly relative to momentum.

For $Pr > 1$ (e.g., oils), momentum diffusivity dominates, and the velocity boundary layer is thicker than the thermal boundary layer.

14.2.3 Trade-offs: heat transfer, pressure loss, and energy consumption

Optimizing flow dynamics and heat transfer systems requires balancing critical trade-offs like heat transfer efficiency, pressure loss, and energy consumption. These trade-offs are crucial in applications such as heating, ventilation, and air conditioning (HVAC) systems, catalytic reactors, and photovoltaic-thermal systems. Effective optimization enhances performance, reduces costs, and minimizes environmental impact.

For instance, in catalytic reactors, advancements in 3D-printed baffled logpile structures have improved heat transfer rates by four to six times compared to conventional packed beds, maintaining similar pressure drops through careful design optimization [1]. Similarly, the use of carbon nanotube nanofluids in photovoltaic-thermal systems improves thermal efficiency but increases pressure drops, requiring careful trade-off management [2]. Also, in passenger vehicles, aerodynamic efficiency is crucial for reducing fuel consumption and carbon emissions. The design of the vehicle, including features like spoilers and diffusers, can significantly impact drag and lift forces. A reduction in drag coefficient by 16% and lift coefficient by 232% was observed in certain vehicle configurations, demonstrating the potential for aerodynamic improvements to enhance energy efficiency [3].

Energy optimization is vital in HVAC systems, where hybrid controllers like DP-NARX-RL use reinforcement learning to reduce energy consumption while maintaining comfort [4]. Pressure optimization in water supply systems reduces energy usage and water leakage, offering a simple yet impactful strategy [5]. Integrating active insulation with HVAC controls further enhances energy efficiency, balancing energy use and peak demand through multi-objective optimization frameworks [6]. Modern computing techniques, such as dynamic core scaling combined with dynamic voltage and frequency scaling, optimize energy consumption and computational performance, demonstrating versatility in high-performance applications [7].

Emerging technologies like 3D printing and nanofluids, combined with advanced control strategies such as reinforcement learning, enable dynamic system optimization. These techniques address challenges of cost, complexity, and sustainability, driving innovation and efficiency in engineering systems. Harnessing these advancements, engineers can design optimized solutions that balance performance, efficiency, and sustainability.

14.3 Optimization techniques: classical approaches

Classical optimization techniques are fundamental in engineering, offering systematic approaches to improve performance and efficiency in systems such as fluid dynamics and heat transfer. These methods optimize variables like shape, flow rates, and temperature distributions by applying mathematical principles to solve constrained and unconstrained problems. Among these, gradient-based methods, Lagrangian methods, and variational methods stand out for their effectiveness in addressing specific optimization challenges. While classical approaches are particularly well-suited for continuous and differentiable functions, they often face difficulties in handling complex, nonlinear, and real-world systems. The following sections will discuss these three methods in detail, highlighting their principles, applications, and limitations.

14.3.1 Gradient-based methods and their applications

Gradient-based optimization methods are fundamental tools in engineering, providing systematic approaches to optimize complex systems. These methods are employed in various contexts, from improving heat exchanger designs to estimating discretization errors in computational fluid dynamics (CFD). This method leverages the gradient of an objective function, which describes how the function changes with respect to its variables. In the domains of flow dynamics and heat transfer, gradient-based methods play a thoughtful part in enhancing system performance, efficiency, and sustainability. They are particularly effective in problems requiring iterative refinements, such as shape optimization, flow control, and thermal management.

The main principle behind gradient-based methods is the use of gradients to guide the optimization process toward the desired outcome. The update rule for these methods is generally expressed as

$$x_{k+1} = x_k - \alpha \nabla f(x_k)$$

where α is the step size, and $\nabla f(x_k)$ represents the gradient of the objective function at the current iteration. This formula ensures that the optimization process moves in the direction of steepest descent, gradually converging to a local or global optimum [8–10].

The gradient-based methods are broadly categorized into implicit and explicit approaches, each serving distinct purposes in optimization tasks. Implicit gradient methods address nested optimization problems by treating the inner-level problem as a fixed equilibrium while allowing the outer-level problem to iteratively update based on the computed equilibrium. This approach simplifies complex, coupled problems often encountered in engineering systems. On the other hand, explicit gradient methods compute gradients at each step directly, enabling dynamic updates to optimize the objective function. These methods are widely utilized in fluid dynamics and thermal systems due to their straightforward implementation and adaptability.

Convergence is a key consideration in gradient-based methods. The process is typically guided by criteria such as the norm of the gradient ($\| \nabla f(x_k) \|$) becoming sufficiently small or the change in the objective function between iterations drops below a threshold. Advanced variants, such as momentum-based methods, further enhance convergence rates by incorporating past gradient information to smooth oscillations and accelerate optimization.

The success of gradient-based methods heavily depends on accurate gradient computation, which is achieved using numerical techniques such as the finite element method (FEM) and the finite volume method (FVM). FEM is particularly effective in discretizing complex geometries and solving partial differential equations that govern fluid flow and heat transfer. By approximating the solution over discrete elements, FEM provides precise gradient calculations for irregular domains. FVM, in contrast, focuses on the conservation of physical quantities, such as mass and energy, across control volumes. This method ensures robust optimization in CFD, particularly for systems with varying boundary conditions and turbulent flows.

Both FEM and FVM enable gradient evaluation by solving the governing equations of fluid flow and heat transfer. For instance, in CFD, the Navier-Stokes equations for momentum conservation are often solved alongside the energy equation.

Navier-Stokes equation for momentum conservation:

$$\rho\left(\frac{\partial\vec{v}}{\partial t}+(\vec{v}\cdot\nabla)\vec{v}\right)=-\nabla p+\mu\nabla^2\vec{v}+\vec{f}$$

Energy equation for heat transfer:

$$\rho c_p\frac{\partial T}{\partial t}+\nabla\cdot(\rho c_p\vec{v}T)=\nabla\cdot(k\nabla T)+S$$

where $\vec{v}$ is velocity, p is pressure, T is temperature, k is thermal conductivity, and S represents any internal heat source. These equations provide the foundation for calculating gradients that drive optimization

Applications of gradient-based methods extend to a variety of real-world problems, such as wing shape optimization in aerospace, heat exchanger design in energy systems, and flow control in manufacturing processes. Through the use of FEM and FVM, these methods effectively handle challenges such as irregular geometries, turbulent behavior, and complex boundary interactions. The integration of advanced techniques, such as adaptive step sizes and preconditioning, further enhances their scalability and efficiency in large-scale systems.

Gradient-based methods are employed to optimize the shape of aerodynamic surfaces, such as wings and fuselages, to enhance performance metrics like lift-to-drag ratio, fuel efficiency, and overall aerodynamic efficiency. The gradient-based methods utilize sensitivity information to iteratively refine designs, aiming to minimize objectives such as drag or maximize lift. The process involves computing gradients of performance metrics with respect to design variables, guiding modifications to the aerodynamic shape. Techniques like the adjoint method are often employed to efficiently calculate these gradients, even for complex configurations. This approach enables the optimization of components like airfoils and wings, leading to improved aerodynamic efficiency [11, 12]. In structural engineering, gradient-based methods are applied to optimize parameters such as weight, strength, and stiffness. By evaluating the gradient of an objective function such as structural compliancewith respect to design variables, engineers can determine the most effective directions for design modifications. This process facilitates the development of structures that meet performance criteria while minimizing material usage and cost. Gradient-based algorithms are particularly effective in handling large-scale structural optimization problems, providing precise and efficient solutions [13, 14]. Figure 14.5 shows the various processes followed in this method, represented as a flowchart.

The advancements in gradient-based methods, including the development of hybrid algorithms like momentum-based or adaptive techniques have significantly expanded their utility. These methods provide dynamic and efficient optimization solutions, ensuring they remain integral to modern engineering practice.

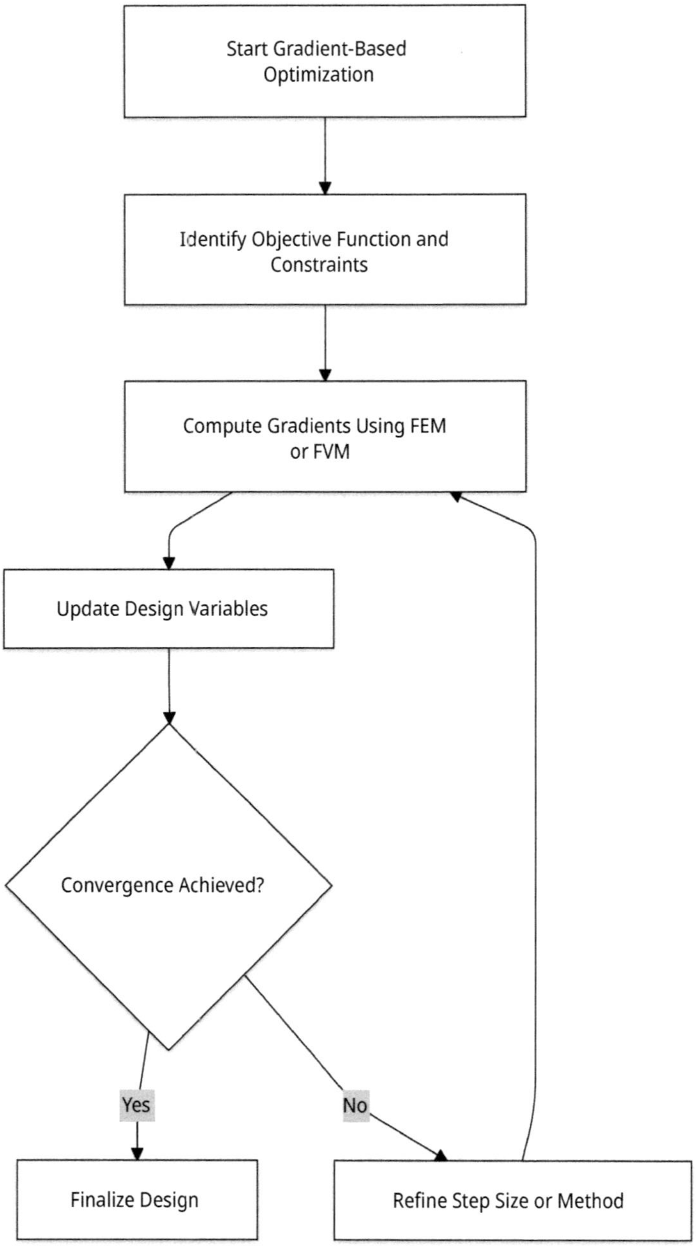

Figure 14.5: Gradient-based optimization process flowchart.

14.3.2 Lagrangian methods and their applications

Lagrangian methods are powerful optimization tools that address problems involving constraints by introducing auxiliary variables known as Lagrange multipliers. These methods transform constrained optimization problems into unconstrained ones, making them more tractable for analysis and solution. In the context of flow dynamics and heat transfer, Lagrangian methods are particularly useful for incorporating physical constraints, such as conservation laws, directly into the optimization framework.

The fundamental principle of Lagrangian methods is the construction of a Lagrangian function, which combines the original objective function $f(x)$ with the constraints $g(x)$ using Lagrange multipliers λ [15–17]:

$$\mathcal{L}(x,\lambda) = f(x) + \lambda^T g(x)$$

where $g(x)$ represents equality constraints of the form $g(x) = 0$. The optimization process involves finding x and λ that satisfy the necessary conditions for optimality:

$$\nabla_x \mathcal{L}(x,\lambda) = 0, \quad g(x) = 0$$

In heat transfer and fluid dynamics, Lagrangian methods are widely applied to optimize systems while maintaining key constraints. For instance, in the design of heat exchangers, the constraints may include conservation of energy and adherence to specific flow rates. Similarly, in fluid flow optimization, the method ensures that solutions satisfy mass conservation and momentum balance.

Lagrangian methods also play a critical role in minimizing entropy production or energy dissipation in fluid and thermal systems. For example, in optimizing channel flows for convective heat transfer, the Lagrangian framework allows engineers to derive equations that minimize viscous dissipation while ensuring temperature and flow rate constraints are met.

Although highly effective, Lagrangian methods can be computationally intensive for complex, nonlinear systems. Solving the Lagrangian equations often requires iterative techniques and advanced numerical methods, such as the FEM or FVM, to discretize and solve the governing equations. Figure 14.6, shows the flow chart process of the Lagrangian methods.

Lagrangian methods are widely employed in heat transfer modelling, optimization, and performance evaluation. These methods allow for precise tracking of individual particles within a continuous medium, making them ideal for studying heat exchange in gas-particle mixtures. For example, the Eulerian-Lagrangian framework combines particle-scale equations to predict effective thermal properties and temperature distributions, providing detailed insights into macro-scale heat transfer behavior. In heat exchanger design, Lagrangian methods help optimize temperature distribution to minimize thermal resistance or enhance heat dissipation. Taking advantage of variational principles, engineers can determine optimal configurations that im-

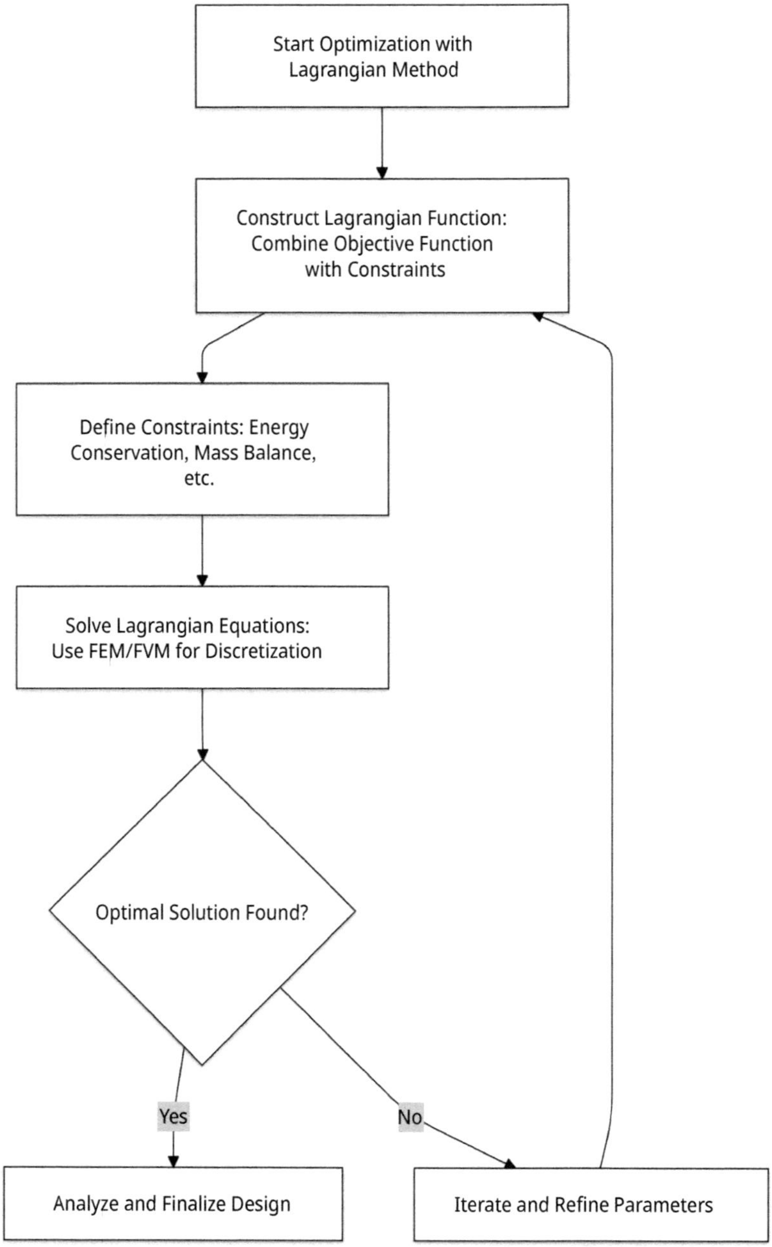

Figure 14.6: Flowchart of Lagrangian methods.

prove efficiency while maintaining structural constraints. Additionally, Lagrangian tracer particles are used to evaluate heat transfer efficiency in turbulent flows, offering valuable insights into local heat transfer characteristics and informing the optimization of thermal system designs [18, 19]. The Lagrangian methods effectively determine velocity fields that minimize energy dissipation and drag. These approaches define functionals based on energy integrals, leading to governing equations for optimal flow conditions. Additionally, Lagrangian methods are highly efficient in modeling particle transport, tracking individual particle trajectories to accurately predict concentration distributions. This capability is essential for designing effective ventilation systems and analyzing pollutant dispersion. Variational methods, built upon Lagrangian principles, are further applied to optimize turbulent flow systems, deriving equations that enhance heat transfer performance and guide systems toward thermal efficiency [20].

14.3.3 Variational methods and their applications

Variational methods provide a rigorous mathematical framework for solving optimization problems by finding stationary points of a functional. These methods are widely employed in flow dynamics and heat transfer to derive equations that optimize specific performance criteria, such as minimizing drag, maximizing lift, or optimizing heat distribution.

The core idea behind variational methods is to define a functional $J[u]$, which is a mapping from a function space to real numbers, and to find a function $u(x)$ that makes $J[u]$ stationary (either a minimum or maximum). The necessary conditions for stationarity are obtained through the Euler-Lagrange equation [21, 22]:

$$\frac{\partial}{\partial x}\left(\frac{\partial L}{\partial(\partial u/\partial x)}\right) - \frac{\partial L}{\partial u} = 0$$

where L is the Lagrangian density of the system, and $u(x)$ represents the function to be optimized.

In heat transfer problems, variational methods are used to derive optimal temperature distributions that minimize thermal resistance or maximize heat dissipation. For instance, the optimization of heat transfer in a fin system can be framed as a variational problem, where the objective is to minimize the total heat loss while maintaining structural constraints. Similarly, in fluid dynamics, variational methods assist in solving flow optimization problems, such as determining velocity fields that minimize energy dissipation or drag. By defining the functional as an energy integral, the Euler-Lagrange equation provides the governing equations for the optimal flow field. An example application is optimizing turbulent flow systems to enhance heat transfer. Variational methods help derive the reverse-temperature equation and volumetric force sources that guide the system toward optimal thermal performance. Figure 14.7, flowchart highlights the iterative process of refining parameters to achieve an optimal solution.

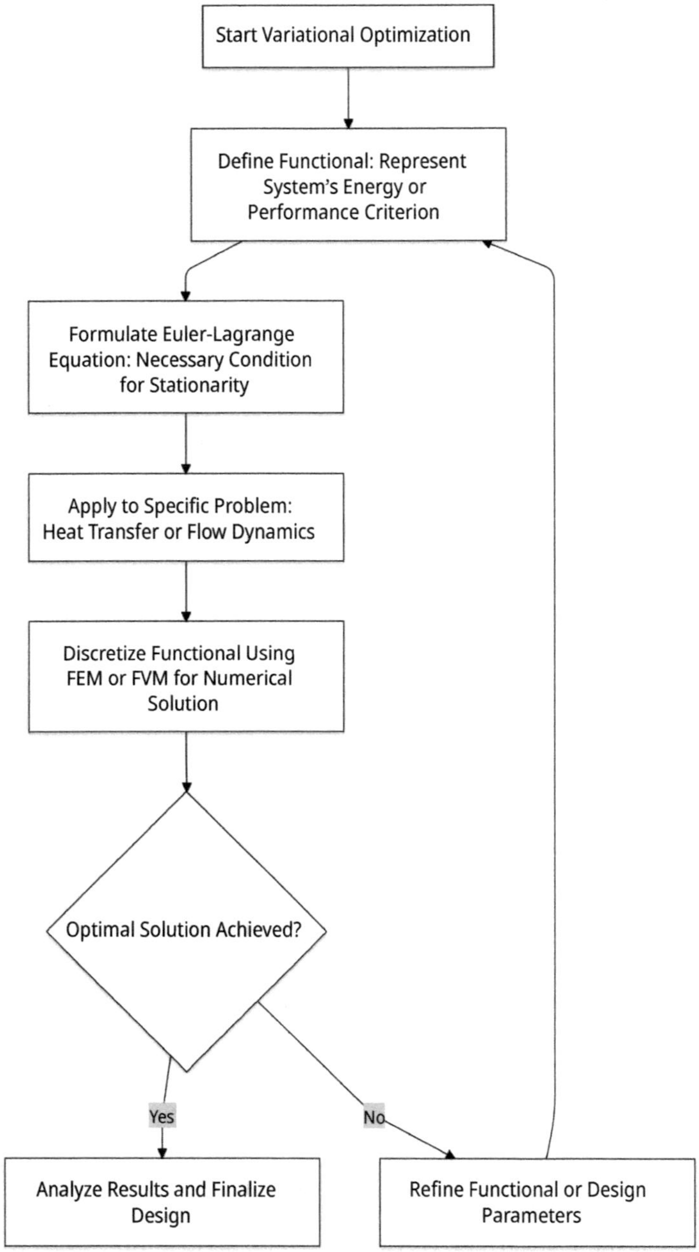

Figure 14.7: Flowchart showing variational methods in heat transfer and flow optimization.

This approach involves formulating a functional representing the system's thermal energy and applying the Euler-Lagrange equation to find the temperature distribution that optimizes the desired thermal performance. For instance, in designing heat exchangers, variational methods help determine configurations that achieve efficient heat transfer while adhering to physical constraints [23]. Also, in laminar forced convection scenarios, variational methods help identify optimal velocity fields that maximize overall heat transfer coefficients while minimizing viscous dissipation. The integration of geometrical design considerations with flow optimization allows for enhanced performance in systems such as evaporators, where adjustments to tube shapes can significantly impact heat transfer efficiency [24].

While variational methods are mathematically rigorous and provide elegant solutions, their implementation in complex systems often requires numerical techniques like FEM or FVM. These methods discretize the functional and solve the resulting equations iteratively, enabling engineers to handle irregular geometries, turbulent flows, and multi-physics interactions effectively.

14.3.4 Comparison of classical methods

The three methods discussed above gradient-based, Lagrangian, and Variational methods serve as fundamental approaches in the optimization of engineering systems. Gradient-based methods focus on leveraging derivatives for iterative improvements, Lagrangian methods effectively handle constrained optimization, and variational methods derive governing equations by identifying stationary points of functionals. Together, these methods provide robust tools for addressing diverse challenges in flow dynamics and heat transfer.

Table 14.1 provides a comparative overview of these methods, highlighting their descriptions, applications, advantages, and limitations [25–28].

14.4 Modern optimization techniques

Modern optimization techniques have revolutionized problem-solving in engineering, offering innovative approaches to address complex, multi-objective challenges. These methods leverage advanced algorithms to explore vast solution spaces, making them ideal for applications in flow dynamics and heat transfer. Among the many modern optimization techniques available, this chapter focuses on three widely used methods: genetic algorithms (GAs), particle swarm optimization (PSO), and simulated annealing (SA).

Table 14.1: Comparison of optimization methods for flow and heat transfer analysis.

Method	Description	Applications	Advantages	Limitations
Gradient-based methods	Optimization techniques that utilize gradients of the objective function to find local minima or maxima.	1. Aerodynamic shape optimization 2. Structural optimization 3. Multi-objective optimization in engineering	1. Efficient for smooth and differentiable problems 2. Can handle large-scale problems effectively	1. May converge to local minima 2. Requires gradient computation, which can be costly for complex functions
Lagrangian methods	Techniques that transform constrained optimization problems into unconstrained ones using Lagrange multipliers.	1. Heat exchanger design 2. Fluid flow optimization 3. Minimizing entropy production in thermal systems	1. Direct incorporation of constraints 2. Provides a systematic approach to handle multiple constraints	1. Computationally intensive for non-linear systems 2. Requires iterative numerical methods for solutions
Variational methods	Mathematical framework for finding stationary points of functional, often leading to differential equations.	1. Optimal temperature distribution in heat transfer 2. Flow optimization problems 3. Turbulent flow systems	1. Provides rigorous mathematical solutions 2. Can derive governing equations for complex systems	1. Implementation can be mathematically intensive 2. Often requires numerical methods for complex geometries

14.4.1 Genetic algorithms for multi-objective problems with applications

The evolutionary concepts of natural selection and genetics serve as the foundation for GAs), which are adaptable heuristic search engines. They work especially well at resolving complicated optimization issues with several competing goals, as those in heat transfer and flow dynamics. In these fields, GAs makes it easier to investigate a wide range of viable solutions, making it possible to determine the best trade-offs between conflicting goals. GAs seeks to approximate the Pareto front of a collection of nondominated solutions in multi-objective optimization, where no single solution is better than any other across all goals. Decision-makers are presented with a range of ideal options using this method, each of which reflects a unique balance between the goals. A popular GA variation for multi-objective optimization, the nondominated sorting genetic algorithm II (NSGA-II) is renowned for its effectiveness in solving challenging issues [29].

GAs operate by initializing a random population of applicant solutions, each evaluated based on objective functions, $z(x) = \{z1(x), z2(x), \ldots, z_k(x)\}$, where dominance criteria are applied to determine the fitness. A solution x is said to dominate another solution y if $z_i(x) \leq z_i(y)$ for all objectives i, and $z_j(x) < z_j(y)$ for at least one objective j. Solutions are ranked into Pareto fronts using this dominance criterion, with higher-ranking solutions selected for reproduction. Choice mechanisms such as contest selection are used to choose parent solutions, which then undergo crossover and mutation to generate offspring. During crossover, offspring are generated by blending parent solutions using the formula

$$x_{\text{offspring}} = \alpha x_{\text{parent1}} + (1 - \alpha) x_{\text{parent2}}$$

where α is a blending parameter between 0 and 1. Mutation introduces diversity into the population by altering solutions with a random factor:

$$x_{\text{mutated}} = x + \sigma \mathcal{N}(0,1)$$

where σ is the mutation rate and $\mathcal{N}$ (0,1) represents a standard normal distribution. To maintain diversity on the Pareto front, GAs calculates the crowding distance for each solution, defined as

$$d(x) = \sum_{i=1}^{k} \frac{z_i^{\max} - z_i^{\min}}{z_i(x+1) - z_i(x-1)}$$

This metric ensures evenly spaced solutions along the Pareto front, preventing premature convergence and maintaining a diverse solution set. The iterative process continues until a predefined stopping criterion, such as the number of generations or convergence to an optimal front, is met.

GAs is excellent at solving nonlinear, nondifferentiable problems and efficiently generate a wide range of solutions that reflect compromises between competing goals. They are widely used in optimizing complex systems like flow dynamics and heat transfer due to their robustness and adaptability. With the aid of these mechanisms, GAs provides engineers with powerful tools for solving multi-objective optimization problems with real-world complexities [30]. Figure 14.8 shows the flowchart steps in the GA process, from initializing a population and evaluating fitness to generating offspring through crossover and mutation,ultimately converging to a Pareto-optimal front.

In the design of tube bundles subjected to cross-flow, GAs have been utilized to optimize configurations for enhanced heat transfer and reduced pressure drop. By employing a multi-objective GA (MOGA), researchers generated a Pareto front representing the trade-offs between average heat flux and pressure drop. The technique for order preference by similarity to an ideal solution (TOPSIS) was then applied to

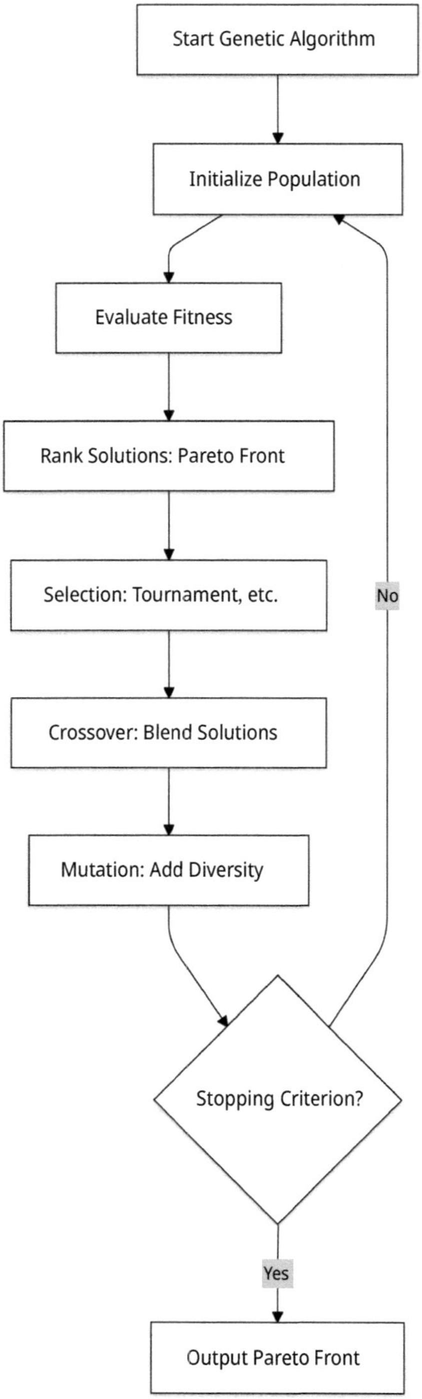

Figure 14.8: Genetic algorithm optimization flowchart.

select the optimal configuration from the Pareto-optimal solutions. This approach led to significant improvements in performance metrics, with optimal average heat flux values ranging from 2708.27 to 3641.25 W/m^2 and pressure drops between 380.32 and 1117.74 Pa [31]. Additionally, GAs have been used to optimize plate-fin heat exchangers, which are essential components of cooling systems for automobile engines. The nondominated sorting GA II (NSGA-II) and CFD simulations were combined to optimize parameters including fluid flow rates and fin angle. Objective functions were modeled using support vector machine regression (SVR), and TOPSIS made it easier to choose the optimal compromise option from the Pareto front. With an oil pressure drop of 13.63 kPa, an oil outlet temperature of 65.11 °C, and a heat transfer rate of 9.79 kW, the optimized design demonstrated a well-balanced improvement in the heat exchanger's performance [32].

14.4.2 Particle swarm optimization (PSO) methods and their applications

The collective actions of fish schools and flocks of birds serve as the inspiration for particle swarm optimization (PSO), a potent population-based metaheuristic. By iteratively enhancing a candidate solution according to fitness metric, it is frequently employed to address optimization problems. PSO offers an effective way to address challenging, multi-objective optimization issues in the domains of flow dynamics and heat transfer, such as maximizing heat-transfer efficiency or minimizing drag.

The fundamental working principle of PSO involves updating the positions and velocities of particles in a swarm to explore the solution space. The position of each particle is updated using the equation:

$$x_i(t+1) = x_i(t) + v_i(t+1)$$

where x_i represents the position of the ith particle, and $v_i(t+1)$ is its velocity at the next iteration. The velocity of each particle is influenced by three components, its previous velocity, its best-known position (personal best), and the global best position in the swarm. The velocity update equation is given by

$$v_i(t+1) = \omega v_i(t) + c_1 r_1 (p_i - x_i(t)) + c_2 r_2 (g - x_i(t))$$

where p_i is the personal best position, g is the global best position, ω is the inertia weight, and c_1,c_2 are cognitive and social coefficients, respectively. The random factors r_1 and r_2 ensure stochastic exploration.

The PSO process begins with the random initialization of particle positions and velocities within a predefined search space. Each particle's fitness is evaluated based on the objectives of the problem, such as enhancing thermal performance in a heat exchanger or reducing energy loss in turbulent flows. The personal best p_i and global best g solutions are then updated based on the evaluated fitness values. Particles are

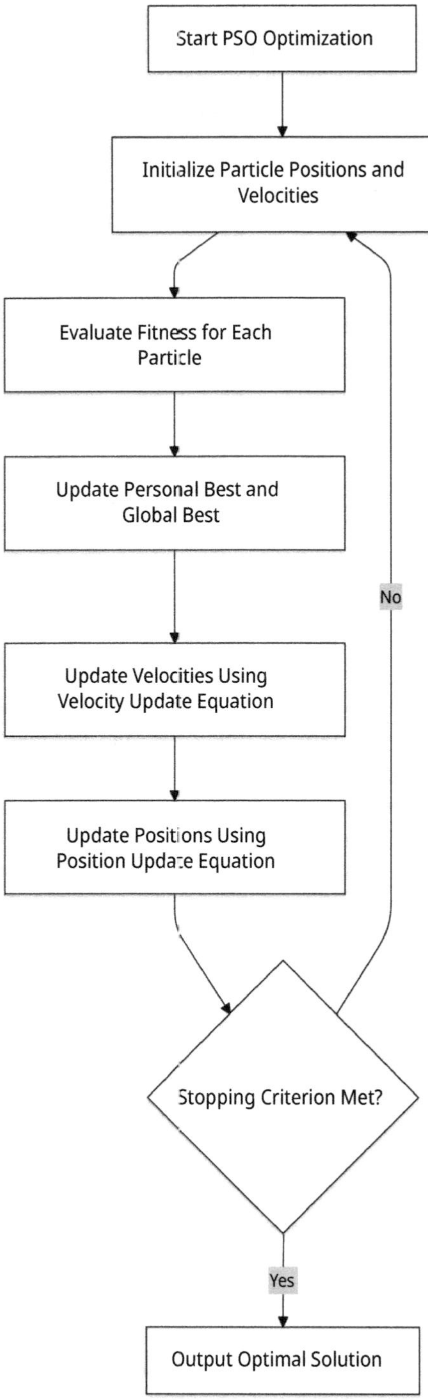

Figure 14.9: Particle swarm optimizationoptimization process flowchart.

iteratively directed towards optimal solutions using the velocity and position update equations until a stopping requirement, like a fitness threshold or a limit number of iterations, is satisfied [33].The iterative steps of the PSO process, from initializing particles and assessing fitness to updating velocities and locations depending on personal and global bests, are depicted in the flowchart in Figure 14.9. An ideal solution is obtained by repeating the process until the stopping requirements are satisfied.

PSO has been utilized to estimate unknown parameters in inverse heat transfer scenarios. For instance, Felde et al. applied PSO to determine the heat transfer coefficient on cylindrical surfaces, achieving accurate estimations with reduced computational time [34]. In the context of heat exchanger networks, PSO has been employed to optimize cleaning schedules, balancing maintenance costs with heat recovery efficiency. A study demonstrated that PSO could achieve significant savings by determining optimal cleaning intervals [35]. PSO has been applied to optimize the geometric parameters of vapor chambers, which are critical in thermal management systems. The study utilized PSO to minimize temperature rise by adjusting the widths of vapor chamber regions and wick porosity, leading to improved thermal performance [36].

14.4.3 Simulated annealing (SA) methods and their applications

Inspired by the annealing process in metallurgy, which involves heating and gradually cooling materials to produce a stable crystalline structure, simulated annealing (SA) is a probabilistic optimization technique. It works especially well for resolving complicated optimization issues with several local optima, such those in heat transfer and flow dynamics. SA is a flexible and reliable optimization technique that finds global optima by exploring the solution space and simulating the physical annealing process.

The core principle of SA is based on an iterative process that evaluates candidate solutions and probabilistically accepts worse solutions to escape local minima. At each iteration, a neighboring solution is generated, and the change in the objective function is calculated as $\Delta E = E(s') - E(s)$, where s is the current solution, s' is the new solution, and E represents the energy or cost function. The acceptance of a worse solution is governed by the probability:

$$P = \exp\left(-\frac{\Delta E}{T}\right)$$

where T is the current temperature, which decreases over time. The cooling schedule, typically defined as $T_{k+1} = \alpha T_k$, where α is a cooling rate between 0 and 1, controls the gradual reduction of the temperature, balancing exploration, and exploitation.

SA begins with the random initialization of a solution and a high initial temperature. Through iterative updates, the algorithm generates neighboring solutions by making small changes to the current solution and evaluates their fitness based on the

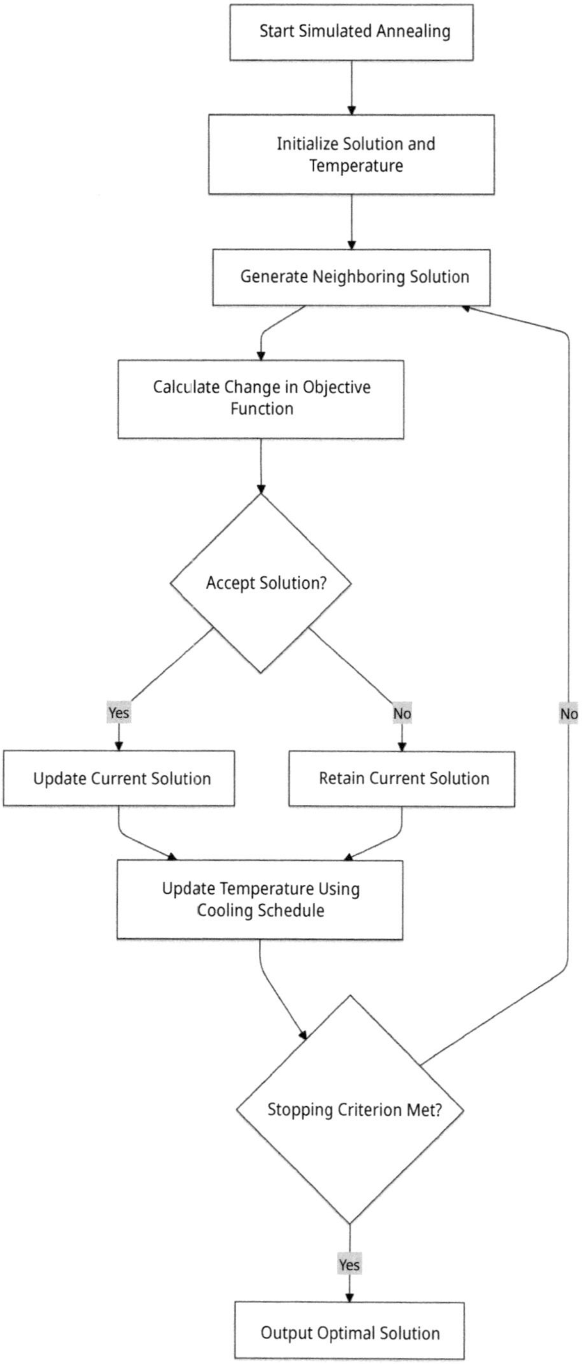

Figure 14.10: Simulated annealing optimization flowchart.

objective function. By accepting suboptimal solutions with a probability that decreases as the temperature cools, SA avoids being trapped in local minima. The process continues until the system converges, either by reaching a maximum number of iterations or when the temperature becomes negligible [37]. Figure 14.10 shows the flowchart representing the iterative steps of the SA process, from initializing a solution to updating based on the cooling schedule until the stopping criterion is met.

SA has been applied to optimize flow dynamics within systems such as chemical reactors and fluid transport networks. By simulating the cooling process, SA can effectively navigate through complex solution spaces to find configurations that enhance flow efficiency and reduce resistance, which is critical in chemical engineering and related fields [38]. SA is utilized in the operation and maintenance of heat transfer units, particularly in managing scale formation, which adversely affects heat transfer efficiency. By formulating the problem as an optimization challenge that considers various constraints and discontinuities, SA helps identify optimal strategies for maintaining effective heat transfer rates [39, 40]. This application is crucial in industries where heat exchangers are integral to process efficiency.

14.4.4 Comparison of modern optimization techniques

Modern optimization techniques such as GAs, PSO, and SA have proven to be highly effective in addressing complex, multi-objective problems in flow dynamics and heat transfer. These methods offer unique strengths, allowing engineers to explore diverse solution spaces, balance trade-offs, and optimize performance under challenging constraints. Table 14.2 provides a comparative overview of these techniques, highlighting their descriptions, applications, advantages, and limitations.

14.5 Artificial intelligence in flow and heat transfer optimization

AI has significantly advanced the optimization of flow dynamics and heat transfer by addressing complex nonlinear and multi-objective problems. Techniques such as neural networks and genetic algorithms enable the development of data-driven models that improve thermal system efficiency by predicting performance and streamlining simulations. These approaches also facilitate the analysis of turbulent flows, uncovering patterns to enhance design optimization and boost heat transfer efficiency.

AI has transformed how nonlinear and multi-objective problems are addressed in flow dynamics and heat transfer. Approaches like the adaptive multi-objective optimization method using surrogate models have proven effective in reducing computational costs for problems like wind turbine airfoil optimization while maintaining ac-

Table 14.2: Comparison of modern optimization techniques.

Method	Description	Applications	Advantages	Limitations
Genetic algorithms (GAs)	Heuristic search inspired by natural selection; uses crossover, mutation, and selection to evolve solutions.	1. Optimizing heat exchanger designs 2. Tube bundle optimization 3. Multi-objective thermal system designs	1. Handles nonlinear, nondifferentiable problems 2. Explores diverse solutions 3. Robust to noise	1. Computationally expensive 2. May converge slowly to global optima 3. Relies on parameter tuning
Particle swarm optimization (PSO)	A population-based optimization technique modeled after the collective movement of birds and fish, adjusting positions and velocities iteratively.	1. Inverse heat transfer problems 2. Vapor chamber design 3. Flow control in turbulent systems	1. Simple to implement 2. Balances exploration and exploitation 3. Efficient for high-dimensional problems	1. Prone to premature convergence 2. Requires proper balance of parameters for inertia and acceleration
Simulated annealing (SA)	A stochastic approach influenced by the annealing process in metallurgy, which explores potential solutions by occasionally accepting suboptimal ones to escape local optima.	1. Flow optimization in chemical reactors 2. Scale management in heat exchangers 3. Fluid transport networks	1. Escapes local minima 2. Handles discrete and continuous variables 3. Simple and flexible	1. Slow convergence 2. Depends heavily on the cooling schedule 3. Computational cost for large problems

curacy. Similarly, multi-objective evolutionary algorithms combined with variable-fidelity surrogates have improved aerodynamic designs, reducing high-fidelity simulation reliance by up to 30% while achieving better convergence and solution quality [41, 42]. Generative adversarial networks (GANs) have been used to optimize coolant channel designs in fuel cells, achieving significant improvements in thermal and structural performance by 20.44% and 9.25%, respectively while drastically reducing computation time [43].

Data-driven models are increasingly integrated into thermal systems to optimize flow dynamics and heat transfer. Techniques like R-LSTM-ED models enhance accuracy in temperature predictions for two-phase immersion cooling by up to 25%, while machine learning methods improve flow regime identification and optimize gas turbine surrogate models, reducing errors by 20% [44, 45]. Hybrid approaches such as artificial neural networks combined with particle swarm optimization optimize fin geometries for thermal energy storage, achieving a 43% improvement in energy storage per unit mass [46]. Digital twinning further enhances thermal system control by predicting heat transfer coefficients and regulating transient thermal systems without requiring retraining [47]. Advanced methodologies like PINN-GCN models provide high-resolution predictions for steady-state fluid flow at speeds 1,000 times faster than traditional simulations, offering high accuracy even with limited training data [48].

AI also offers remarkable capabilities in scrutinizing turbulent flow behaviors, enabling the modeling of complex flow and heat transfer phenomena. Studies using triangular vortex generators demonstrated enhanced heat transfer and a significant increase in the Nusselt number, highlighting their efficacy in optimizing turbulent flows [49]. AI methods such as artificial neural networks predict heat transfer in hybrid nanofluid systems with remarkable precision, outperforming traditional approaches [50]. In solar collector systems, particle swarm optimization combined with neural networks effectively handles the complexities of fluid flow and heat transfer, demonstrating the power of soft computing methods in optimizing complex dynamics [51]. Machine learning models like multi-layer perceptron (MLP) and long short-term memory (LSTM) networks efficiently predict turbulent coherent structures, reducing computational costs and improving the understanding of flow behaviors [52].

While AI-driven methods offer significant advancements, challenges remain in balancing computational efficiency with accuracy and ensuring robustness across diverse scenarios. The integration of AI with traditional methods continues to hold promise for driving innovations in flow dynamics and heat transfer optimization, enhancing both system performance and efficiency.

14.6 Challenges in optimizing flow dynamics and heat transfer

Optimizing flow dynamics and heat transfer is a complex task that requires balancing multiple factors, including physical principles, computational techniques, and design parameters. The challenge lies in improving heat-transfer efficiency while effectively managing flow behavior across various systems, such as heat exchangers, hybrid heat-electricity systems, and thermal management solutions for batteries. Achieving this balance calls for innovative strategies in design optimization, advanced computational modeling, and precise control methods.

Optimizing flow dynamics and heat transfer involves balancing conflicting objectives, such as enhancing heat transfer while minimizing pressure drop. Advanced models like neural networks adds complexity, requiring precise data and tuning. Sensitivity to parameters like nanoparticle volume and Reynolds number further complicates optimization. Mult objective approaches, such as genetic algorithms, demand balancing multiple goals with sophisticated techniques. Additionally, varying material and fluid properties, along with the need for accurate empirical correlations, add layers of complexity to the process [53]. Understanding the intricate relationship between protrusion radius and distance in micro-channels poses significant complexity. While thermal performance initially improves with these parameters, it exhibits fluctuating trends, marked by local maxima and minima. Optimization results highlight that flow friction has minimal impact on thermal performance, with heat transfer emerging as the dominant factor. However, the study overlooks the influence of varying fluid properties or alternative working fluids, which could substantially alter flow dynamics and thermal efficiency. Moreover, the use of genetic algorithms for optimization faces potential limitations in handling abrupt flow changes, such as vortex formation or flow separation, which may compromise the accuracy and reliability of the optimization results [54].

PSO is a widely used algorithm for addressing optimization problems in flow dynamics and heat transfer, yet it faces notable challenges. Dynamic environments with time-varying parameters demand continuous adaptation, which traditional PSO struggles to handle effectively. Multi-objective optimization adds complexity, requiring the algorithm to balance conflicting goals like minimizing energy use while maximizing heat-transfer efficiency. Additionally, the presence of complex constraints necessitates modifications to ensure feasible solutions throughout the process. Maintaining population diversity is critical to prevent premature convergence to suboptimal solutions, particularly in dynamic settings. Computational efficiency is another hurdle, as these problems often involve high-dimensional spaces or real-time requirements, necessitating strategies like cooperative agents or memory-based tracking to enhance performance. Scalability poses a final challenge, as PSO must handle large-scale problems effectively, often requiring parallelization or advanced optimization techniques to maintain performance [55]. Also, this technique includes addressing dynamic environments, where changing parameters over time can lead to suboptimal solutions unless adaptive techniques are employed. Balancing conflicting objectives, such as maximizing heat transfer efficiency while minimizing energy consumption, often necessitates modifications to standard PSO to generate diverse and effective solutions. Complex constraints present another challenge, requiring adaptations to ensure feasibility throughout the optimization process. Maintaining population diversity is critical to avoid premature convergence, with strategies like knowledge sharing among particles improving exploration of the solution space. Additionally, computational efficiency is a concern, particularly for high-dimensional problems or real-time applications. Enhancements like cooperative agents and memory-based tracking can reduce computational demands by focusing efforts on promising areas. Scalability also becomes critical for large-scale problems, requiring parallelization or advanced techniques to sustain performance [56]. Adapting PSO to diverse problems remains a significant challenge, requiring flexible operators and integration of new procedures to address varied optimization scenarios effectively. While PSO's decentralized nature supports parallelization, implementing it efficiently without compromising performance is complex. The algorithm's simplicity is strength, but balancing it with advanced modifications for specific applications poses difficulties. Additionally, staying updated with the numerous PSO variants and evolving research directions demands continuous effort from researchers [57].

Challenges in using SA for geometric optimization in heat transfer include the lack of universally recommended cooling schedules, making it difficult to generalize results across different scenarios. The study reveals that while hybrid cooling schedules like BoltzExp, ConstExp1, and ConstExp2 improve performance, there is still limited understanding of their effectiveness in optimizing complex geometries. Additionally, optimizing intricate configurations, such as double-T shaped cavities, demands significant computational resources, highlighting the need for further research to refine cooling schedules and enhance the reliability of simulated annealing in diverse geometrical optimization problems [58]. Further, implementing simulated annealing requires precise parameter tuning to ensure efficiency. Developing accurate yet computationally efficient surrogate models for complex systems is another hurdle. Integrating CFD with optimization techniques adds complexity, demanding seamless data utilization and precise modeling of multiphysics interactions like fluid flow and heat transfer. High computational resource demands and difficulties in validating results against real-world data further complicate the optimization process [59]. However, the hybrid approaches, such as combining simulated annealing with sequential quadratic programming, aim to address these challenges adopting efficient local refinement methods and trust-region frameworks. These methods enhance solution accuracy and ensure convergence to optimal designs, even in the presence of metamodel inaccuracies, by integrating adaptive ensemble strategies that improve prediction reliability and optimization outcomes [60].

14.7 Conclusions

The optimization of flow dynamics and heat transfer has emerged as a cornerstone of modern engineering, addressing challenges that span energy efficiency, sustainability, and cost-effectiveness. This chapter provides a comprehensive exploration of classical and modern optimization techniques, showcasing their relevance and applicability in advancing thermal systems and fluid flow performance. Classical methods, including gradient-based, Lagrangian, and variational approaches, remain invaluable for their systematic and mathematically rigorous frameworks, particularly in scenarios involving smooth and continuous functions. Despite their limitations in handling complex and nonlinear systems, these methods serve as a foundation for optimization in engineering.

Modern techniques, such as GAs, PSO, and SA, demonstrate remarkable versatility in solving multi-objective and computationally intensive problems. Quantitative results show significant improvements: GAs applied to tube bundles achieved optimal heat flux values between 2,708.27 and 3,641.25 W/m^2 while maintaining pressure drops between 380.32 and 1,117.74 Pa. NSGA-II optimization of plate-fin heat exchangers resulted in an oil pressure drop of 13.63 kPa with heat transfer rates of 9.79 kW. The integration of AI-driven methods enhanced predictive and computational capabilities, with R-LSTM-ED models improving temperature prediction accuracy by up to 25% in two-phase immersion cooling, while machine learning methods reduced errors in gas turbine surrogate models by 20%.

Advanced optimization approaches have demonstrated substantial performance gains across multiple applications. Hybrid ANN-PSO methods optimizing fin geometries for thermal energy storage achieved a 43% improvement in energy storage per unit mass. GAN-based optimization of coolant channels in fuel cells improved thermal and structural performance by 20.44% and 9.25%, respectively, while dramatically reducing computation time. In aerodynamic design, variable-fidelity surrogate models reduced high-fidelity simulation reliance by up to 30% while achieving better convergence and solution quality. PINN-GCN models now provide high-resolution predictions for steady-state fluid flow at speeds 1,000 times faster than traditional simulations.

While significant progress has been achieved, challenges persist, including balancing computational efficiency and accuracy, addressing the high-dimensional nature of optimization problems, and managing trade-offs between competing objectives. Hybrid methodologies, combining traditional approaches with modern algorithms and AI-driven techniques, show great promise in overcoming these hurdles. The integration of physics-based models, adaptive frameworks, and data-driven insights further refines optimization processes, paving the way for future advancements.

Ultimately, the continuous evolution of optimization methodologies underscores their importance in achieving sustainable and efficient engineering solutions. By utilizing these techniques, engineers and researchers can address emerging challenges and drive innovation in the fields of flow dynamics and heat transfer, ensuring their critical role in advancing energy systems, environmental sustainability, and industrial applications.

References

[1] Rosseau L. R., Van Lanen T., Roghair I., and van Sint Annaland M. (2024 Feb 15). Experimental demonstration of the heat transfer – Pressure drop trade-off in 3D printed baffled logpile structures. Chemical Engineering Journal, 482, 149092.

[2] Hormozi Moghaddam M., and Karami M. (2022 Mar). Heat transfer and pressure drop through mono and hybrid nanofluid-based photovoltaic-thermal systems. Energy Science & Engineering, 10(3), 918–931.

[3] Piratla S. (2023 Oct 31). Evaluating the aerodynamics of different passenger vehicle configurations. Authorea Preprints.

[4] Dawood S. M., Hatami A., and Homod R. Z. (2022 Nov 2). Trade-off decisions in a novel deep reinforcement learning for energy savings in HVAC systems. Journal of Building Performance Simulation, 15(6), 809–831.

[5] Ćipranić I., Jevrić M., Radulović M., and Sekulić G. (2021 Feb 1). The energy–pressure nexus in the water supply system. Water Supply, 21(1), 344–355.

[6] Sepehri A., and Pavlak G. (2023 Oct 21). Tradeoffs in optimization of Active Insulation Systems and HVAC: A case study in residential buildings. Science and Technology for the Built Environment, 29(9), 927–935.

[7] Zhang W., Zhang H., and Lach J. (2020 Oct 7). Extending performance-energy trade-offs via dynamic core scaling. IEEE Transactions on Computers, 70(11), 1875–1886.

[8] Hoffman M., (2022). T M. Foundations of the discrete gradient method. arXiv preprint arXiv:1805.06444v5. (Accessed March 17, 2025, at https://arxiv.org/abs/1805.06444)

[9] Chau M., Fu M. C., Qu H., and Ryzhov I. O. (2014 Dec 7). Simulation optimization: A tutorial overview and recent developments in gradient-based methods. In Proceedings of the Winter Simulation Conference 2014, IEEE, pp. 21–35.

[10] Daoud M. S., Shehab M., Al-Mimi H. M., Abualigah L., Zitar R. A., and Shambour M. K. (2023 May). Gradient-based optimizer (GBO): A review, theory, variants, and applications. Archives of Computational Methods in Engineering, 30(4), 2431–2449.

[11] Li J., Du X., and Martins J. R. (2022 Oct 1). Machine learning in aerodynamic shape optimization. Progress in Aerospace Sciences, 134, 100849.

[12] Wu H., Chen R., Lou J., You Y., Huang L., Xu M., and Ruan Y. (2024 May 1). A gradient aerodynamic optimization method based on deep learning. Physics of Fluids, 36(5), 057117.

[13] Issa M., and Mostafa Y. (2022 Jun 5). Gradient-based optimizer for structural optimization problems. In Integrating Meta-Heuristics and Machine Learning for Real-world Optimization Problems, Cham: Springer International Publishing, pp. 461–480.

[14] Dababneh O., Kipouros T., and Whidborne J. F. (2018 Jan 4). Application of an efficient gradient-based optimization strategy for aircraft wing structures. Aerospace, 5(1), 3.

[15] Baeldung. (2024). Simulated Annealing Explained. (Accessed March 17, 2025, at https://www.baeldung.com/cs/simulated-annealing)

[16] Lucht, P. (2016). The Method of Lagrange Multipliers. Rimrock Digital Technology. Retrieved from https://user.xmission.com/çrimrock/Documents/The%20Method%20of%20Lagrange%20Multipliers.pdf

[17] Shivaraju N. (2014). Interpretation of the Lagrange Multipliers. International Journal of Research and Analytical Reviews (IJRAR), 1(1), 341–347.

[18] Giacomini M., Désidéri J. A., and Duvigneau R. Comparison of multiobjective gradient-based methods for structural shape optimization (Doctoral dissertation, INRIA).

[19] Pratt J., Busse A., and Müller W. C. (2020 Aug 3). Lagrangian statistics of heat transfer in homogeneous turbulence driven by Boussinesq convection. Fluids, 5(3), 127.

[20] Zhang Z., and Chen Q. (2007 Aug 1). Comparison of the Eulerian and Lagrangian methods for predicting particle transport in enclosed spaces. Atmospheric Environment, 41(25), 5236–5248.

[21] Gelfand I. M., Fomin S. V., and Silverman R. A. (2000). Calculus of Variations (Unabridged Repr. Ed.), Silverman R. A. editor Mineola, New York: Dover Publications, pp. 3.

[22] Delfour M. C., and Zolésio J. P. (2011 Jan 1). Shapes and geometries: Metrics, analysis, differential calculus, and optimization. Society for Industrial and Applied Mathematics.

[23] Nigam S. D., and Agrawal H. C. (1960 Jan 1). A variational principle for convection of heat. Journal of Mathematics and Mechanics, 9(6) 869–883.

[24] Zhao T., Liu D., He K. L., Chen X., and Chen Q. (2020 Aug 9). An integrated three-level synergetic and reliable optimization method considering heat transfer process, component, and system. Energies, 13(16), 4112.

[25] Hottois R., Châtel A., Coussement G., Debruyn T., and Verstraete T. (2023 Mar 1). Comparing gradient-free and gradient-based multi-objective optimization methodologies on the VKI-LS89 turbine vane test case. Journal of Turbomachinery, 145(3), 031001.

[26] Theljani A., and Chen K. (2019 Apr 1). An augmented Lagrangian method for solving a new variational model based on gradients similarity measures and high order regularization for multimodality registration. Inverse Problems and Imaging, 13(2), 309–335.
[27] Iusem A. N., and Nasri M. (2010 Jan). Augmented Lagrangian methods for variational inequality problems. RAIRO-Operations Research, 44(1), 5–25.
[28] University of Cambridge. Lectures on fluid dynamics and mechanics. Cambridge. (Accessed March 17, 2025, at https://www.damtp.cam.ac.uk/user/reh10/lectures/nst-mmii-chapter1.pdf)
[29] Selçuklu S. B. (2024 Jul 17). Multi-objective genetic algorithms. In Handbook of Formal Optimization, Singapore: Springer Nature Singapore, pp. 1007–1044.
[30] Konak A., Coit D. W., and Smith A. E. (2006 Sep 1). Multi-objective optimization using genetic algorithms: A tutorial. Reliability Engineering & System Safety, 91(9), 992–1007.
[31] Ge Y., Xin F., Pan Y., Liu Z., and Liu W. (2019 Jul 1). Optimal arrangement design of a tube bundle in cross-flow using computational fluid dynamics and multi-objective genetic algorithm. Journal of Heat Transfer, 141(7), 071801.
[32] Li S., Deng Z., Liu J., and Liu D. (2022 Nov 20). Multi-objective optimization of plate-fin heat exchangers via non-dominated sequencing genetic algorithm (NSGA-II). Applied Sciences, 12(22), 11792.
[33] Marini F., and Walczak B. (2015 Dec 15). Particle swarm optimization (PSO). A tutorial. Chemometrics and Intelligent Laboratory Systems, 149, 153–165.
[34] Yatim H. M., Mohd-Ghazali N., Mohamad M., Pamitran A. S., and Novianto S. (2023 May 24). Two-phase heat transfer microchannel system identification with Particle Swarm Optimization (PSO) approach. International Journal of Air-Conditioning and Refrigeration, 31(1), 13.
[35] Biyanto T. R., Suganda S. W., Susatio Y., and Justiono H. (2015 Dec 2). Cleaning schedule optimization of heat exchanger networks using particle swarm optimization. arXiv preprint arXiv:1512.00883.
[36] Hassan H., and Harmand S. (2020 Feb 1). Optimization of the width of vapor chamber regions by using particle swarm optimization method. Journal of Heat Transfer, 142(2), 021901.
[37] Baeldung. (2024). Simulated Annealing Explained. (Accessed March 17, 2025, at https://www.baeldung.com/cs/simulated-annealing)
[38] Raoufi M. H., Farasat A., and Mohammadifard M. (2015 Sep). Application of simulated annealing optimization algorithm to optimal operation of intelligent well completions in an offshore oil reservoir. Journal of Petroleum Exploration and Production Technology, 5, 327–338.
[39] Giri J. M. (2023). Simulated annealing and its applications to mechanical engineering: A review. International Journal of Innovative Research in Computer Science and Technology, 11, 9–14.
[40] Batzias F. A., and Marcoulaki E. C. Heat Transfer Optimization Using a Simulated Annealing Algorithm.
[41] Younis A., and Dong Z. (2024 Jan 1). Adaptive surrogate assisted multi-objective optimization approach for highly nonlinear and complex engineering design problems. Applied Soft Computing, 150, 111065.
[42] Zhang Y., Wang Z., and Han Z. H. (2024 Aug 1). A parallel variable-fidelity algorithm for efficient constrained multi-objective aerodynamic design optimization. Physics of Fluids, 36(8), 087135.
[43] Wang H., Wang Z., Qu Z., and Zhang J. (2023 Dec 15). Deep-learning accelerating topology optimization of three-dimensional coolant channels for flow and heat transfer in a proton exchange membrane fuel cell. Applied Energy, 352, 121889.
[44] Feng Y., Winter J., Adams N. A., and Schranner F. S. (2024 Aug 1). A general multi-objective Bayesian optimization framework for the design of hybrid schemes towards adaptive complex flow simulations. Journal of Computational Physics, 510, 113088.
[45] Ransing R. S. (2024 Sep 2). Guest editorial: Data-driven methods for heat transfer and fluid flow. International Journal of Numerical Methods for Heat & Fluid Flow, 34(8), 2833–2835.

[46] Tavakoli A., Hashemi J., Najafian M., and Ebrahimi A. (2023 Nov 1). Physics-based modelling and data-driven optimisation of a latent heat thermal energy storage system with corrugated fins. Renewable Energy, 217, 119200.

[47] Di Meglio A., Massarotti N., and Nithiarasu P. (2024 Jul 16). A physics-driven and machine learning-based digital twinning approach to transient thermal systems. International Journal of Numerical Methods for Heat & Fluid Flow, 34(6), 2229–2256.

[48] Peng J. Z., Hua Y., Aubry N., Chen Z. H., Mei M., and Wu W. T. (2024 Jun 1). Data and physics-driven modeling for fluid flow with a physics-informed graph convolutional neural network. Ocean Engineering, 301, 117551.

[49] Shah M. A., Pan K., Ibrahim M., Saeed T., and Ibrahim S. (2022 Jun 1). Numerical study and optimization of important parameters of non-Newtonian nanofluid turbulent flow in a channel with triangular vortex generators using artificial neural network. Engineering Analysis with Boundary Elements, 139, 94–108.

[50] Mishra A., Rawat S. K., Yaseen M., and Pant M. (2023 Nov 1). Development of machine learning algorithm for assessment of heat transfer of ternary hybrid nanofluid flow towards three different geometries: Case of artificial neural network. Heliyon, 9(11), e21453.

[51] Rawat S. K., Yaseen M., Pant M., Ujarari C. S., Joshi D. K., Chaube S., Negi A. S., and Kumar M. (2023 Nov 1). Designing soft computing algorithms to study heat transfer simulation of ternary hybrid nanofluid flow between parallel plates in a parabolic trough solar collector: Case of artificial neural network and particle swarm optimization. International Communications in Heat and Mass Transfer, 148, 107011.

[52] Aguilar-Fuertes J. J., Noguero-Rodríguez F., Jaen Ruiz J. C., García-raffi L. M., and Hoyas S. (2021 Feb 13). Tracking turbulent coherent structures by means of neural networks. Energies, 14(4), 984.

[53] Beigzadeh R., and Eiamsa-ard S. (2020). Genetic algorithm multiobjective optimization of a thermal system with three heat transfer enhancement characteristics. Journal of Enhanced Heat Transfer, 27 (2), 123–141.

[54] Yang K., Zheng Z., and Zhang D. (2019). Genetic algorithm-based heat transfer enhancement technique of a protruded micro-channel. Thermal Science, 23(Suppl. 3), 727–735.

[55] Kouka N., Fdhila R., Hussain A., and Alimi A. M. (2020 Jul 19). Dynamic multi objective particle swarm optimization with cooperative agents. In 2020 IEEE Congress on Evolutionary Computation (CEC), IEEE, pp. 1–8.

[56] Dos Santos Coelho L., Guerra F. A., Pasquim B., and Mariani V. C. (2013 Sep 20). Chaotic quantum-behaved particle swarm optimization approach applied to inverse heat transfer problem. In International Conference on Evolutionary Computation Theory and Applications, vol. 2, SCITEPRESS, pp. 97–102.

[57] Parsopoulos K. E. Particle Swarm Methods. 2018.

[58] Gonzales G. V., dos Santos ED, Emmendorfer L. R., Isoldi L. A., Rocha L. A., and Estrada E. D. (2015 Aug 31). A comparative study of simulated annealing with different cooling schedules for geometric optimization of a heat transfer problem according to constructal design. Scientia Plena, 11(8), 081321.

[59] Maakala V., Järvinen M., and Vuorinen V. (2018 Oct 1). Optimizing the heat transfer performance of the recovery boiler superheaters using simulated annealing, surrogate modeling, and computational fluid dynamics. Energy, 160, 361–377.

[60] Tenne, Y. (2017 Apr 26). A simulated annealing based optimization algorithm. In H. Peyvandi (Ed.), Computational optimization in engineering (Chapter 3). IntechOpen.

S. Senthilraja*, R. Gangadevi, M. Baskaran, N. Saravanan

Chapter 15
Advanced optimization methods in flow dynamics

Abstract: In order to ensure safety, enhance efficiency, and preserve functionality, it is imperative that modern technology and industry implement efficient heat transfer systems. The efficacy of electronics, energy, transportation, and industrial processes is directly influenced by advancements in heat transfer materials and technologies. The efficacy and effectiveness of heat transfer in modern equipment are influenced by a variety of factors, including the thermal conductivity of the material, the contact surface area, the mode of heat transfer, and the properties of the fluid. Therefore, the optimisation of various parameters in heat transfer systems has become increasingly important in recent years. Many advanced optimisation techniques, including genetic algorithm (GA), artificial bee colony (ABC), cuckoo search algorithm (CSA), firefly algorithm (FFA), ant colony optimisation (ACO) and particle swarm optimisation (PSO) have been employed for the design optimisation of heat transfer systems over the past two decades. This chapter offers a concise overview of the various optimisation techniques that have been employed in various heat transfer systems over the years. The information presented in this chapter assists young researchers in determining the parameters of the systems using a variety of optimisation techniques prior to executing them in real-time applications.

Keywords: Optimization techniques, Genetic algorithm, Ant colony optimization, Firefly algorithm, cuckoo search algorithm

15.1 Introduction

Heat transfer is a significant concept used in many engineering applications, denoting the flow of thermal energy from one location to another due to temperature variations. A device used for transferring heat and circulating the heat transfer fluid inside

---

***Corresponding author: S. Senthilraja,** Department of Mechatronics Engineering, SRM Institute of Science and Technology, Kattankulathur, Tamil Nadu, 603203, India
R. Gangadevi, Department of Mechatronics Engineering, SRM Institute of Science and Technology, Kattankulathur, Tamil Nadu, 603203 India
M. Baskaran, Department of Mechatronics Engineering, K.S.Rangasamy College of Technology, Tiruchengode, Tamil Nadu 637215 India
N. Saravanan, Department of Mechanical Engineering, Erode Sengunthar Engineering College, Thudupathi, Perundurai, Erode, Tamil Nadu, 638057 India

https://doi.org/10.1515/9783111661674-015

systems is crucial in modern applications, ensuring optimal performance, energy efficiency, and reliability. The regulation of thermal energy and fluid dynamics can be performed by these devices, which include heat exchangers, pumps, and fans. This allows for the maintenance of the ideal temperatures and flow rates. Effective heat transmission is essential in sectors such as automotive, aircraft, and manufacturing. In the automobile industry, efficient engine cooling systems depend on excellent heat transfer to keep engine temperature within safe parameters. In aircraft applications, effective thermal management assures that sensitive electrical components function optimally without overheating. The effectiveness of a heat transfer system is a significant element controlled by numerous major parameters, including the type of material, properties of the heat transfer fluid, surface area, and temperature differences. Furthermore, factors such as natural convection, fluid flow dynamics, and pressure differentials significantly impact the efficacy of heat transfer mechanisms. Therefore, it is crucial to understand and optimise these parameters to improve heat transfer efficiency and achieve energy conservation goals in thermal engineering systems successfully.

Optimization is an important principle in computer science and mathematics that is essential for identifying the best solution from a range of possibilities. This complex process depends on algorithms, which are analytical techniques intended to assist in determining the optimal output. Optimization often entails an iterative approach in which the algorithm progressively improves the solution, hence improving outcomes with each iteration. Optimisation is crucial in several domains like engineering, finance, logistics, and machine learning. In engineering, optimisation is used to build structures that are both resilient and economical. Numerous methodologies may be used to enhance procedures effectively. The classification of optimization techniques used in different applications is presented in Figure 15.1.

The linear optimisation method addresses the linear correlations among variables. The nonlinear optimisation approach employs the nonlinear interactions among the variables. The dynamic optimisation approach used for systems with time-dependent variables. Deterministic methods provide exact answers derived from mathematical models, such as gradient descent and the simplex approach. Generic algorithms and simulated annealing are examples of stochastic methods, which are used to develop solutions that are close to optimum. Single-Objective Optimisation emphasises on enhancing one variable, such as maximum heat transfer efficiency. Multi-Objective Optimisation addresses several conflicting objectives, including cost, efficiency, and environmental effect. Static Optimisation requires constant circumstances throughout the whole procedure. Real-Time optimisation dynamically modifies solutions in response to fluctuating inputs or situations.

Optimising heat transfer processes using complex algorithms is a high-tech approach with many important applications, such as heat exchangers in power plants. One of the key goals of the optimisation process is to minimise the amount of energy that is used while simultaneously maximising the efficiency of heat transmission. This

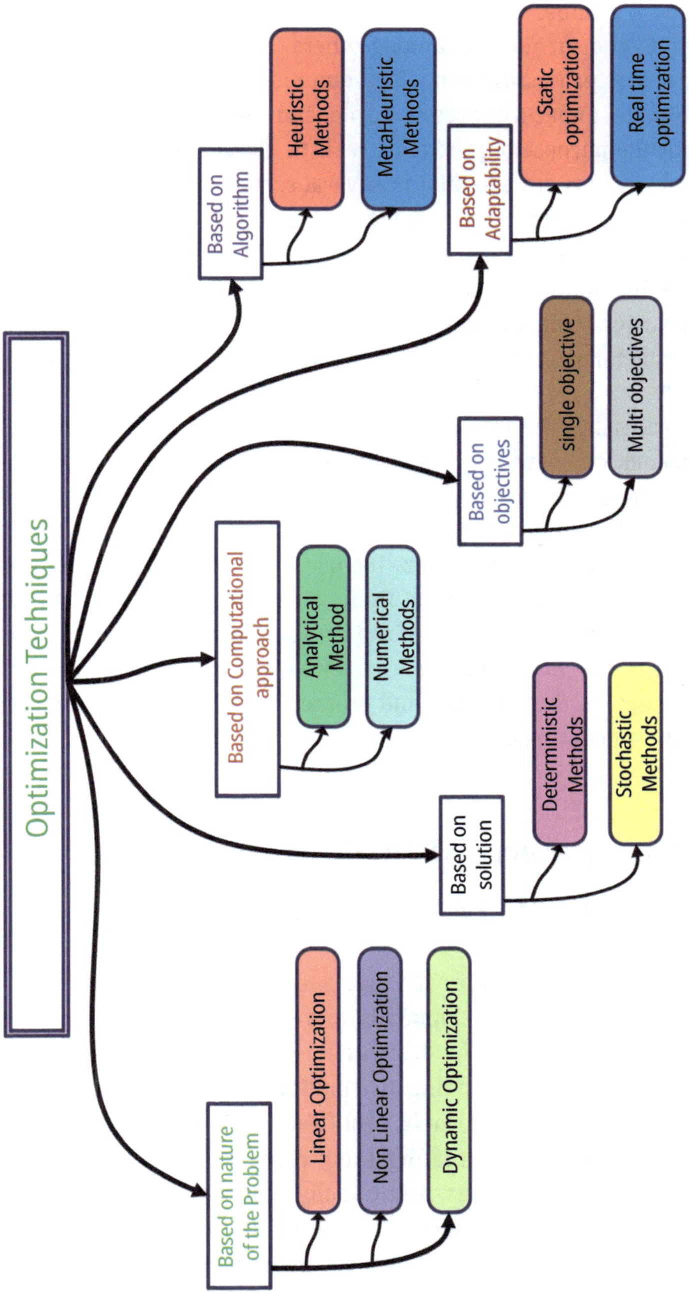

Figure 15.1: Classification of optimization techniques.

process is influenced by a number of parameters, including thermal conductivity, fluid flow characteristics, and turbulence, among others. An investigation into the complexities of heat transfer processes in topology optimisation necessitates the investigation of convection, the impacts of pressure drop, and the use of computational fluid dynamics (CFD) for the purpose of producing realistic simulations. Due to many benefits, optimisation algorithms are essential in several sectors, transforming problem-solving methodologies and improving efficiency. Optimisation algorithms consistently promote innovation and enhancement across several domains by enabling data-driven decision-making and improving operational results. In order to improve the efficiency of heat transfer, a large number of academics have created a variety of optimisation methods, which have then been used in a variety of heat transfer applications.

This chapter is structured in the following order: Section 15.2 presents information relevant to the systematic review of various optimisation strategies in heat transfer applications.The information on the application of Genetic Algorithm in different heat transfer systems is included in Section 15.3. Section 15.4 outlines the details about the Ant Colony Optimization methods to optimize the parameters in heat transfer system. Section 15.5 discusses the details and use of firefly algorithm in the heat transfer devices. Section 15.6 covers the fundamentals of the cuckoo search algorithm and its application in various heat transfer systems. The conclusion section provides a summary of this chapter, along with the problems and prospective applications of optimisation approaches in future heat transfer systems.

15.2 Bibliometric analysis of Optimization techniques in heat transfer applications

Bibliometric analysis is a quantitative approach that is used for the purpose of evaluating and analysing the tendencies, developments, and effects of published papers within a particular subject or across many disciplines. Bibliometric analysis is extensively used in research assessment, funding distribution, and policy formulation, assisting stakeholders in detecting rising trends, prolific authors, or knowledge deficiencies. It assists organisations in evaluating their research performance and benchmarking it against worldwide norms. Additionally, it assists organisations in evaluating their research performance and contrasting it with international norms. Bibliometric approaches use bibliographic data from academic publications, including citation counts, co-authorship networks, and keyword analysis, to provide significant insights into the structure and dynamics of research activity. This approach depends on many indicators, including as publication counts, citation rates, and journal impact factors.

In order to carry out the bibliometric study, the necessary collection of articles was collected from Scopus on the 30th of December, 2024. As more number of articles found and widely used by numerous researchers, the Scopus was selected as a search engine to analyze the scientific publications. Due to the increasing quantity of papers discovered and extensively used by many academics, Scopus was chosen as the search engine for analysing scientific publications in this study. The detailed step by step procedure involved in this bibliometric study is shown in Figure 15.2.

The words such as "Optimisation," "Heat Transfer," and "Flow devices" were used, and they were recognised in the titles, abstracts, and keywords of the papers. Around 1748 papers were gathered from 1980 to 2025, and the details of year wise article publications is presented in Figure 15.3. It can be found that the quantity of papers about optimisation in heat transfer applications has been steadily rising since 1980. According to the Scopus database, around 1,748 papers were published during this time in various esteemed journals. From 1980 to 2012, fewer than fifty articles have been published each year. Since then, the rate of publication has slowly gone up until 2014 and a peak of publishing occurred in 2014, with 268 publications followed by 146 publications in the year 2023, 128 in the year 2022, 110 in the year 2021 and 103 in the year 2020.

There is a fluctuation in the number of citations for each year and article, as can be seen when looking at the article mean citations. The year 2008 had the greatest mean citation per article count of 33.00, followed by the year 2005 with 30.63, the year 2011 with 29.44, the year 2020 with 27.09, and the year 2004 with 27.06 citations. Similarly, the mean citation per year reached its highest point of 5.42 in the year 2020, and then it gradually declined to 5.05 in the year 2021. During the years 2022, 2023, and 2019, the numbers dropped even lower, reaching 4.63, 3.77, and 3.25, respectively. The findings of this research clearly suggest that the potential for optimisation in heat transfer applications has significantly expanded in recent years.

15.3 Application of genetic algorithm in heat transfer and flow devices

A genetic algorithm (GA) is an optimisation approach that employs the concepts of both genetics and natural selection to address optimisation problems. GA imitate biological processes, including reproduction, transformation, and natural selection, to repeatedly enhance a population of viable solutions to a specific issue. Selection, crossover, and mutation are the three primary biological activities that this method incorporates into its implementation. As part of the selection process, the chromosomes of the appropriate parents are selected based on the fitness ratings. Two distinct parental chromosomes fuse together to generate progeny, inheriting the optimal features of both parents during the crossover phase. Mutation provides minor, ran-

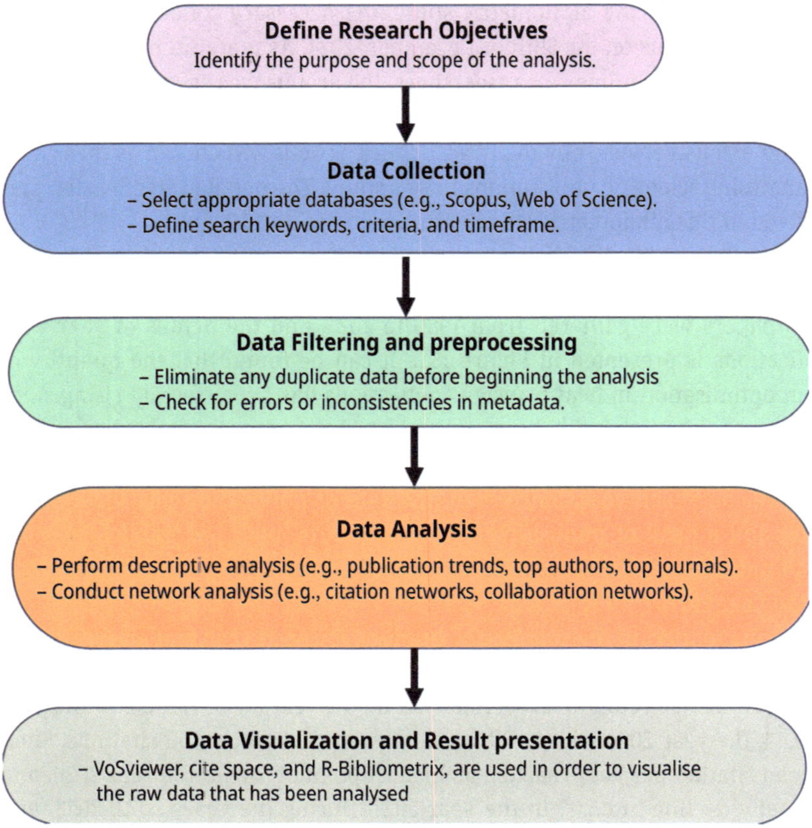

Figure 15.2: Detailed steps involved in bibliometric analysis.

dom alterations in a chromosome, facilitating genetic variety and averting premature convergence to local optima. The merits and demerits of GA technique is presented in Table 15.1. The details of the steps involved in GA technique is illustrated in Figure 15.4. As a result of its many benefits, the GA plays a significant part in the process of optimising the performance of a variety of real-time systems.

He et al. [1] optimised the operational parameters and efficiency of a heat pipe incorporated cooling system with the GA approach for various loads and the surrounding temperature. The results revealed that the greatest fluctuation in cooling capacity is 1.37%, but the largest deviation of the other parameter is 12.87%. Liu et al. [2] introduced an innovative strategy to enhance design and operational variables using response surface methodology (RSM) and multi-objective genetic algorithm (MOGA). According to the findings, the diameter of the spiral had the most important impact, followed by the pitch and the fluid velocity. These findings also help designers construct ground heat exchangers that are highly efficient.

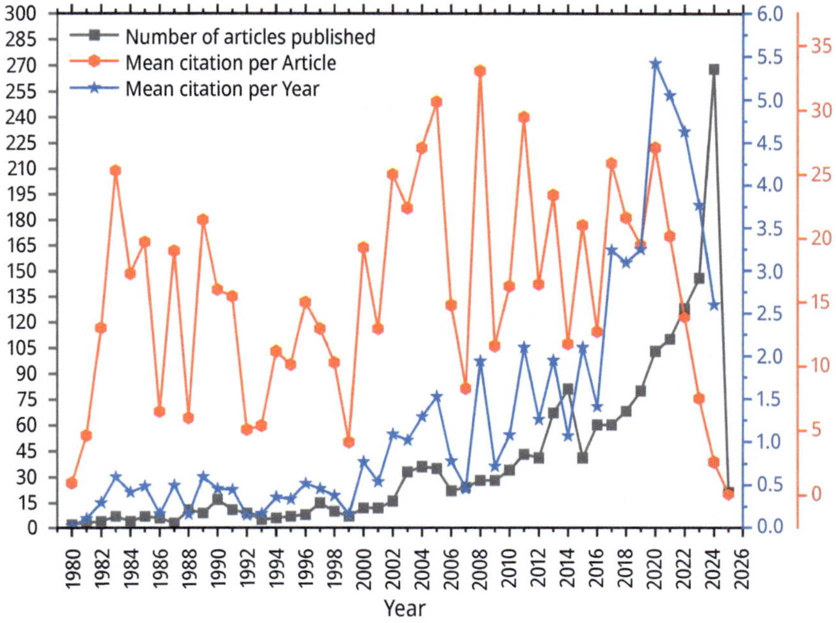

Figure 15.3: The details of year wise article publications and mean citations.

Table 15.1: Merits and Demerits of GA technique.

S. No	Merits	Demerits
1	**Global Search Capability**: GAs explore a broad search space, reducing the likelihood of getting trapped in local optima.	**Computationally Expensive**: Evaluating the fitness of individuals in large populations can be time-consuming.
2	**No Gradient Information Needed**: GAs do not rely on derivative information, making them suitable for non-differentiable, discontinuous, or complex objective functions.	**Parameter Sensitivity**: Performance depends on the careful tuning of parameters such as population size, mutation rate, and crossover rate.
3	**Versatility**: Applicable to a wide range of problems, including optimization, machine learning, and creative tasks.	**Premature Convergence**: The algorithm can converge to suboptimal solutions if diversity is lost early.
4	**Parallelism**: GAs can be easily parallelized, allowing faster computations on modern multi-core systems.	**Lack of Guarantee for Optimality**: GAs do not always guarantee finding the exact global optimum.

Table 15.1 (continued)

S. No	Merits	Demerits
5	**Flexibility**: Can handle complex, multi-modal, and multi-objective problems effectively.	**Fitness Function Dependency**: The quality of results heavily depends on the design of the fitness function.
6	**Stochastic Nature**: Randomness in GAs allows exploration of unexpected solutions, which can lead to innovative outcomes.	**Requires Large Populations**: A small population size may lead to insufficient exploration, while a large size increases computational costs.
7	**Adaptive Approach**: GAs can dynamically adapt to changes in the problem space or constraints.	**Solution Encoding Challenges**: Designing an effective chromosome representation can be complex for some problems.
8	**Avoids Assumptions**: Does not require prior knowledge or assumptions about the problem's mathematical properties.	**Slower Convergence**: Compared to gradient-based methods, GAs can be slower in reaching solutions, especially in high-dimensional problems.

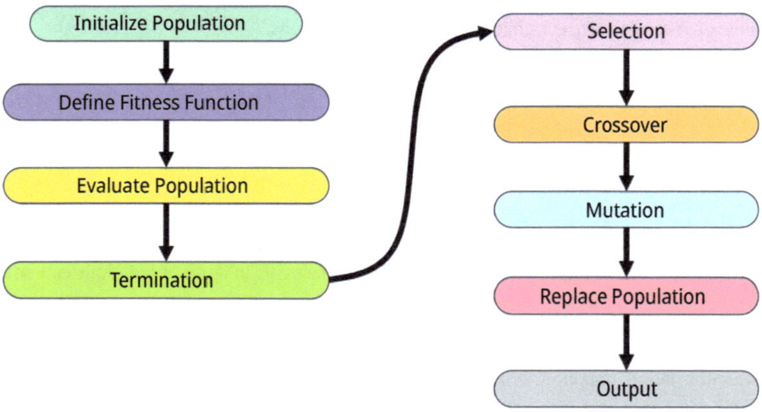

Figure 15.4: Steps involved in Genetic algorithm.

Shen et al. [3] have developed a numerical model using the GA technique to optimise the various variables of heat storage systems based on objective functions and boundary conditions. It was observed that the Phase Change Material (PCM) mass has a significant impact on the variables and thermal performance. Deep learning and multi-objective genetic algorithms were used by Guo et al. [4] in order to carry out the modelling and optimisation of the heat pipe cooled battery thermal management system. Due to the greater heat production that is produced by the development of solid electrolyte interphase, it has been noticed that the thermal management system that is based on heat pipes is unable to regulate the temperature of the aged battery pack. Using a genetic algorithm, Peng et al. [5] developed a novel model with the goal of lowering the amount of electricity that is used by an air conditioning system that is based on a heat pump. They discovered that the functioning of the AC system may be achieved by modifying the frequency of the compressor, as well as the frequency of the fans both inside and outside the building.

15.4 Ant colony optimization and its applications

Ant colony optimisation, also known as ACO, is one of the other optimisation techniques that is used to improve the efficiency of heat transfer and flow devices by optimising various parameters of the system. ACO is a method that imitates the behaviour of ants in order to determine the shortest route from their nests to food sources. Marco Dorigo, a scientist, came up with the idea for the ACO approach in the nineteenth century by drawing inspiration from behavioural characteristics of social insects. Pheromones are chemical compounds laid down by ants when they move from a place to another. Utilising this, the other ants are adhering to the route to arrive at the destination. This input mechanism enables the ant colony to settle on an optimum or near-optimal route, demonstrating a decentralised but remarkably efficient process for making choices. The steps need to follow in ACO optimization problems is illustrated in Figure 15.5. The advantages and disadvantages of ACO is presented in Figure 15.6.

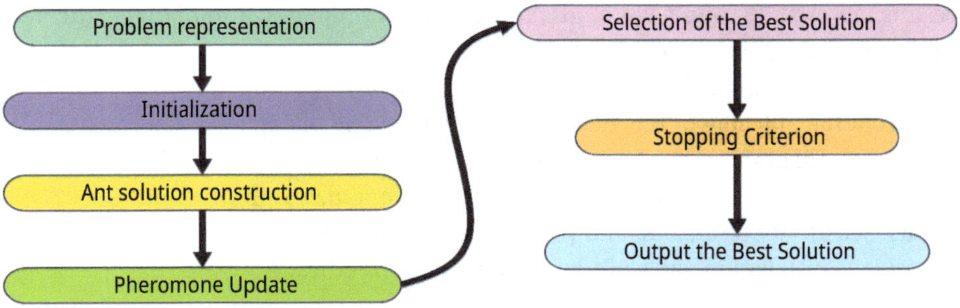

Figure 15.5: Steps need to follow for ACO optimization.

ACO significantly improves the efficiency and performance of heat transfer systems via the optimisation of their design, operation, and energy consumption. Heat transfer systems, including heat exchangers, cooling systems, and thermal energy storage units, include intricate processes necessitating meticulous monitoring and optimisation to reduce energy losses and enhance thermal efficiency. Due to several benefits,

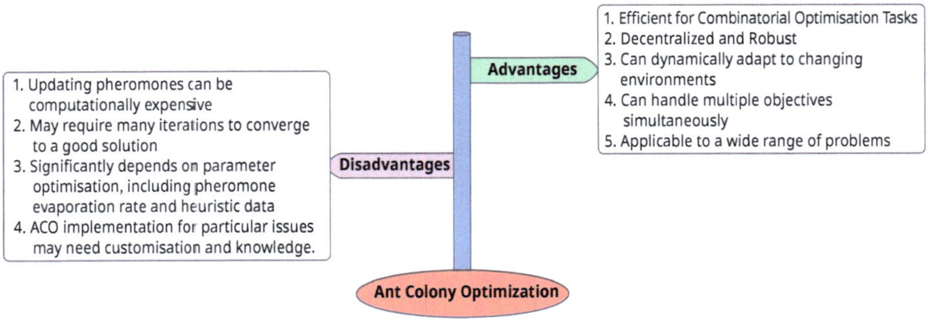

Figure 15.6: Advantages and disadvantages of ACO technique.

ACO methods have been extensively used in various heat transfer applications in recent years.

Gawronska et al. [6] used the ACO approach to rebuild the heat transfer coefficient in relation to the continuous boundary condition. From the obtained results, the authors suggested that the ACO technique may improve industrial processes, perform data analysis, and support data-driven decision-making. The assessment of the heat transfer rate in a tube heat exchanger employing nanofluid and ribs through diverse optimisation methodologies, including the Harris Hawks Optimiser (HHO), Grey Wolf Optimiser (GWO), Whale Optimisation Algorithm (WOA), Artificial Bee Colony (ABC), Ant Colony Optimisation (ACO), Ant Lion Optimiser (ALO), Biogeography-Based Optimisation (BBO), and Dragonfly Algorithm (DA), was conducted by Bahiraei et al. [7], revealing that the ALO algorithm is the most effective method for achieving optimal results.Liu et al. [8] and Qian et al. [9] used the ACO approach to optimise the heat dissipation and dimensions of systems, observing that ACO may successfully enhance the heat dissipation performance of electronic devices.

From the outcome of the above literature, it can be understood that the ACO is a potent and adaptable optimisation method with several benefits, including as flexibility, resilience, and relevance to diverse issues. Nonetheless, its computing requirements, sensitivity to factors, and propensity for stagnation underscore the need for meticulous installation and calibration to optimise its efficacy.

15.5 Applications of firefly algorithm

In 2008, researcher Xin-She Yang, inspired by the behaviour of fireflies, devised an algorithm known as the Firefly Algorithm (FA). Commonly, fireflies transmit light signals and communicate with one another. The algorithm works based on the concept of attraction, whereby fireflies are attracted to more luminous counterparts. In FA, brightness represents the efficacy of a solution inside optimisation issues. A firefly with a superior solution produces a more intense "light," prompting others to approach it in pursuit of optimum solutions. The Firefly Algorithm has been used in several optimisation challenges throughout engineering, research, and commercial sectors. In engineering design, it is used for structural optimisation, feature selection, and parameter modification. In data science and artificial intelligence, FA has shown efficacy in grouping, classification, and neural network training. Moreover, it has been used in addressing practical issues such as scheduling, resource allocation, and network optimisation. The schematic representation of the firefly algorithm is presented in in Figure 15.7. The steps need to follow for firefly algorithm is given in Figure 15.8.

Over the last several decades, many academics have devised various firefly algorithms, optimised parameters for diverse heat transfer and flow systems, and improved system efficiency. Mohanty [10] used the firefly algorithm to optimise the parameters of a shell and tube heat exchanger and evaluated its performance and economic aspects. In this research, the method was used in two distinct scenarios, yielding data that indicated a reduction in operational costs by 77% and overall costs by 29%.

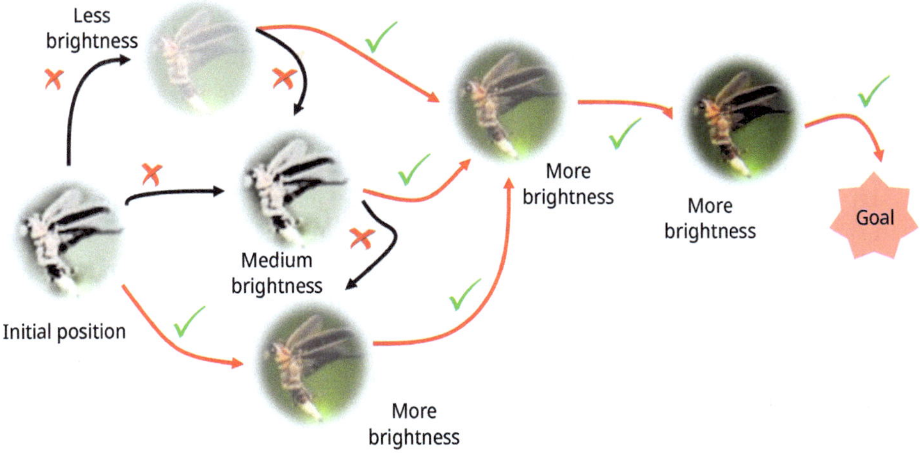

Figure 15.7: Schematic representation of firefly algorithm.

Wang et al. [11] evaluated the interfacial heat transfer coefficient during gas cooling using the firefly algorithm and finite element approach, demonstrating that the findings correlated with those obtained experimentally and via the finite element method. Despite the fact that the group scale fireflies in the ZFA are not less than 100, just twenty fireflies in the ZGFA are able to guarantee that all of the fireflies will travel to the ideal location owing to the global optimisation factor. Singh et al. [12] developed an algorithm to assess the heating efficiency of the photovoltaic thermal greenhouse

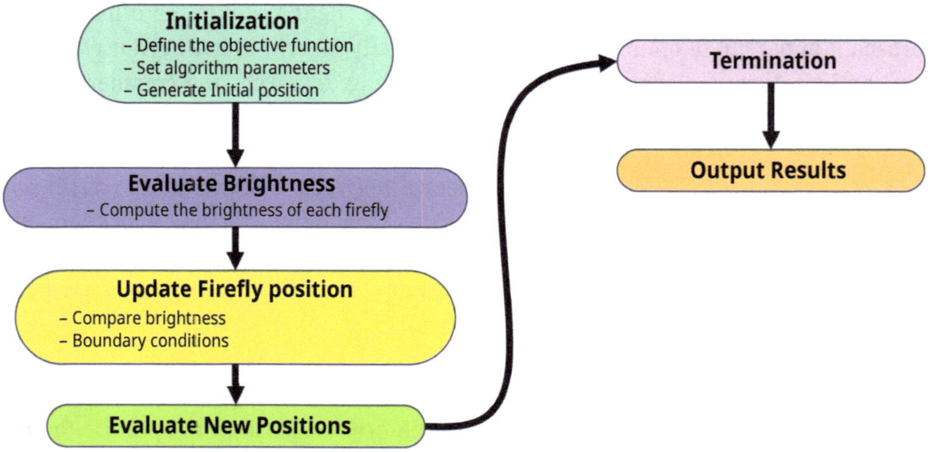

Figure 15.8: Procedure for firefly algorithm.

system using a self-adaptive firefly neural network model, and determined that the experimental findings closely aligned with the model output. Petinrin et al. [13] examined entropy production and optimised the shell and tube heat exchanger inside the preheat train of a crude oil distillation plant. The sensitivity research findings on the choice factors indicated a significant decrease in entropy production rate and a rise in pumping power when the mass flow rate escalates for all variables. The selection of tube diameter and quantity significantly influenced the variations in entropy production rate and pumping power.

Liu et al. [14], Amoozad Khalili et al. [15], Zieba Falama [16], and Hamidia [17] used several firefly algorithms to optimise the parameters of heat transfer and flow systems, observing a significant improvement in performance. The aforementioned literature clearly demonstrates that the firefly method has the potential for application across several disciplines since it was developed in 2008. Furthermore, the developmental domains of this algorithm are very dynamic, as novel applications emerge almost every day.

15.6 Cuckoo search algorithm and its applications

Cuckoo Search algorithm (CSA) is an another swarm intelligence algorithm that was invented by Xin-She Yang and Suash Deb in 2009, and it is an optimisation strategy that is inspired by nature. It was motivated by the brood parasitism shown by some cuckoo species that deposit their eggs in the nests of other kinds of birds. The method is based on the observation of cuckoo behaviours and their reproductive optimisation tactics. The method incorporates this intriguing natural behaviour with the principle of Lévy flights, a stochastic process defined by step lengths according to a Lévy distribution. This combination enables the CSA to effectively navigate and use the search space, rendering it very proficient in addressing intricate optimisation challenges. Cuckoos enhance the survival of their progeny by deliberately depositing their eggs in nests with greater survival probabilities. This behaviour resembles the pursuit of optimum solutions within a specified issue domain. The host bird's response to finding foreign eggs includes a survival element that simulates the natural selection process, hence augmenting the algorithm's exploratory potential. By using these ideas, the CS algorithm achieves a balance between exploration (global search) and exploitation (local refining). The important milestones of CSA is presented in Table 15.2.The Pros and Cons of CSA is depicted in Figure 15.9.

Heat transfer issues can include several variables, non-linear correlations, and limitations that make conventional optimisation techniques less efficient or computationally prohibitive. CSA is becoming recognised as a successful tool for tackling intricate optimisation problems in heat transfer applications. Enhancing these systems is crucial for improving efficiency, reducing energy consumption, and lowering operating expenses. Conventional optimisation techniques may encounter difficulties with these constraints, particularly when the problem exhibits a multi-modal landscape including several local optima.

Table 15.2: Milestones of Cuckoo Search Algorithm.

Year	Milestone	Details
2009	Development of Cuckoo Search Algorithm	– Introduced by **Xin-She Yang** and **Suash Deb**. – Inspired by cuckoo brood parasitism and Lévy flights.
2010	Initial Applications and Benchmarking	– Applied to standard test functions for optimization. – Demonstrated efficiency over other algorithms like Genetic Algorithms (GA) and Particle Swarm Optimization (PSO).
2011	Widespread Adoption in Engineering Optimization	– Used in engineering design problems such as structure optimization and resource allocation.
2012	Hybridization with Other Algorithms	– First hybrid approaches combining CS with methods like GA and PSO for enhanced performance.
2013	Extensions to Handle Constrained Optimization Problems	– Researchers adapted the algorithm to solve constrained problems more effectively.
2014	Multi-objective Optimization	– Extended to handle multi-objective optimization tasks, further broadening its applicability.

Table 15.2 (continued)

Year	Milestone	Details
2015	Applications in Machine Learning	– Began integration into neural network training, feature selection, and clustering problems.
2016	Advances in Parameter Adaptation	– Focus on dynamic parameter tuning to improve algorithm efficiency and robustness.
2018	Integration with Big Data and Cloud Computing	– Used in large-scale optimization problems in data-intensive and distributed computing environments.
2020	Development of Variants for Real-time and Dynamic Optimization	– Introduced real-time CS variants for adaptive and dynamic environments.
2022	Applications in Renewable Energy and Sustainability	– Widely applied in energy management, smart grids, and renewable energy systems optimization.
2023	Enhanced Hybrid Approaches and AI Integration	– Hybridized with AI techniques like deep learning for complex decision-making processes.

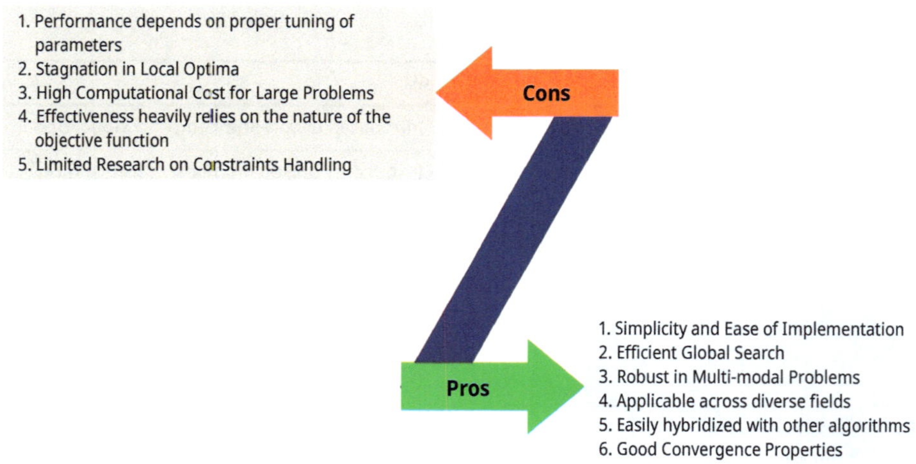

Figure 15.9: Pros and Cons of Cuckoo Search Algorithm.

As a result of the many benefits it offers, the importance of CSA in applications involving heat transfer has significantly expanded during the last several years. A few of the findings from the study conducted on the application of CSA optimisation to heat transfer and flow systems are provided in this section of the section. The steps followed to optimize the parameters of heat transfer systems and flow devices is shown

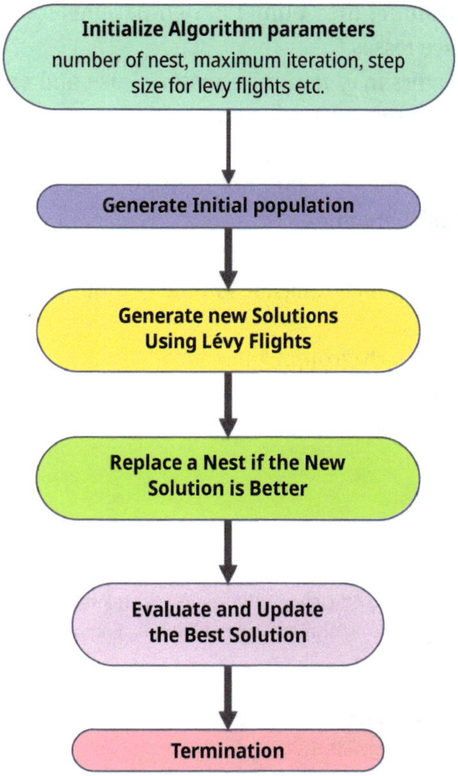

Figure 15.10: Steps need to follow for cuckoo search algorithm.

in Figure 15.10. Asadi et al. [18] optimised the design parameters of shell and tube heat exchangers with the CSA approach for total yearly cost and discovered that around 77% of energy was conserved. They also observed that costs were lowered by about 9.4% and 13.1% when employing PSO and GA procedures, respectively.CSA was used by Ranjan et al. [19] in order to optimise the performance of split elliptical shaped fins in order to increase the rate of heat transfer. The researchers observed that the peak value of the Nusselt number was attained as 0.125 and 0 at transverse offset and longitudinal offset respectively, with a numerical value equivalent to 0.97% accuracy.

15.7 Conclusions

This chapter included a thorough examination of diverse optimisation approaches and their applications across various heat transfer systems. The main results of this investigation are summarised and provided below.

1. Optimisation approaches assist in determining the optimal design parameters to enhance heat transfer efficiency or reduce losses.
2. Optimising design and operating parameters may minimise material use and energy consumption, hence decreasing production and operational costs.
3. Optimisation techniques, including genetic algorithms, particle swarm optimisation, and Cuckoo Search, combine well with computational tools such as Finite Element Analysis and Computational Fluid Dynamics.
4. Despite its many potential advantages, the use of optimisation algorithms in heat transfer devices encounters some challenges, including substantial computational expenses and environmental considerations.
5. Regardless of advancements by many research groups, more investigation is required regarding the utilisation and development of novel optimisation approaches and their applications in heat transfer systems.

References

[1] He Z., Xi H., Ding T., Wang J., and Li Z. (2021 Jan 5). Energy efficiency optimization of an integrated heat pipe cooling system in data center based on genetic algorithm. Applied Thermal Engineering, 182, 115800.

[2] Liu Q., Tao Y., Shi L., Zhou T., Huang Y., Peng Y., Wang Y., and Tu J. (2023 Feb 25). Parametric optimization of a spiral ground heat exchanger by response surface methodology and multi-objective genetic algorithm. Applied Thermal Engineering, 221, 119824.

[3] Shen Y., Liu Y., Liu S., and Mazhar A. R. (2022 Feb 1). A dynamic method to optimize cascaded latent heat storage systems with a genetic algorithm: A case study of cylindrical concentric heat exchanger. International Journal of Heat and Mass Transfer, 183, 122051.

[4] Guo Z., Wang Y., Zhao S., Zhao T., and Ni M. (2023 Jun 15). Modeling and optimization of micro heat pipe cooling battery thermal management system via deep learning and multi-objective genetic algorithms. International Journal of Heat and Mass Transfer, 207, 124024.

[5] Peng Q., Yang G., and Du Q. (2023 Dec). Energy-saving optimization of frequency-variable heat pump air conditioning system for electric vehicles based on a genetic algorithm. International Journal of Automotive Technology, 24(6), 1543–1551.

[6] Gawronska E., Zych M., Dyja R., and Domek G. (2023 Sep 15). Using artificial intelligence algorithms to reconstruct the heat transfer coefficient during heat conduction modeling. Scientific Reports, 13(1), 15343.

[7] Bahiraei M., Foong L. K., Hosseini S., and Mazaheri N. (2021 Mar 1). Predicting heat transfer rate of a ribbed triple-tube heat exchanger working with nanofluid using neural network enhanced by advanced optimization algorithms. Powder Technology, 381, 459–476.

[8] Liu Y., Wang K., and Yao D. (2024 Apr 26). Optimization of heat dissipation performance of electronic device integration modules based on intelligent algorithms. In 2024 Third International Conference on Distributed Computing and Electrical Circuits and Electronics (ICDCECE), IEEE, pp. 01–05.

[9] Qian Y., Hou F., Fan J., Lv Q., Fan X., and Zhang G. (2021 May 14). Design of a fan-out panel-level SiC MOSFET power module using ant colony optimization-back propagation neural network. IEEE Transactions on Electron Devices, 68(7), 3460–3467.

[10] Mohanty D. K. (2016 Apr 1). Application of firefly algorithm for design optimization of a shell and tube heat exchanger from economic point of view. International Journal of Thermal Sciences, 102, 228–238.
[11] Wang X., Li H., He L., Li Z., and Wang Z. (2019 Sep 1). Estimated temperature-dependent interfacial heat transfer coefficient during gas cooling based on firefly algorithm and finite element method. Heat and Mass Transfer, 55(9), 2545–2558.
[12] Singh G., Jain V. K., and Singh A. (2018 Nov). Adaptive network architecture and firefly algorithm for biogas heating model aided by photovoltaic thermal greenhouse system. Energy & Environment, 29(7), 1073–1097.
[13] Petinrin M. O., Bello-Ochende T., Dare A. A., and Oyewola M. O. (2018 Dec 25). Entropy generation minimisation of shell-and-tube heat exchanger in crude oil preheat train using firefly algorithm. Applied Thermal Engineering, 145, 264–276.
[14] Liu G., Zhang B., Zhang Y., and Guo C. (2020 May 3). Modeling of double-layer triangular microchannel heat sink based on thermal resistance network and multivariate structural optimization using firefly algorithm. Numerical Heat Transfer, Part B: Fundamentals, 77(5), 417–428.
[15] Amoozad Khalili H., Salehi G., Momeni S. M., and Eshagh Nimvari M. (2021 Jun 1). Multi objective optimization of heat pipe using firefly algorithm. Journal of Industrial Strategic Management, 6(2), 93–102.
[16] Zieba Falama R., Bakari H., and Dumbrava V. (2021 Dec). Double-objective optimization-based firefly algorithm of a stand-alone photovoltaic/water pumping system for water supply in rural and remote areas: A case study. Journal of Electrical Systems and Information Technology, 8, 1–25.
[17] Hamidia F., Abbadi A., and Boucherit M. S. (2019). PV/pumping system based on firefly optimizing algorithm. In Renewable Energy for Smart and Sustainable Cities: Artificial Intelligence in Renewable Energetic Systems 2, Springer International Publishing, pp. 171–177.
[18] Asadi M., Song Y., Sunden B., and Xie G. (2014 Dec 5). Economic optimization design of shell-and-tube heat exchangers by a cuckoo-search-algorithm. Applied Thermal Engineering, 73(1), 1032–1040.
[19] Ranjan A., Das R., Pal S., Majumder A., and Deb M. (2021 Jun 1). Use of cuckoo search algorithm for performance evaluation of split elliptic shaped fins for enhanced rate of heat transfer. Journal of Heat Transfer, 143(6), 063301.

Index

https://doi.org/10.1515/9783111661674-016

De Gruyter Series in Advanced Mechanical Engineering

Already published in the series

Volume 15
Arbind Prasad, Manoj Kumar Sinha, J. Paulo Davim (Eds.)
High-Performance Metallic Biomaterials: Surface Modification and Coating of Implants, 2025
ISBN 978-3-11-157137-9, e-ISBN (PDF) 978-3-11-157142-3,
e-ISBN (EPUB) 978-3-11-157212-3

Volume 14
Ranjith Rajamanickam, Amit Sharma, Dhivya Ranjith, J. Paulo Davim (Eds.)
Machine Learning for Medical Applications:

Volume I: Computational Drug Discovery, Bioimaging, Smart Biomaterials, 2025
ISBN 978-3-11-150318-9, e-ISBN (PDF) 978-3-11-150320-2,
e-ISBN (EPUB) 978-3-11-150374-5

Volume II: Computer Vision, Image Processing, Disease Detection, 2025
ISBN 978-3-11-914782-8, e-ISBN (PDF) 978-3-11-220519-8,
e-ISBN (EPUB) 978-3-11-220529-7

Volume 13
Raj Kumar Arya, George Verros, J. Paulo Davim (Eds.)
Compressors and Blowers: Maintenance, Practical Guidance, Energy-Efficient Technologies, 2025
ISBN 978-3-11-136335-6, e-ISBN (PDF) 978-3-11-136395-0,
e-ISBN (EPUB) 978-3-11-136417-9

Volume 12
Nishant Kumar Singh, Rajesh Kumar Verma, Virendra Kumar,
J. Paulo Davim (Eds.)
Machining and Tribology of Advanced Materials: From Coatings, Lubrications, Surface Treatments to Modeling and Simulation, 2025
ISBN 978-3-11-137642-4, e-ISBN (PDF) 978-3-11-137729-2,
e-ISBN (EPUB) 978-3-11-137772-8

Volume 11
Kanak Kalita, Ranjan Kumar Ghadai, J. Paulo Davim (Eds.)
Microfabrication and Nanofabrication: Precision Manufacturing, 2024
ISBN 978-3-11-133543-8, e-ISBN (PDF) 978-3-11-133553-7,
e-ISBN (EPUB) 978-3-11-133559-9

Volume 10
Kanak Singh, Ranvijay Kumar, Vinay Kumar and J. Paulo Davim (Eds.)
3D Printing: Polymer, Metal and Gel Based Additive Manufacturing, 2024
ISBN 978-3-11-108061-1, e-ISBN (PDF) 978-3-11-108163-2
e-ISBN (EPUB) 978-3-11-108206-6

Volume 9
Raj Kumar Arya, Anurag Kumar Tiwari, George Verros, Prashant Malik and J. Paulo Davim (Eds.)
Sustainable Hydrogen Energy: Production, Storage & Transportation, 2024
ISBN 978-3-11-124641-3, e-ISBN (PDF) 978-3-11-124647-5
e-ISBN (EPUB) 978-3-11-124728-1

Volume 8
Tatjana Sibalija and J. Paulo Davim (Eds.)
Soft Computing in Smart Manufacturing: Solutions toward Industry 5.0, 2022
ISBN 978-3-11-069317-1, e-ISBN (PDF) 978-3-11-069322-5
e-ISBN (EPUB) 978-3-11-069325-6

Volume 7
J. Paulo Davim (Ed.)
Nonconventional Machining, 2023
ISBN 978-3-11-058410-3, e-ISBN (PDF) 978-3-11-058447-9
e-ISBN (EPUB) 978-3-11-058415-8

Volume 6
J. Paulo Davim (Ed.)
Manufacturing in Digital Industries, 2020
ISBN 978-3-11-057507-1, e-ISBN (PDF) 978-3-11-057542-2,
e-ISBN (EPUB) 978-3-11-057526-2

Volume 5
Kaushik Kumar, Divya Zindani, J. Paulo Davim
Rapid Prototyping, Rapid Tooling and Reverse Engineering, 2020
ISBN 978-3-11-066324-2, e-ISBN (PDF) 978-3-11-066490-4,
e-ISBN (EPUB) 978-3-11-066345-7

Volume 4
J. Paulo Davim (Ed.)
Additive and Subtractive Manufacturing, 2019
ISBN 978-3-11-054816-7, e-ISBN (PDF) 978-3-11-054977-5,
e-ISBN (EPUB) 978-3-11-054833-4

Volume 3
J. Paulo Davim (Ed.)
Drilling Technology, 2017
ISBN 978-3-11-047863-1, e-ISBN 978-3-11-048120-4,
e-ISBN (EPUB) 978-3-11-047871-6

Volume 2
J. Paulo Davim (Ed.)
Progress in Green Tribology, 2017
ISBN 978-3-11-037272-4, e-ISBN 978-3-11-036705-8,
e-ISBN (EPUB) 978-3-11-039252-4

Volume 1
J. Paulo Davim (Ed.)
Metal Cutting Technology, 2016
ISBN 978-3-11-044942-6, e-ISBN 978-3-11-045174-0,
e-ISBN (EPUB) 978-3-11-044947-1, Set-ISBN 978-3-11-045175-7